Handbook of Smart Actuators and Smart Sensors

Handbook of Smart Actuators and Smart Sensors

Edited by **Bob Tucker**

NY RESEARCH PRESS
New York

Published by NY Research Press,
23 West, 55th Street, Suite 816,
New York, NY 10019, USA
www.nyresearchpress.com

Handbook of Smart Actuators and Smart Sensors
Edited by Bob Tucker

International Standard Book Number: 978-1-63238-279-5 (Hardback)

Printed in China.

Contents

Preface

I am honored to present to you this unique book which encompasses the most up-to-date data in the field. I was extremely pleased to get this opportunity of editing the work of experts from across the globe. I have also written papers in this field and researched the various aspects revolving around the progress of the discipline. I have tried to unify my knowledge along with that of stalwarts from every corner of the world, to produce a text which not only benefits the readers but also facilitates the growth of the field.

Smart actuators and sensors are used in diverse electric applications. This book is a comprehensive summary of recent technical researches in the field of smart actuators and sensors. It is a compilation of researches accomplished by experts from around the globe. The book will allow the reader to appreciate diverse points of view, engineering techniques and tools that must be considered together when designing and realizing smart actuation and sensing systems. It covers significant issues regarding SMA-based systems and smart polymers.

Finally, I would like to thank all the contributing authors for their valuable time and contributions. This book would not have been possible without their efforts. I would also like to thank my friends and family for their constant support.

Editor

Section 1

SMA-Based Systems

Chapter 1

New Developments on Mini/Micro Shape Memory Actuators

Adelaide Nespoli, Carlo Alberto Biffi, Riccardo Casati, Francesca Passaretti, Ausonio Tuissi and Elena Villa

Additional information is available at the end of the chapter

1. Introduction

Shape Memory Alloys (SMAs) are functional materials, characterized by some attractive features, such as pseudo-elasticity (PE) and shape memory effect (SME) [1]. The latter consist in the capability of the material to recover high deformation values by heating and can be considered very suitable for actuation applications.

In the actuation field, the most common shape memory alloy is quasi-equiatomic NiTi system, commercially known as Nitinol®. Ti-rich NiTi compounds show characteristic transformation temperatures higher than the room one and they can recover high values of deformation. Moreover, these intermetallic compounds are widely used since they exhibit high thermal and mechanical cycling stability. Other NiTi-based alloys are also employed for this kind of applications. In particular, NiTiCu, with Cu substituting Ni in the 3-10at% range, or NiTiCo system are used respectively when the application requires narrow thermal hysteresis or high stiffness.

In Figure 1, the comparison between differential scanning calorimetry (DSC) data derived from NiTi wire (200 μm in diameter) in fully annealed condition and after aging (500ºC for 10 minutes) are reported. As it can be seen, the fully annealed NiTi alloy goes into a one-step transformation both during cooling and heating. In this case, shape memory effect occurs by a martensitic transformation (MT) between a low-temperature monoclinic structure, B19′, called martensite, and a high-temperature body-centered cubic parent phase, B2, called austenite. When the material is aged at specific temperatures, this transition may occur in two steps in association with a trigonal phase, called R-phase [1].

During heating, the phase transformation, named inverse MT, is defined by austenitic start (As) and finish (Af) temperatures. Similarly, during cooling direct MT is defined by martensitic start (Ms) and finish (Mf) temperatures, respectively.

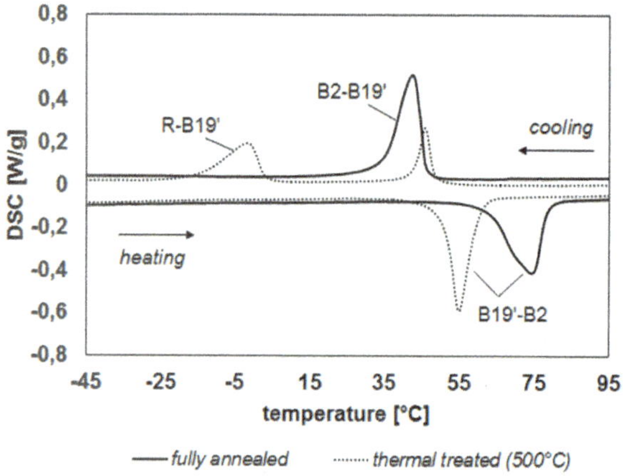

Figure 1. DSC scan of 200μm NiTi wire in the fully annealed state and after aging (500°C for 10 minutes).

Then, the deformation induced by a state of stress on a SMA element can be recovered heating the material over the austenitic characteristic temperatures: the shape recovery starts at As and finishes at Af.

The transition temperatures are strongly influenced by the applied load, since it stabilizes the martensitic structure causing an increase of all transition temperatures. This behavior of SMA is well described by the Clausius-Clapeyron law [1-5].

Furthermore, SMA actuators generally need a training to stabilize their fatigue properties in terms of transformation temperatures and strain recovery. For example, a possible procedure consists in the thermal cycling under a constant load. Figure 2 depicts the evolution of the temperature-stroke relationship during the first thermo-mechanical cycles. It is seen that the stabilization of the material behavior can be reached after few thermal loops.

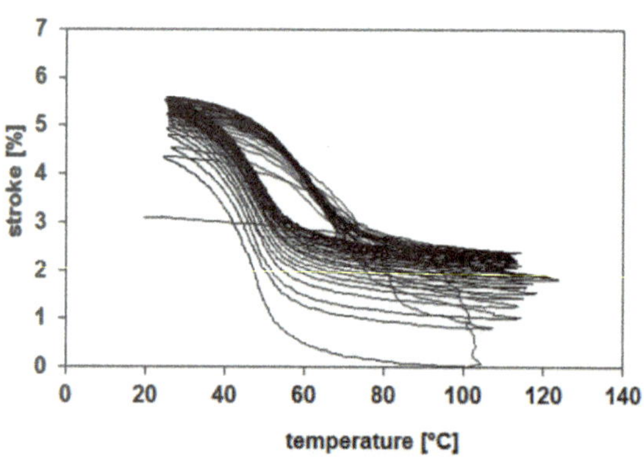

Figure 2. Example of thermo-mechanical hysteresis typical of a SMA material.

Since SMA can generate mechanical work, they are employed as actuators. The functional properties of SMA element used in actuators strictly depend on several factors, such as the chemical composition and the thermo-mechanical history[1,4,5].

Nowadays, the tendency is to produce ever smaller actuators because of the continued miniaturization of all the industrial products. In general, a mini-actuator may be composed by only a SMA element or by a mini-modular device, in which the SMA element is embedded and acts as the core and active part.

In this chapter, the fundamental operating parameters, which affect the performance of shape memory thin wires, are reviewed. Then, the influence of electrical heating conditions performed by different waveforms on functional fatigue of NiTi micro-wires is also reported.

When the SMA element is embedded into a mini-modular mechanical device, its shape could be a serious problem as spaces are very restricted. To solve this drawbacks, a new SMA conformation suitable for the mini and micro scales is presented: it consists of a planar wavy formed NiTi wire, called snake-like arrangement. Currently, this configuration of the SMA wire is principally exploited in the textile and medical domains [6-8]. n the micro scale, original works are reported by Mineta et al. [7-8], who produced different SMA snake elements by means of electrochemical etching for obtaining bending motion of active catheters. Khol et al. [9] used the micro snake geometry to activate a microgripper system. Moreover, Leester-Schadel et al. [10] adopted laser technology to produce micro SMA snake actuators and then used the batch fabrication process to obtain more articulated samples. In this chapter the main mechanical performance of this unusual geometry and its exploitation in a mini-modular mechanical device are presented.

Another topic of this work is related to a current research on fabrication and characterization of SMA micro-snakes, by means of laser technology and following polishing processes. The reason why this non-traditional production technology has been chosen for this purpose is due to its flexibility in the machining of small features, high productivity and repeatability [11-12]. Considering the direction of miniaturization of the products during the recent years [13], in the last part of the chapter the rescaling of SMA mini-snakes down to micro-scale is presented. The attention is also focused on the evaluation of the capability of laser micro-cutting industrial process in the fabrication of micro-snakes from NiTi SMA sheets [14-17]. The evaluation of the thermo-mechanical properties of the produced actuator in correspondence of the different fabrications steps is then presented and discussed in order to describe the behavior of the micro-snake-like element for actuation.

2. SMA linear actuators

The first case study is the application of a simple straight wire as base element for actuation systems [18]. The most important functional properties, which characterize shape memory wire for actuators, are the capability to recover high strain values, the characteristic transformation temperatures and the thermo-mechanical cycling stability [1]. These characteristics of SMA can be tailored acting on chemical composition, thermal and mechanical processing, shape setting treatment and training [1, 20-25].

In recent years, the good ductility of NiTi intermetallic has been exploited for the production of few tens micron wires [21, 26, 27].

Thanks to their small dimensions, such thin wires permit to reach very short cooling time but, obviously, they can be employed just in all those applications in which high loads are not required. The fabrication of these products has led to the miniaturization of SMA components for actuators that are nowadays employed for new kind of devices such as optical image stabilizer and autofocus for small camera [26]

During working life, the SMA wire is subjected to thermo-mechanical cycling and accumulates plastic deformation, elongating irreversibly and reducing its diameter (see Figure 3). This behavior has to be considered in actuator design. Another important point to consider is the unsteadiness of the functional properties of shape memory wires during the working life. The plastic deformation induced during the thermo-mechanical cycling leads to a change of microstructure of SMA wire. Lattice defects, dislocations and nano-scaled precipitates are introduced into the matrix and they cause significant changes in transformation temperature and in the capability of the material to recover a deformation [28-30], as depicted in Figure 4.

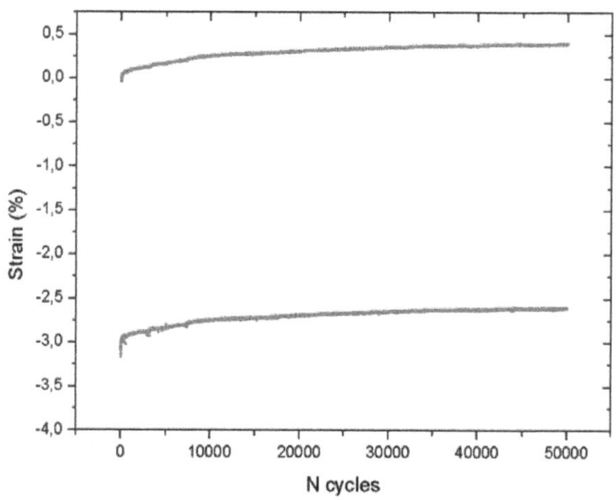

Figure 3. Thermal cycling of a 80μm wire under constant stress (200MPa). The heating was performed by electrical pulse, and stroke was limited at 3%.

In most cases, SMA wires used as actuator, are heated by means of an electrical pulse (Joule effect) and cooled by natural air convection. So, it is very easy to reach short actuation time (heating time) but the reset time (cooling time) cannot be substantially controlled. Slight improvement in the reduction of reset time is achieved positioning the shape memory wire horizontally [19], as shown in Figure 5. In this figure, it can be noticed that after heating (at

200s) the strain is maintained to two constant values for both the two configurations till time is near to 230s and 250s for horizontal and vertical position respectively. After these points, the two wires start the cooling process which results to be faster for the horizontally positioned wire than the vertical one. As an example, it can be seen that at 275s the horizontally positioned wire recovers half of its maximum strain while at the same time the vertical positioned wire has just started the cooling route.

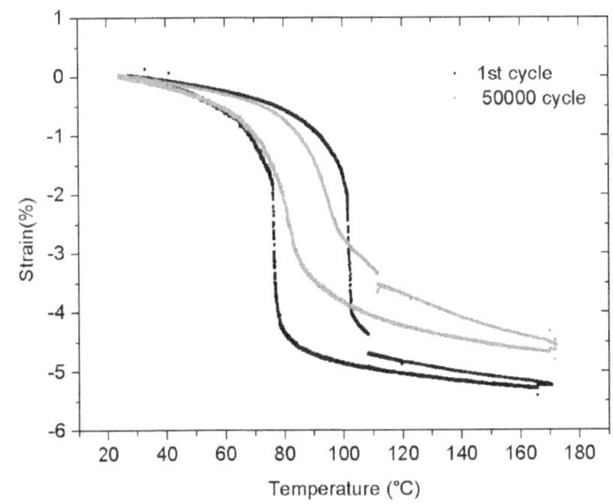

Figure 4. Thermal loop under constant stress (200MPa). The heating was performed by a thermal chamber. The test was carried out on a 80μm wire before and after $5x10^4$ thermal cycles under constant stress (200MPa).

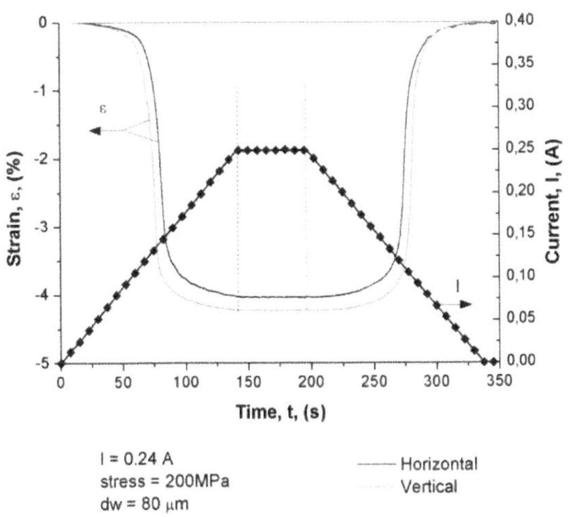

Figure 5. Comparison between the mechanical performance of horizontal and vertical configuration of 80μm SMA wire used as actuator [19].

As previously stated, the heating process of a SMA actuator can be easily obtained by Joule effect. In this case, it has to be considered that the shape of the electrical pulse used for the actuation strongly affects the functional properties of SMA [27]. As an example, Figure 6 shows the time required by the wires to recover the 3% of deformation (actuation time) when it is heated by two different current pulses. At the 1st cycle ramp and step electrical pulse employed to heat two 80μm wires were designed to have the same electrical efficiency but the actuation time is 400ms by step current pulse while it is 623ms by ramp. After $5{\cdot}10^3$ cycles the wire heated by step pulse employs 618ms, so 218ms more than the time employed at the first cycle. As opposite, the ramp pulse leads to an increase of just 22ms. After $5{\cdot}10^4$ cycles the actuation time related to the wire heated by step is even higher than the one related to wire heated by ramp. This behavior leads to a drastic decrease of the step heating method efficiency (from 0.91% to 0.55%), vice versa the efficiency of the ramp heating method does not change so much (from 0.91% to 0.75%).

Then, in order to achieve fatigue performance acceptable for the specific device, the right electrical pulse has to be chosen considering the number of cycles that the SMA actuator has been designed to work.

Recently, the effect of drawing procedure on functional fatigue of thin NiTi wires has also been investigated [21]. Basically, 80μm NiTi wires were produced through two different drawing procedures reaching the same final cold working level before shape setting. These two processes differ for the number of drawing steps carried out to reach a certain cold working level before each heat treatment. After the last thermal treatment the specimen that underwent to a less number of drawing steps shows a narrow thermal hysteresis, even after thermo-mechanical cycling (see Figure 7). It means that adopting a severe drawing procedure leads to improve the actuation performance.

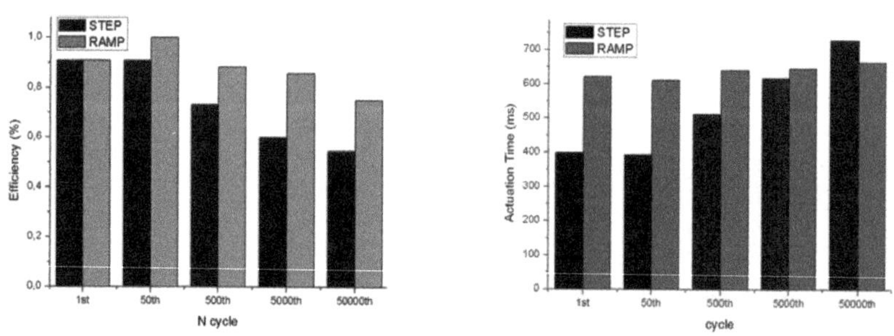

Figure 6. Actuation time tests performed by electrical heating (on the left) and by efficiencies (on the right).

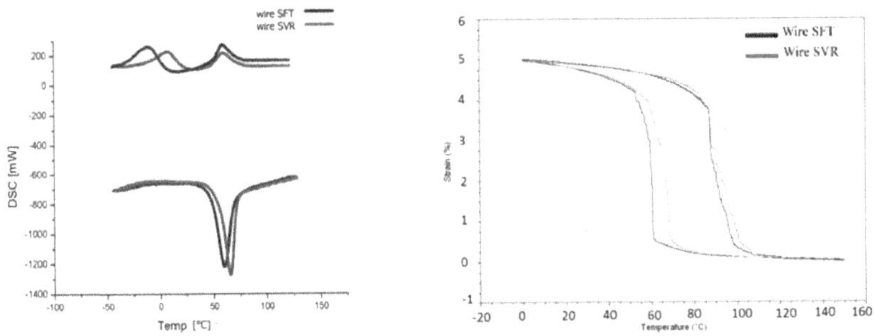

Figure 7. DSC scans, on the left, and thermal loop under constant tensile stress, on the right, of two 80μm NiTi wires obtained by a severe (SVR) and soft (SFT) drawing procedure.

3. Snake-like SMA actuators

The main focus of miniaturization is to develop devices with high mechanical performance in very limited spaces. In the SMA field, the mini-actuator may be composed by only the SMA element or by a mini-modular device in which the SMA element is embedded and acts as the core part. In the latter case, some issues typical of SMA miniaturization should be taken into a serious consideration: first of all, the fixing points of the SMA element and the electrical contacts should be adequately designed to prevent abrupt rupture, and then the use of a wire geometry generally implies the employment of additional mechanical parts which may affect the long run performance of the SMA element.

The mechanical work resulting from the SMA shape recovery is used to produce linear or rotational motion in smart actuators. To attain high strokes in limited space, designers developed several miniature devices which include SMA elements with a straight or a helical spring shape [31]. These two shapes present two opposite mechanical behaviors: the straight conformation allows high forces but smart wire arrangements should be employed in order to achieve also high strokes. As an example, Jansen et al. used two sets of seven pulleys to arrange a 1000mm long SMA wire with a diameter of 0.200mm, and attain 3% of stroke under a constant load of about 130MPa [32]. The use of additional mechanical components may affect the overall performance of the SMA element, leading to friction which could influence the final functional parameters of the device. As opposite, the helical spring shape guarantees high strokes without the use of any further mechanical part but fairly small forces can be attained especially when embedded in miniature devices.

Hence, straight and helical spring conformations represent two opposite limits in the force versus stroke plane. In order to answer both to the need of stroke and force between these two limits, new SMA geometries should be studied.

In a recent work, a new SMA conformation has been proposed [33]; it consists in a planar wavy formed NiTi wire, named snake-like arrangement. To understand the mechanical performance of this unusual conformation as respect to other shapes, three NiTi samples with different geometry, but with identical elemental composition, thermal history and power consumption were tested by means of a standard tensile test. In this investigation the NiTi wire was formed to have the straight conformation, the helical spring and the snake-like arrangements, and austenite and martensite mechanical responses were assessed separately. As can be seen in Figure 8 the snake-like shape has a mechanical behavior between the wire and the helical spring ones.

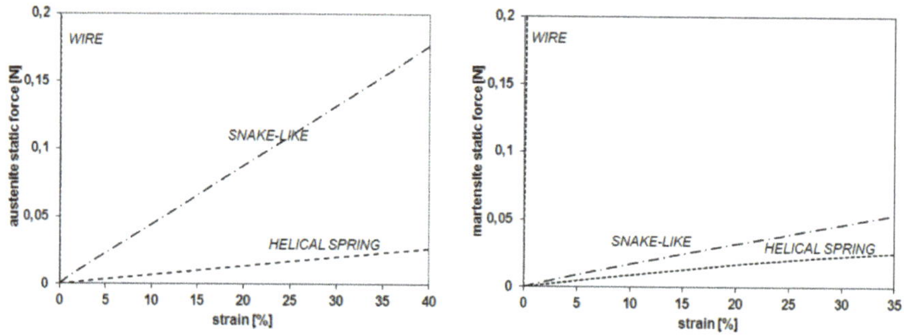

Figure 8. Comparison between austenite and martensite mechanical behavior of different NiTi wire geometries for actuators.

The snake-like conformation is characterized by four main parameters: the number of curvatures, N, the curvature radius, R, the distance between two consecutive curvatures, D, and the height, H. A snake-like specimen can be prepared by constraining a SMA wire in the snake-like arrangement by using a drilled aluminum bar and a series of nails; heating at high temperature is finally used to fix the shape. Figure 9 reports a snake-like sample prepared from a commercial NiTi wire having the diameter of 0.2mm, formed at 500°C during ten minutes and quenched in water at room temperature, with N=4, H=5.27mm±0.07mm, R=0.54mm±0.03mm and D=0.82mm±0.06mm.

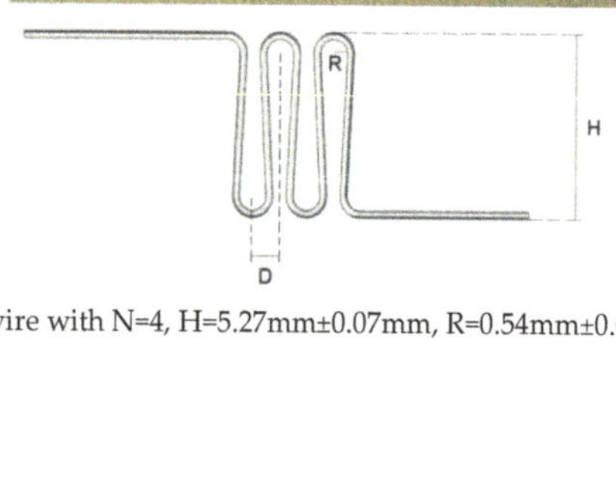

Figure 9. Snake-like NiTi wire with N=4, H=5.27mm±0.07mm, R=0.54mm±0.03mm and D=0.82mm±0.06mm.

As previously stated, during the ten minutes formation process the curved parts of the snake element are in contact with metal devices. This fact does not modify the calorimetric properties of the SMA element in all its length, as can be seen in Figure 10 where the overlapping of the DSC scans relative to the straight and the curved segments is clearly visible.

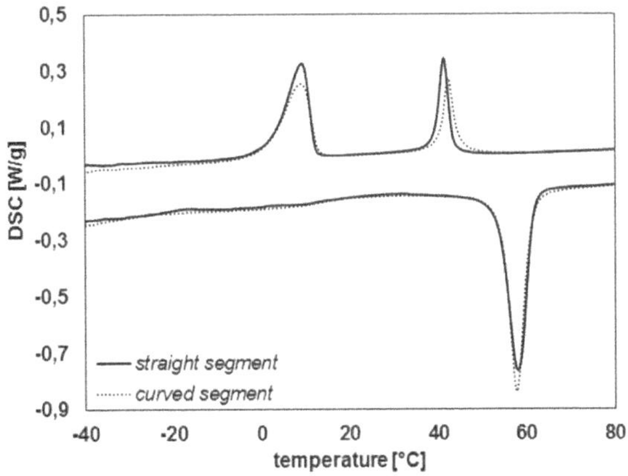

Figure 10. DSC scans of the straight and the curved part of the snake-like NiTi wire.

For small deformation values (see Figure 11), this snake-like shape shows an almost complete recovery of the deformation both for austenite and for martensite.

The relative higher output stroke performed by the snake-like NiTi wire as respect to the straight wire conformation is substantially due to the bend/unbend movement of the curvatures. In order to avoid plastic deformation and to achieve simultaneously a significant output stroke, the snake sample, formed as previously explained, should be stressed with a load near 0.1N, as forces lower and higher than 0.1N do not guarantee a proper functioning of this snake element: in fact, lower forces do not induce martensite formation, as opposite higher forces cause plastic deformations. As an example, in Figure 12 the curves derived from five thermal cycles of the snake-like NiTi sample under two constant loads, 0.1N and 0.2N is depicted. By comparing the two trends, it can be seen that in both cases the snake sample lose part of its shape recovery during cycling; this is much more visible for the specimen stressed with 0.2N.

Fatigue life was studied under the constant load of 0.1N; the snake-like sample was electrically activated with 0.6A with a maximum power consumption of 1W and a work frequency of 0.1Hz. Fracture takes place at the apex of that curvature most far from the applied load and from Figure 13 it can be observed that it occurs near the $7{,}5\text{x}10^4$ cycle. The fractured section was analyzed through SEM observation, see the inset of Figure 13; it can be seen a region with the striation of the crack propagation departing from a nucleation site placed on the inner surface of the snake and a region characterized by dimples, typical of ductile materials.

After the cycling test, the DSC scan shows a feeble shifting of the transformation temperatures towards higher values; a better definition of the rhombohedral phase during heating is also visible (see Figure 14).

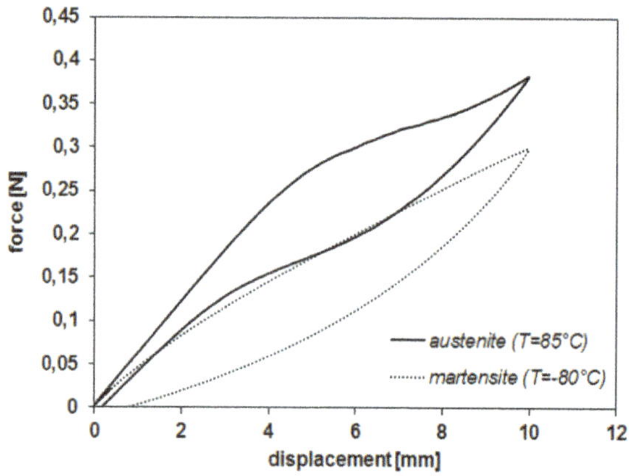

Figure 11. Force versus displacement of austenite and martensite of the snake-like NiTi wire.

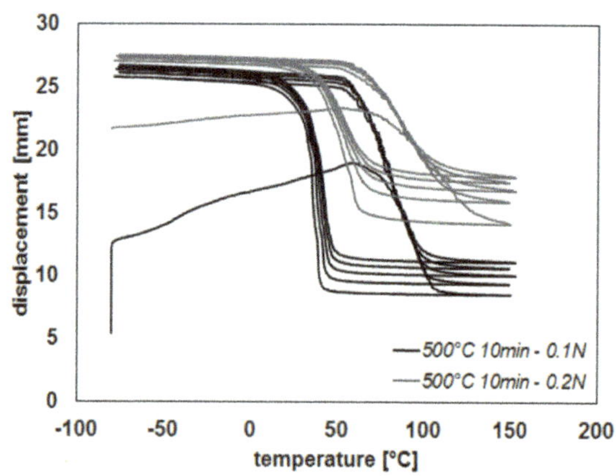

Figure 12. Hysteresis curves of the snake-like NiTi sample treated at 500°C for 10 minutes under 0.1N and 0.2N constant load.

When embedded in mini-modular device, the snake-like wire does not lose its mechanical performance. As an example, in Figure 15 a miniature rotational actuator composed by two mutual antagonist NiTi snake-like wire is reported [34]. The device exerts a stroke of about 120° under a constant torque of 0.1Nmm, as shown in Figure 16; it is activated by 0.6A with a maximum power consumption of 1W and a work frequency of 0.05Hz. Due to the particular construction, plastic deformation of the SMA element is prevented as the SMA elements never exceed 6mm of deformation.

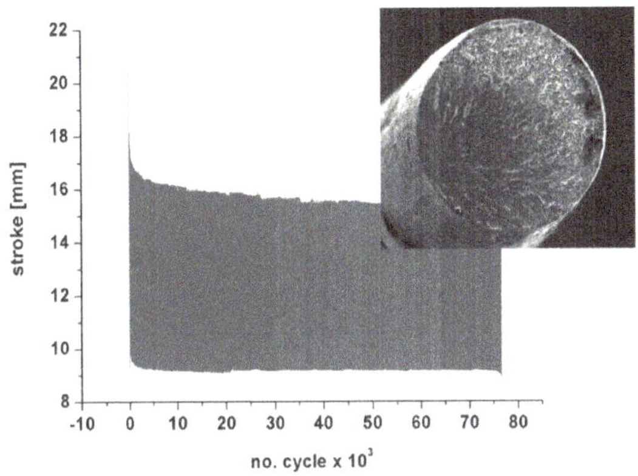

Figure 13. Snake-like NiTi sample fatigue test result under 0.1N load.

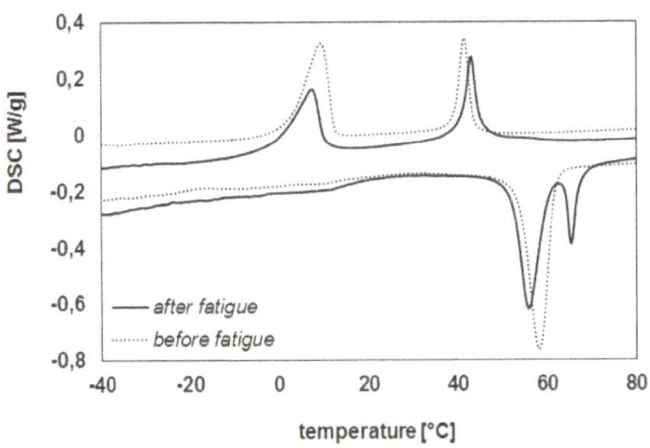

Figure 14. Snake-like NiTi sample DSC scan before and after the fatigue test

Figure 15. Mini rotational actuator activated by two antagonists snake-like NiTi wires.

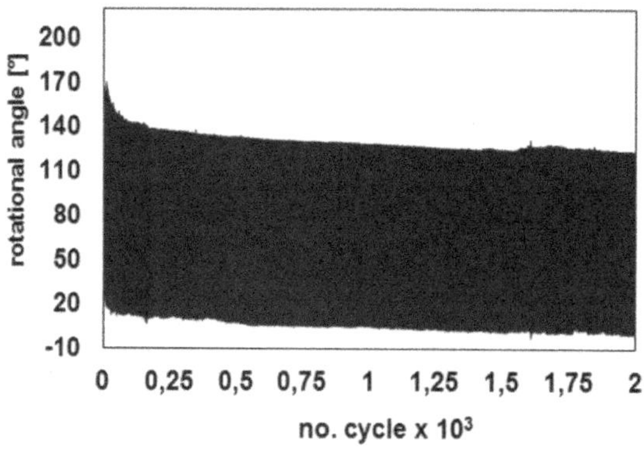

Figure 16. Fatigue test results of the mini rotational actuator under 0.1Nmm constant torque.

3.1. Micro snake-like shape memory alloy elements

In the last years the trend of miniaturization of components and products has been evidently introduced in several industrial fields, such as biomedical, electronics, aerospace and mechanics [35-36]. SMAs, as it already happens in biomedical applications [14,15,37-40], can be machined using laser machining with appreciated and high qualitative results [12]. The decision to investigate this method for the production of SMA actuators is considered an interesting topic because it shows some relevant advantages in comparison with other technologies, such as high productivity, high quality and repeatability, when applied in the industrial world.

Hence, the snake geometry can be easily scale down to the micro scale by the employment of laser technology with appreciated and high qualitative results. Because laser material processing is a thermal process, in which an heat source is used to remove a certain volume of material from the work piece, some thermal damages are evidently obtained, such as melted material, heat affected zone and oxides. To remove these defects, chemical etching and electro-chemical polishing have to be performed on the snake SMA samples [37,41]. Figure 17 depicts SEM images of the snake NiTi sample after laser, chemical etching and electro-chemical processes; top and bottom surfaces as well as snake sample section are shown. It can be seen that laser damages are mainly visible on the bottom surface and that the chemical etching remove the great part of defects. Moreover, a significant loss of mass and consequent reduction of the dimensions of the micro-snake can be evidently observed from SEM pictures due to the material removal of the post-processing. The snake NiTi sample geometry and calorimetric properties change after each process, as can be seen in Figure 18 and in Figure 19 respectively. In particular, the geometrical dimension are referred to a micro-snake obtained starting from a 120μm NiTi sheet. As concern the DSC data, it can be observed that the laser machining produces a refinement of the rhombohedral phase peak both on cooling and heating while the martensite peak barely change as respect to the unworked 120mm thick sheet, being almost spread and invisible. The laser process also yields a conspicuous diminution of the rhombohedral phase transformation temperatures and of the martensite starting temperature while the other characteristic transformation temperatures do not have an analogous variation being almost unchanged. Chemical etching and electro-polishing steps do not visibly alter any phase transformation temperature both on cooling and on heating, even if some little change due probably to the hydrogen introduction in the chemical process, is visible [45,46].

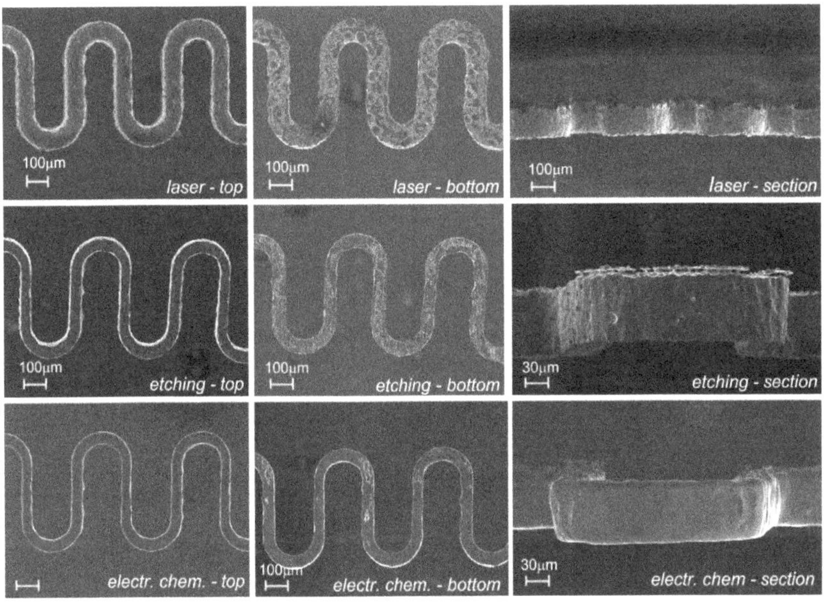

Figure 17. SEM images of the snake NiTi sample surfaces and section after laser, chemical etching and electro-polishing processes.

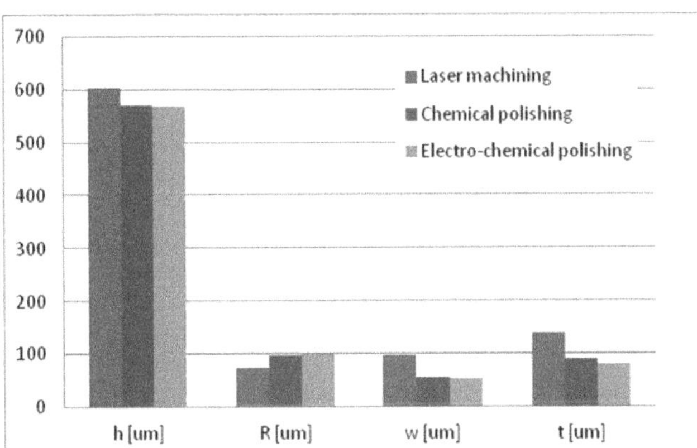

Figure 18. Variation of the geometrical parameters of the micro-snake during the fabrication process (h: high, R: radius of curvatures, w: width, t: thickness).

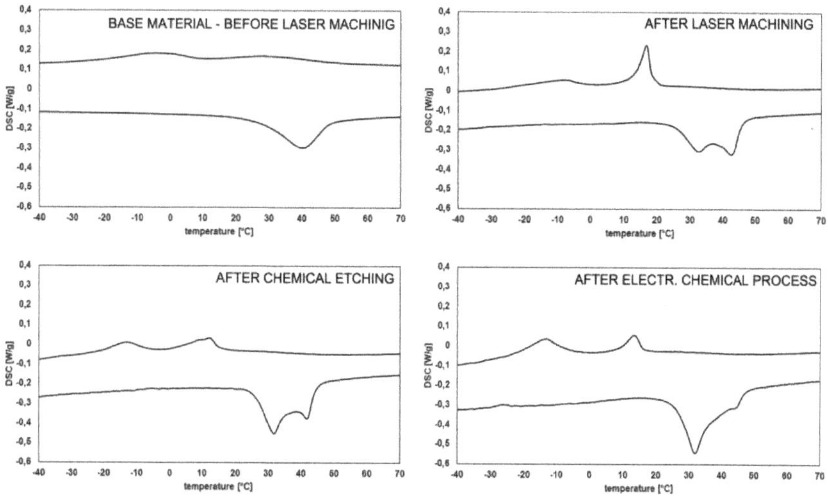

Figure 19. DSC scans before laser machining and in correspondence of each fabrication process.

The fabrication process previously described can be used to produce snake SMA sample of different length. As an example, Figure 20 depicts the SEM image of a micro-snake NiTi sample with 19 curvatures (i.e. 3.8mm of length) obtained starting from a 120μm NiTi sheet treated at 400°C for 15 minutes and quenched in water.

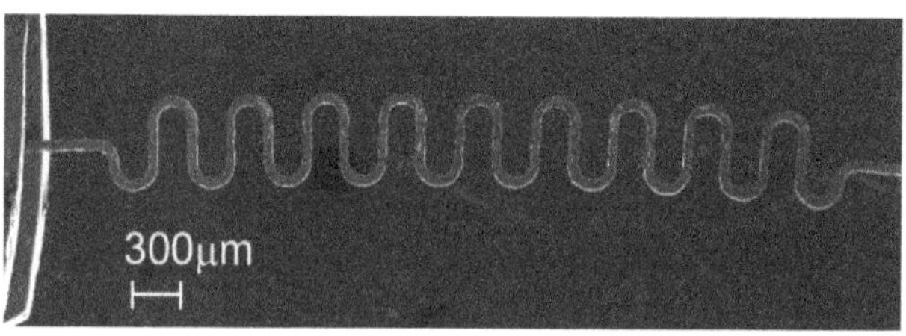

Figure 20. SEM image of a micro-snake NiTi sample obtained starting from a 120μm NiTi sheet.

Strain-recovery and thermo-mechanical cycling tests demonstrate the high performance of this micro-snake sample. As an example, under the constant load of 16mN, it reaches about 40% of recovery with a cycling stabilization within few hundreds of cycles, when heated by electrical current (130mA) with a maximum power consumption of 0.3W and a frequency of 0.15Hz.

4. Conclusions

In this work, new developments on SMA actuators in the mini/micro scale are presented: the attention was focused on straight NiTi thin wires and mini/micro snake-like NiTi samples.

Several working parameters have proved to strongly affect the functional and fatigue properties of SMA actuator, such as the electrical pulse performed to heat a wire, the different drawing procedures carried out to produce the wire or the wire arrangement into the device.

Then, the functional parameters of a snake-like NiTi wire demonstrate that this kind of geometry may be a good alternative to the most common geometries. It can generate a shape recovery in a very large range of deformation comparing to the straight SMA wire. This point is very beneficial for practical uses since the new proposed SMA conformation just occupies small space to generate the same displacement as the straight one, and it demonstrates to maintain its general mechanical performance when embedded in a mini-modular mechanical device. Finally, the presented fabrication process adopted for the manufacturing of micro-actuators looks to be very attractive (laser process can be performed on a large family of materials with high repeatability). The thermo-mechanical behavior of the micro-snake actuator can be considered interesting in the range of low applied load values with a good cycling stability.

Author details

Adelaide Nespoli*, Carlo Alberto Biffi, Riccardo Casati,
Francesca Passaretti, Ausonio Tuissi and Elena Villa
Consiglio Nazionale delle Ricerche, Istituto per l'Energetica e le Interfasi (CNR-IENI), Lecco, Italy

Acknowledgement

Authors would like to thank Prof. Barbara Previtali, head of Sitec-Laboratory for Industrial Laser Applications of Mechanical Department of Politecnico di Milano, for her precious support on the laser processing.

5. References

[1] Kazuhiro Otsuka, Clarence M Wayman (1998) ShapeMemoryMaterials. Cambridge University Press. 284 p.
[2] Hiroyasu Funakubo (1984) Shape memory alloys. London: Gordon and Breach Science Publishers.
[3] Miller DA, Lagoudas DC (2001) Influence of cold work and heat treatment on the shape memory effect and plastic strain development of NiTi. Mater. sci. eng. A. 308:161–75.
[4] Fukuda T, Kakeshita T, Kitayama M, Saburi T (1995) Effect of aging on martensitic transformation in a shape memory Ti–40.5Ni–10Cu alloy. J. phys. IV. 5:717–22.
[5] Degeratu S, Rotaru P, ManoleaGh, Manolea HO, Rotaru A (2009) Thermal characteristics of Ni–Ti SMA (shape memory alloy) actuators. J. therm. anal. calorim. 97:695–700.
[6] Villa E, Arnaboldi S, Tuissi A, Giacomelli M, Turco E (2009) Mechanical analysis of hybrid textile composites with NiTi wires. J. mater. eng. perform. 18(5–6):517–21.

* Corresponding Author

[7] Mineta T, Mitsui T, Watanabe Y, Kobayashi S, Haga Y, Esashi M (2001) Batch fabricated flat meandering shape memory alloy actuator for active catheter. Sens. actuators A. 88:112–20.

[8] Mineta T, Mitsui T, Watanabe Y, Kobayashi S, Haga Y, Esashi M (2002) An active guide with shape memory alloy bending actuator fabricated by room temperature process. Sens.actuators A. 97–98:632–7.

[9] Kohl M, Krevet B, Just E (2002) SMA microgripper system. Sens. and actuators A. 97-98:646-652.

[10] Leester-Schadel M, Hoxhold B, Lesche C, Demming S, Buttgenbach S (2008) Micro actuators on the basis of thin SMA foils. Microsyst. technol.14:697-704.

[11] John F.Ready, Dave F. Larson (2001) LIA Handbook of Laser Materials Processing. Laser Institute of America, Magnolia Publishing, Inc. p715.

[12] Dubey AK, Yadava V (2008) Laser beam machining—A review. Int. J. mac. tools manuf. 48:609–628.

[13] Gower MC (2000) Industrial applications of laser micromachining. Opt. Express 7:56-57

[14] Li C, Nikumb S, Wong F (2006) An optimal process of femtosecond laser cutting of NiTi shape memory alloy for fabrication of miniature devices. Opt. and lasers in eng. 44:1078–1087

[15] Momma C, Knop U, Nolte S (1999) Laser cutting of slotted tube coronary stents –state of the art and future developments. Prog. in biomed. res. 39-44.

[16] Buttgenbach S, Butefish S, Leester-Schadel M, Wogersien A (2001) Shape Memory Microactuators. Microsys. technol. 7:165-170.

[17] Meijer J, Du K, Gillner A, Hoffmann D, Kovalenko VS, Masuzawa T, Ostendorf A, Poprawe R, Schulz W (2002) Laser machining by short and ultrashort pulses, state of the art and new opportunities in the age of the photons, CIRP Ann. Manufact. Technol. 51:531-550.

[18] Saikrishna CN, Ramaiah KV, Allam Prabhu S, Bhaumik SK (2009) On stability of NiTi wire during thermo-mechanical cycling. Bull. mater. sci. 32:343–352

[19] Zanotti C, Giuliani P, Arnaboldi S, Tuissi A (2010) Analysis of wire position and operating conditions of functioning of NiTi wires for shape memory actuators. Proc. SMST 2010. 688-696.

[20] Frenzel J, George EP, Dlouhy A, Somsen CH, Wagner MFX, Eggeler G (2010) Influence of Ni on martensitic phase transformations in NiTi shape memory alloys. Actamater. 58:3444–3458.

[21] Casati R, Tuissi A, Belochapkine S, Dickinson C, Tofail SAM Thin NiTi wires with reduced thermal hysteresis for shape memory actuators. doi:10.1142/S1793604 7125000991250009

[22] Miller DA, Lagoudas DC (2001) Influence of cold work and heat treatment on the shape memory effect and plastic strain development of NiTi. Mat. Sci. Eng. A. 308:161-175.

[23] Todoroki T, Tamura H (1987) Effect of heat treatment after cold working on the phase transformation in TiNi alloy. Tran. of the jpn. Inst. of met. 28:83-94.

[24] Frick CP, Ortega AM, Tyber J, Maksound AEM, Maier HJ, Liu Y, Gall K (2005) Thermal processing of polycrystalline NiTi shape memory alloys. Mat. sci. eng. A. 405:34-49

[25] Contardo L, Guenin G (1990) Training and two way memory effect in Cu-Zn-Al alloy. Acta metall. mater. 38:1267-1272.

[26] http://www.cambridgemechatronics.com/

[27] Casati R, Tuissi A (2011) Effect of current pulses on fatigue of thin NiTi wires for shape memory actuators. Proc. SMST 2011. Submitted

[28] Wagner MFX, Dey SR, Gugel H, Frenzel J, Somsen Ch, Eggeler G (2010) Effect of low-temperature precipitation on the transformation characteristics of Ni-rich NiTi shape memory alloys during thermal cycling. Intermet. 18:1172-1179.

[29] Simon T, Kroger A, Somsen C, Dlouhy A, Eggeler G (2010) On the multiplication of dislocations during martensitic transformation in NiTi shape memory alloys. Acta mater. 58:1850-1860.

[30] Ibarra A, San Juan J, Bocanegra EH, Nò ML (2007) Evolution of microstructure and thermomechanical properties during superelastic compression cycling in Cu-Al-Ni single crystal. Acta Mater. 55:4789-4798.

[31] Nespoli A, Besseghini S, Pittaccio S, Villa E, Viscuso S (2010) The high potential of shape memory alloys in developing miniature mechanical devices: A review on shape memory alloy mini-actuators. Sens. and actuators A.158:149–160

[32] Jansen S, Breidert J, Welp EG (2004) Positioning actuator based on shape memory wires. Actuators 2004. 9th International Conference on New Actuators.

[33] Nespoli A, Villa E, Besseghini S (2011) Thermo-mechanical properties of snake-like NiTi wires and their use in miniature devices. J. therm.anal calorim. DOI 10.1007/s10973-011-1324-0

[34] Nespoli A, Bassani E, Besseghini S, Villa E (2010) Rotational mini-actuator activated by two antagonist shape memory alloy wires. Phis. Procedia 10:182-188

[35] Hsu T (2002) Miniaturization – A paradigm shift in advanced manufacturing and education. Proc. IEEE/ASME

[36] Feldmann K, Franke J, Schüßler F (2010) Development of micro assembly processes for further miniaturization in electronics production. CIRP Annals –Manufact. technol. 59:1–4

[37] Kathuria YP (2005) Laser microprocessing of metallic stent for medical therapy. J. of Mater. Process. Technol. 170:545-550

[38] Meng H, Liao J, Zhou Y, Zhang Q (2009) Laser micro-processing of cardiovascular stent with fiber laser cutting system.Opt. And laser technol. 41:300– 302

[39] Previtali B, Arnaboldi S, Bassani P, Biffi CA, Lecis N, Tuissi A, Carnevale M, LoConte A (2010) Microcutting of NiTiCu alloy with pulsed fiber laser.ISBN 978-0-7918-3877-8, Order No. I844CD

[40] BiffiC A, Bassani P, Tuissi A, Carnevale M, Lecis N, LoConte A, Previtali B (2011) Flexural Vibration Suppression of Glass Fiber/CuZnAl SMA Composite. Funct. Mater. Lett., DOI 10.1142/S1793604712500142

[41] Raval A, Choubey A, Engineer C, Kothwala D (2004) Development and assessment of 316LVM cardiovascular stents. Mater.sci. and eng. A.386:331–343.

[42] Kudesia SS, Solana P, Rodden W, Hand DP, Jones J (2005) Appropriate Regimes of Laser Drilling Models Containing Melt Eject Mechanisms. Journal of Laser Applications, 14:159-164.

[43] Rao BT, Kaul R, Tiwari P, Nath AK (2005) Inert gas cutting of titanium sheet with pulsed mode CO2 laser. Opt. and Lasers in Eng. 43:1330–1348

[44] Biffi CA, Lecis N, Previtali B, Vedani M, VimercatiG (2010) Fiber laser microdrilling and its effect on material microstructure. The int. j. of adv. manuf. Technol. 51:983-994

[45] Mazzolai FM, Biscarini A, Coluzzi B, Mazzolai G, Villa E, Tuissi A (2007) Low frequency internal friction of hydrogen-free and hydrogen-doped NiTi alloys. Acta mater.55:4243-425

[46] Biscarini A, Coluzzi B, Mazzolai G, Mazzolai FM, Tuissi A (2003) Mechanical spectroscopy of H-free and H-doped NiTiCu shape memory. J. of alloy and compd.356:669-672

Chapter 2

Optimum Mechanical Design of Binary Actuators Based on Shape Memory Alloys

A. Spaggiari, G. Scirè Mammano and E. Dragoni

Additional information is available at the end of the chapter

1. Introduction

Shape memory alloys are smart materials which have the ability to return to a memorized shape when heated. When an SMA is below its transformation temperature (martensitic phase), it has a low yield strength and can be deformed quite easily and behaves like a pseudoplastic solid. When the deformed material is heated above its transformation temperature there is a change in its crystal structure which causes the return to its original shape (austenitic phase). During this transformation the SMA element can generate a net force, behaving like an intrinsic actuator. The most common shape memory material is a nickel and titanium alloy called Nitinol [1]. SMAs have very good electrical and mechanical properties, high corrosion resistance and biocompatibility. When an electric current is injected in the SMA element, it can generate enough heat to cause the phase transformation due to joule effect. Thanks to their unique behaviour shape memory alloys have become a valuable industrial choice in the engineering world. Pseudo-plasticity, superelasticity, and shape memory effect [2] are increasingly used in many applications including actuators, constant-force springs, and adaptive damping systems. While the application of the superelastic effect is quite well established and understood for the manufacturing of medical devices with peculiar properties, the use of the shape memory effect for building solid state actuators is still characterized by a trial-and-error approach. Although the thermo mechanical phenomena behind the behaviour of SMAs are theoretically well known [3] - [4], there is an open challenge for engineering methods to assist the designer in exploiting these alloys for the development of industrial devices.

1.1. Design methodology review for SMA actuators

Many SMAs applications have been studied in recent years. Kuribayashi [5] proposes a rotary joint based on a bending SMA actuator, while, in [6] design and applications of SMA actuators are presented. Microrobots can be developed using SMA as shown in [7] where there is a basic method to design the SMA spring based on a thermo-electromechanical approach. Reynaerts et al. [8] present design considerations concerning the choice of the active element to evaluate SMA actuator efficiency. Lu et al. [9] design a high strain shape memory actuator taking into account pseudoplasticity and compared its performance with traditional actuators. Due to SMA high non linearity, design curves relying on experiments are proposed in [10] to assess the SMA actuator geometry. A comprehensive review of applications of SMAs in the field of mechanical actuation has recently been published in [11]. Jansen et al [12] develop a linear actuator used as a drive module in an angular positioning mini-actuator. This architecture allows both large force and long strokes to be obtained. Strittmatter et al. [13] propose a SMA actuator for the activation of a hydraulic valve, biased by a conventional spring. Bellini et al. [14] propose a linear SMA actuator able to vary the air inflow for internal combustion engines, improving gas combustion and leading to higher efficiency. Haga et al. [15] propos a mini-actuator to be used in Braille displays. Elwaleed et al. [16] develop a SMA beam actuator able to amplify the SMA actuator strain using elastically instable beams. Among the proposed actuator there is lack of simple design instruments to provide basic information to the designer, either due to specific constraints of application or due to the high complexity of the thermomechanical material models used. In order to answer for an analytical design methodology, the author described in several technical publications a set of equations useful for linear [18] and rotary application [19]. The authors developed the design equations both for SMA actuators under a general system of external forces [20] and produced the design formulae to increase the output stroke thanks to negative stiffness compensators [21]. Moreover two peculiar systems were designed and developed: a telescopic actuator [22] and a wire on drum system [23]. The present work reviews and improves the design rules developed by the authors and set them in a coherent formal analytical framework. Design examples are provided to illustrate the step-by-step application of the design optimization procedures for realistic case studies.

1.2. Challenges and issues in SMA actuators design

The three main challenges in SMA actuator design are: obtaining a simple and reliable material model, increasing the stroke of the actuator and finding design equations to guide the engineer in dimensioning the actuator. To design SMA actuators, a material model must describe the mechanical behaviour of the alloy in two temperature ranges: below the temperature M_f, at which the austenite-martensite transformation is finished (OFF or deactivated or cold state) and above the temperature, A_f, at which the martensite-austenite transformation is completed (ON or activated or hot state). In these two conditions, inside the shape memory material there is only one stable crystalline phase and therefore the macroscopic mechanical properties are known.

The authors proposed two simple material models to describe SMA behaviour, both of them describing the mechanical behaviour of a SMA element at high temperatures (austenitic phase) as linear, characterized by the elastic modulus E_A. The first one approximates the martensitic behaviour as a linear one [18] (Figure 1a). This behaviour is by no means obvious, thus an explanation is needed. The typical stress-strain curve of a shape memory material, shown in Figure 1a, has two characteristic paths corresponding to a low temperature (martensitic curve) and to a high temperature (austenitic curve). Both curves consist of an initial elastic portion (OA and OC), followed by a constant-stress plateau. In the practical use of shape memory alloys for making actuators, the materials remain within the linear elastic range at the higher temperature but are strained beyond the elastic limit when cooled to the lower temperature. The maximum strain, ε_{adm} is small enough to ensure the desired fatigue life [24-25] but large enough to maximize the stroke of the actuator for a given amount of material involved. Since this analysis is aimed at dimensioning the binary actuator for the extreme positions, irrespective of the intermediate state, the behaviour of the material is approximated by segments OA′ (martensite) and OC (austenite) in Figure 1a. The reference elastic moduli are E_M and E_A, respectively. This is also true if the shear behaviour of the shape memory material is considered, when the maximum shear strain, γ_{adm}, replaces ε_{adm} and the shear moduli G_M and G_A replace E_M and E_A.

The second material model describes the behaviour at cold temperatures (martensitic phase) with a bilinear law [21] as shown in the stress-strain diagram of Figure 1b. The model is defined by a first leg OD with an elastic modulus E_{MA} and a second leg DE with a gradient E_{MB}. We define ε_g as the deformation of the occurrence of the change of slope. Due to the bilinear stress-strain response, the SMA elements used in the actuators (springs or wires) also have bilinear force-displacement behaviour when disabled. The elastic moduli E_A, E_{MA} and E_{MB} are replaced by the stiffness K_A, K_{MA}, K_{MB}, while the deformation ε_g is replaced by the displacement x_g. Assuming that the geometric changes related to the deformations of the springs do not influence the elastic constant value, a parameter of merit of the SMA material, s_1, can be defined as follows:

$$s_1 = \frac{E_A}{E_M} = \frac{G_A}{G_M} = \frac{k_a}{k_m} \tag{1}$$

This non dimensional group expresses for both models the shape memory capability of the alloy, the larger s_1 the better the material is. The only difference is that in case of the simplest linear model (Figure 1a) the denominator is the secant modulus, while in case of bilinear model (Figure 1b) the denominator is E_{MA} the modulus of the first linear martensitic region. In order to balance the active SMA element authors evaluates the influence of three backup elements: a constant force (Section 3.1) a conventional spring (Section 3.2) and an antagonistic SMA (Section 3.3). Moreover the authors propose two compensator systems in order to increase the stroke of the actuator. The systems are either based on a leverage (rocker-arm, Section 4.2.1) or on articulated mechanisms (double quadrilateral, Section 4.2.2), which can be considered as backup elements with negative stiffness.

2. Design procedure of binary SMA actuators

In this section a design procedure for binary SMA actuators is described and discussed. The system is made up by a generic actuator which moves the output port by means of an elastic system containing an active SMA element and a bias (backup) element.

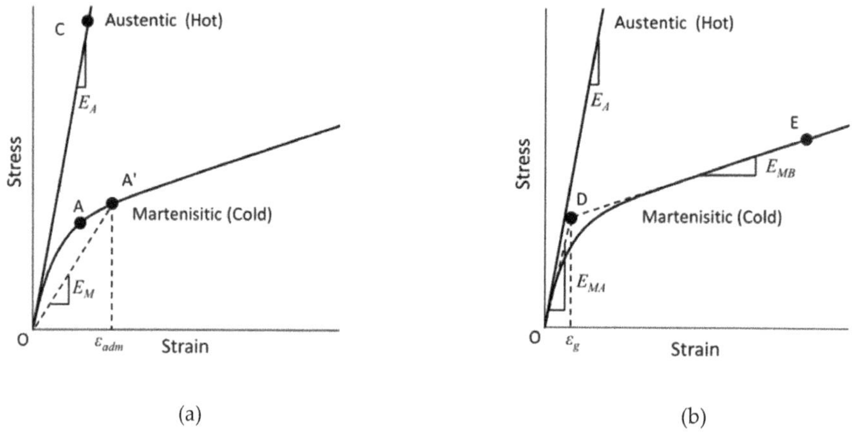

Figure 1. Linear (a) and bilinear (b) material model for martensitic phase of SMA elements

According to the particular means of applying the bias force, the three cases shown in Figure 2 are analyzed:

- a primary SMA spring biased by a constant force (Figure 2a);
- a primary SMA spring biased by a traditional spring (Figure 2b);
- two antagonist SMA springs (when one is hot, the other one is cold, Figure 2c).

Without loss of generality, each spring in Figure 2 is modeled as a traction spring exhibiting a linear force-deflection relationship. While the assumption of linearity is obvious for the traditional spring in Figure 2b, the linear behaviour for the SMA spring is an approximation needed to use the model in Figure 1a.

Each actuator presented in Figure 2 is intended to move the output port E through a total useful stroke Δx when working against an external dissipative force F_F and an external conservative force F_0. The force F_F is always opposite to the velocity of the cursor and is assumed to be constant. For example, if the cursor is subject to dry friction forces characterized by a static value F_S and a dynamic value $F_D < F_S$, the design dissipative force $F_F = \text{MAX}(F_S, F_D) = F_S$ will be adopted for the calculation.

This approach comprises every possible external constant load, as exemplified by the membrane pump shown in Figure 3a. This example represents the most general case of external constant forces the actuator has to deal with. The SMA actuator undergoes the following external loads: two generic dissipative forces F_1 (force during aspiration) and F_2 (force during pumping) and a conservative force F_u due to the gravity force on the piston. The dissipative force F_1 and the force F_u act together against the primary spring, when the piston is moving towards the primary SMA spring (inlet of the fluid). By contrast, the dissipative force F_2 acts on the cursors against the force F_u and the primary spring, when the cursor moves away from the primary SMA spring (outlet of the fluid). This force system can be always reduced to the two forces considered in the method: a conservative one, F_0 and two symmetric dissipative ones, F_F, rearranging the forces as follows:

$$F_0 = F_u + \frac{F_1 - F_2}{2} \quad , \quad F_F = \frac{F_1 + F_2}{2} \tag{2}$$

The direction of the forces depends on the piston speed, v, as shown in Figure 3b-c.

A second non dimensional parameter, s_2, useful in the configurations with two springs, is defined as the ratio between the minimum value assumed by the stiffness of bias spring 2 and the stiffness of active spring 1 in the cold state:

$$s_2 = \frac{K_{2\min}}{K_{1SC}} \tag{3}$$

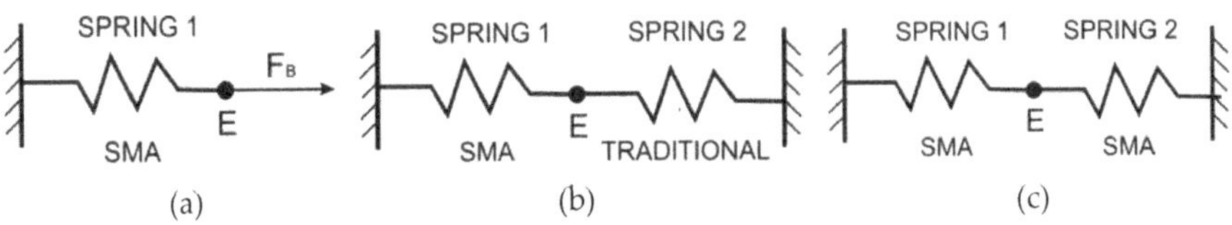

Figure 2. Three cases of the shape memory actuator biased by: a constant force (a), a traditional spring (b), a shape memory spring (c).

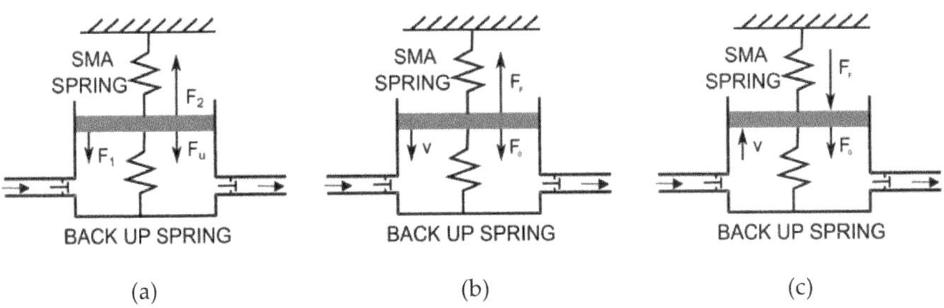

Figure 3. Example of generic conservative and dissipative forces acting on the system (a), equivalent forces when SMA is inactive (b) and when SMA is active (c)

The minimum value assumed by the stiffness of spring 2, K_{2min}, coincides with the only stiffness of spring 2, K_{2C}, if spring 2 is a traditional one, while it coincides with K_{2SC} if spring 2 is an active one. A third dimensionless parameter, s_F, is introduced in order to consider the influence of the dissipative forces in the motion of the SMA actuator. This parameter is defined as the ratio between the dissipative force F_F and the maximum force sustained by the primary spring in the cold state, calculated as the product between the cold spring stiffness, K_{1SC}, and the maximum deflection, L_C-L_{01}, in the cold state:

$$s_F = \frac{F_F}{K_{1SC} \cdot (L_C - L_{01})} \tag{4}$$

The fourth dimensionless parameter, s_0, is introduced in order to consider the influence of the conservative force in the motion of the SMA actuator. This parameter is defined as the ratio between the conservative force F_0 and the maximum force sustained by the primary spring in the cold state, calculated as the product between the cold spring stiffness, K_{1SC}, and the maximum deflection, L_C-L_{01}, in the cold state:

$$s_0 = \frac{F_0}{K_{1SC} \cdot (L_C - L_{01})} \tag{5}$$

The combination of definitions (4) and (5) gives the relationship between s_F and s_0:

$$s_0 = \frac{F_0}{F_F} \cdot s_F \tag{6}$$

2.1. SMA actuator backed up by a constant force

Figure 4a shows the actuator biased by a constant force in three characteristic positions. Figure 4b shows the relationship between the applied force and the spring deflection during the travel of the actuator between these positions.

The procedure can be retrieved from [20]. The useful stroke of the actuator is obtained as:

$$\Delta x = \frac{F_B}{K_{1SC}} \cdot \frac{(s_1 - 1 - 2s_F)}{s_1 \cdot (s_F + 1 - s_0)} \tag{7}$$

The bias force is:

$$F_B = F_F \cdot \frac{s_F + 1 - s_0}{s_F} \tag{8}$$

Thus, the stroke can be cast as:

$$\Delta x = (L_C - L_{01}) \frac{s_1 - 1 - 2s_F}{s_1} \tag{9}$$

Eq. (7) shows that meaningful strokes ($\Delta x > 0$) are only possible if parameter s_F is lower than the following critical value.

$$s_{Fcr} = \frac{s_1 - 1}{2} \tag{10}$$

Moreover, a possible way to choose the value of s_F consists on studying the free length of the active spring, L_{01}, and its wire diameter, d_1, as described below in eq. 14.

2.1.1. *Embodiment of SMA actuator*

If helical springs or wires are considered, two fundamental expressions can be written for the diameter or the wire and for the length of the springs, as a function of the stiffness, K and of the deflection of the SMA element, f.

$$d = m_d \cdot \sqrt{K \cdot f} \tag{11}$$

$$L_0 = m_l \cdot f \tag{12}$$

The constant m_d and m_f depends on the embodiment and can be calculated as shown in the examples in Section 3.1.2 and Section 3.2.2.

A generic expression of the free length, L_{01}, can be written as follows:

$$L_{01} = m_l \cdot \Delta x \cdot \frac{s_1}{s_1 - 2 \cdot s_F - 1} \tag{13}$$

In (13) the free length, L_{01}, depends on the particular embodiment of the spring, but it is always minimized if s_F is chosen as small as possible. The minimum value of s_F can be determined by fixing the maximum allowable wire diameter, d_l, of the SMA element. This maximum value can be determined, for example, using cooling time considerations, the bigger the wire, the slower the cooling, [8], [18].

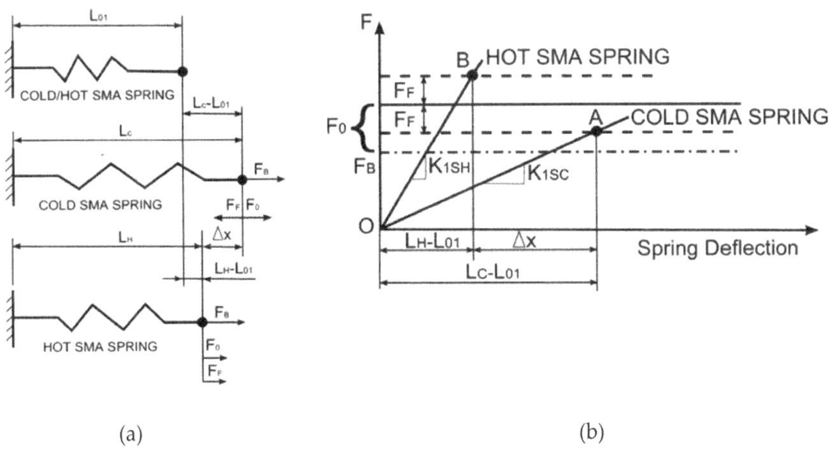

Figure 4. Actuator model (a) and force-deflection diagram (b) of the shape memory actuator biased by a constant force

Combining expression (11) with (1) and (3), the following relationship can be used to determine the minimum value of s_F.

$$d_1 = m_{d1} \cdot \sqrt{\frac{F_F}{s_F}} \tag{14}$$

Where m_{d1} depends only on the embodiment of the SMA primary spring.

2.1.2. *Case study: SMA wire based actuator*

The equations (11) and (12) represent the relationships between stiffness, K, and free length, L_0, either for a spring or a wire.

In particular the stiffness of a SMA actuator in wire form is obtained using the coefficient:

$$m_{dw} = \sqrt{\frac{4}{\pi \cdot E_m \cdot \varepsilon_{adm}}} \tag{15}$$

where it is shown that the limit is given by the maximum strain of the SMA wire in martensitic phase. The free length is obtained from eq. (11) by considering the following coefficient for a SMA wire, which comes out simply from the definition of axial strain in a rod.

$$m_{lw} = 1/\varepsilon_{adm} \tag{16}$$

2.2. SMA actuator backed up by an elastic spring

Figure 5a shows the actuator biased by a traditional spring in three characteristic positions. Figure 5b displays the relationship between the applied force and the deflection of each spring during the thermal operation of the actuator between those positions.

The detailed procedure can be retrieved from [20]. The pre-stretch is

$$p = \left(L_C - L_{01}\right)\frac{s_2 + s_F + 1 - s_0}{s_2} \tag{17}$$

$$\Delta x = p\,\frac{s_2 \cdot \left(s_1 - 2 \cdot s_F - 1\right)}{\left(s_1 + s_2\right) \cdot \left(s_2 + s_F + 1 - s_0\right)} \tag{18}$$

The following alternative expression of the overall travel can be written

$$\Delta x = \left(L_C - L_{01}\right)\frac{s_1 - 1 - 2s_F}{s_1 + s_2} \tag{19}$$

Lastly, Figure 5b shows that the maximum deflection in the cold state of the bias spring amounts to $p-(L_H-L_{01})$. This expression can be written as:

$$p-\left(L_H-L_{01}\right)=\left(L_C-L_{01}\right)\cdot\frac{s_1\cdot\left(s_2+s_F+1\right)-s_2\cdot s_F-s_0\cdot\left(s_1+s_2\right)}{s_2\cdot\left(s_1+s_2\right)} \tag{20}$$

2.2.1. *Embodiment of the SMA actuator*

If helical springs or wires are considered, the two fundamental expressions (11) and (12) can be written for the diameter of the wire and for the length of each spring. Following [20] a generic expression of the total length can be obtained, which depends on the dimensionless parameters s_1, s_2, s_0 and s_F and on the embodiment of the springs.

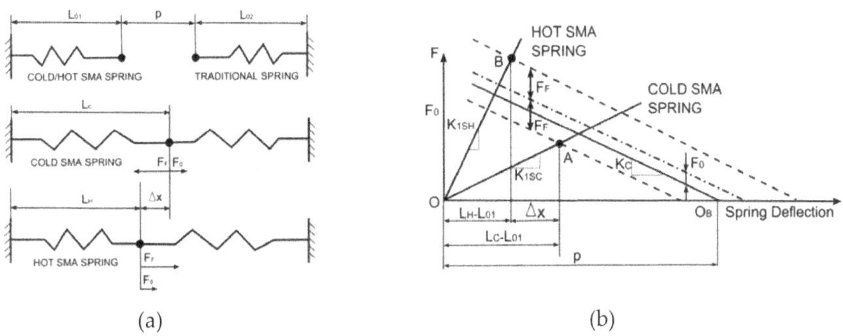

Figure 5. Actuator model (a) and force-deflection diagram (b) of the shape memory actuator biased by a traditional spring

This expression is minimized if s_F is chosen as small as possible and if a precise value of s_2^* is chosen, obtained by equating to zero the derivative of the total length of the actuator. The optimal value s_2^* can be expressed as:

$$s_2^*=\sqrt{\frac{s_1\cdot\left(m_{l2}+1\right)\cdot\left(s_F+1-s_0\right)}{\left(m_{l1}+1\right)}} \tag{21}$$

As a first attempt, it is possible to determine the m_l coefficients considering the desired embodiment of each spring and a value of $C = 7$, because the value of s_2^* is not greatly affected by C.

2.2.2. *Case study: SMA spring based actuator*

The equations (11) and (12) represent the relationships between stiffness, K, and free length, L_0, either for a spring or a wire.

In particular the wire diameter of a helical spring can be cast as follows [26]:

$$d=\sqrt{\frac{4\cdot K\cdot f\cdot K_b}{\pi\cdot G\cdot\gamma_{adm}}} \tag{22}$$

Were K is the stiffness of the spring, f is the deflection of the spring and K_b is the coefficient of Bergstrasser, given by the following relationship:

$$K_b = \frac{4 \cdot C + 2}{4 \cdot C - 3} \tag{23}$$

Thus for a spring the m_{ds} coefficient in expression (11) can be expressed as:

$$m_{ds} = 4\sqrt{\frac{(C+2)}{\pi \cdot G \cdot \gamma_{adm} \cdot (4C-3)}} \tag{24}$$

The external diameter of a helical spring can be calculated using the definition of the spring index ($C = D/d$) and the number of active coils is given by:

$$N = \frac{G \cdot d}{8 \cdot C^3 \cdot K} \tag{25}$$

The fully compressed length can be calculated using expression:

$$L_{fc} = Nd = \frac{G \cdot d^2}{8 \cdot C^3 \cdot K} \tag{26}$$

The free length is:

$$L_0 = 1.15 \cdot Nd + f = 1.15 \cdot \frac{G \cdot f \cdot d^2}{8 \cdot C^3 \cdot F} + f \tag{27}$$

Thus the coefficient to calculate the free length in eq. (12) is:

$$m_{ls} = \frac{1.15 + \pi \cdot \gamma_{adm} \cdot C^2}{\pi \cdot \gamma_{adm} \cdot C^2} \tag{28}$$

2.3. SMA actuator backed up by an antagonist SMA spring

Figure 6a shows the actuator biased by a second SMA spring in three characteristic positions. Figure 6b shows the relationship between the applied force and the deflection of each spring during the thermal operation of the actuator between these positions.

Following the procedure described in [20], the prestretch p can be written:

$$p = (L_C - L_{01}) \cdot \frac{s_1 \cdot s_2 + s_F + 1 - s_0}{s_1 \cdot s_2} \tag{29}$$

The stroke can be cast as:

$$\Delta x = p \cdot \frac{s_2 \cdot \{(s_1 - 1 - s_F) \cdot (s_1 + 1) - s_0 \cdot (s_1 - 1)\}}{(s_1 + s_2) \cdot (s_1 \cdot s_2 + s_F + 1 - s_0)} \tag{30}$$

The following expression for the optimal value s_2^* is found:

$$s_2^* = \sqrt{s_F + 1 - s_0} \tag{31}$$

The following alternative expression of the overall travel can be obtained:

$$\Delta x = \left(L_C - L_{01}\right) \cdot \frac{\left(s_1 - 1 - s_F\right) \cdot \left(s_1 + 1\right) + s_0 \cdot \left(s_1 - 1\right)}{s_1 \cdot \left(s_1 + s_2\right)} \tag{32}$$

Both Equations (30) and (32) demonstrate that meaningful strokes ($\Delta x > 0$) are only possible if parameter s_F is lower than the following critical value.

$$s_{Fcr} = \frac{F_F \cdot \left(s_1 - 1\right)^2}{F_0 \cdot \left(s_1 - 1\right) + F_F \cdot \left(s_1 + 1\right)} \tag{33}$$

Figure 6b shows that the maximum deflection of the bias spring in the cold state amounts to $p - \left(L_H - L_{01}\right)$. This expression can be written as:

$$p - \left(L_H - L_{01}\right) = \left(L_C - L_{01}\right) \cdot \frac{s_1 \cdot s_2 - s_2 \cdot s_F + s_F + 1 - s_0 \cdot \left(s_2 + 1\right)}{s_2 \cdot \left(s_1 + s_2\right)} \tag{34}$$

2.3.1. *Embodiment of SMA antagonist actuator*

If helical springs or wires are considered, the two fundamental expressions (11) and (12) can be written for the diameter of the wire and for the length of each spring. Following [20] a generic expression of the total length can be obtained, which depends on the dimensionless parameters s_1, s_2, s_0 and s_F and on the embodiment of the springs.

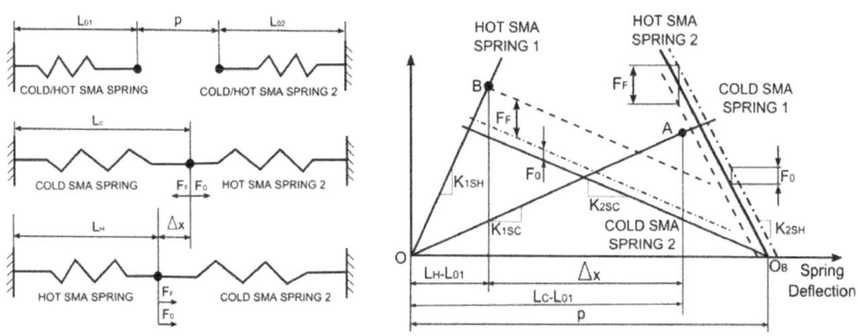

(a) (b)

Figure 6. Actuator model (a) and force-deflection diagram (b) of the agonist antagonist shape memory actuator

This expression is minimized if s_F is chosen as small as possible and if an optimal value for s_2^* is chosen, obtained equating to zero the derivative of the total length of the actuator. The optimal value s_2^* can be expressed as:

$$s_2^* = \sqrt{\frac{(m_{l2}-1)\cdot(s_F+1-s_0)}{(m_{l1}-1)}} \tag{35}$$

Expression (35) reduces to expression (31) if the same spring embodiment is considered for spring 1 and spring 2. As a first attempt, it is possible to determine the m_l coefficients considering the desired embodiment of each spring and a value of $C = 7$, because the value of s_2^* is not greatly affected by C. The minimum value of s_F can be determined by fixing the maximum allowable wire diameter, d_l, of the SMA element. This maximum value can be determined, for example, using cooling time considerations. Relationship (14) can be used to determine the minimum value of s_F.

2.4. Design procedures for a binary SMA actuator

The step by step procedures which guide the designer to apply the above design methods are described in Table 1. The first four steps are the same irrespective of the backup element used in the actuator. From step 5, the procedure is threefold.

These procedures ensure that any actuator biased by a one of the above described back up elements, with the calculated stiffness K_C and containing whatever SMA spring, with the selected material parameter s_1 and the calculated cold stiffness K_{ISC}, satisfies the design problem (useful stroke Δx, design dissipative force F_F and design conservative force F_0) when assembled with the calculated pre-stretch p.

2.4.1. Case study: SMA based swing louver

In this section, the complete design procedure is carried out numerically for the actuation of a swing louver exploited to direct the air flow in domestic air conditioners. An actuator made up of an SMA spring biased by an antagonist SMA spring is designed here as a possible alternative solution to conventional electric motors and linkages. The SMA actuator acts on the louver with a known arm to make the louver swing. The design parameters are: required stroke: 5mm; dissipative force: 5N, conservative force: 2N. The material considered is Nitinol, with the following properties, γ_{adm} = 0.02, to ensure a fatigue limit over 500 thousand cycles, austenitic shear modulus G_a=23000 MPa, equivalent martensitic shear modulus G_m=8000 MPa.

The procedure starts calculating the non dimensional groups from eq. (1), (10), (31) and (5):

$$s_1 = \frac{G_A}{G_M} = \frac{23000}{8000} = 2.875 \tag{36}$$

<table>
<tr><th>Backup: Constant Force (Section 3.1)</th><th>Backup: Traditional spring (Section 3.2)</th><th>Backup: antagonist SMA spring (Section 3.3)</th></tr>
<tr><td colspan="3">—the stroke of the actuator, Δx, the design dissipative force, F_F, and the design conservative force, F_0, are accepted from design specifications;
—a shape memory material for the spring and a maximum strain, ε_{adm} or γ_{adm}, are selected, after which parameter $s_1 = E_A/E_M = G_A/G_M$ is known;
—The spring embodiment is chosen and the parameters ml_1, ml_2 and md_1 are calculated as shown in Section 3.1.2 or 3.2.2. Otherwise, this step is skipped;
—a s_F value lower than the critical value (10) is selected, possibly using (14);</td></tr>
<tr><td>—the value of s_0 is calculated using definition (6);
—the bias force, F_B, is calculated using eq. (8);
—the maximum deflection of the primary spring, L_C-L_{01}, is retrieved from (9);
—the cold stiffness, K_{1SC}, of the spring is calculated from eq. (7);</td><td>—the value of s_0 is calculated using definition (6);
—the lowest affordable value of parameter s_2 is chosen (to exploit the primary spring), or the optimal value (21) is adopted (to conserve overall space). If the parameters ml_1 and ml_2 cannot be determined, expression (31) can be used for a sub-optimal design;
—the maximum deflection of the primary spring, L_C-L_{01}, is retrieved from (19);
—the cold stiffness, K_{1SC}, of the spring is calculated from eq. (4);
—the pre-stretch of the system, p, is calculated from (17);
—the stiffness of the traditional spring, K_C, is calculated from eq. (3);
—the maximum deflection of the traditional spring, p-(L_H- L_{01}), is calculated from (20).</td><td>—a s_F value lower than the critical value (10) is selected. The parameter is chosen as small as possible to reduce the size of the actuator, respecting the constraint of the diameter of the wire, d, given by (14);
—the value of s_0 is calculated using definition (6);
—the lowest affordable value of parameter s_2 is chosen (to exploit the primary spring), alternatively the optimal value (35) is adopted (to conserve overall space). If the parameters ml_1 and ml_2 cannot be determined, expression (31) can be used for a sub-optimal design;
—the maximum deflection of the primary spring, L_C-L_{01}, is retrieved from (32);
—the cold stiffness, K_{1SC}, is calculated from eq. (4);
—the overall pre-stretch of, p, is calculated from (29);
—the stiffness of the secondary SMA spring, K_{2SC}, is calculated from eq. (3);
—the maximum deflection of the secondary SMA spring, p-(L_H- L_{01}) , is calculated from (34).</td></tr>
</table>

Table 1. Step by step procedure for each Backup element considered

$$s_{Fcr} = \frac{s_1 - 1}{2} = \frac{2.875 - 1}{2} = 0.9375 \tag{37}$$

$$s_F = 0.25 \le s_{Fcr} \tag{38}$$

$$s_2 = s_2^* = \sqrt{s_1 \cdot (s_F + 1)} = \sqrt{2.875 \cdot (0.25 + 1)} = 1.896 \tag{39}$$

$$s_0 = \frac{F_0}{F_F} \cdot s_F = \frac{2}{5} \cdot 0.25 = 0.1 \; s_0 = \frac{F_0}{F_F} \cdot s_F = \frac{2}{5} \cdot 0.25 = 0.1 \tag{40}$$

the maximum deflection of the primary spring, L_C-L_{01}, is retrieved from (32);

$$\begin{aligned}(L_C - L_{01}) &= \Delta x \cdot \frac{s_1 \cdot (s_1 + s_2)}{(s_1 - 1 - s_F) \cdot (s_1 + 1) + s_0 \cdot (s_1 - 1)} = \\ &= 5 \cdot \frac{2.875 \cdot (2.875 + 1.896)}{(2.875 - 1 - 0.5) \cdot (2.875 + 1) + 0.1 \cdot (2.875 - 1)} = 12.43\, mm\end{aligned} \tag{41}$$

while eq. (4) allows the SMA spring cold stiffness to be calculated

$$K_{1SC} = \frac{F_F}{s_F \cdot (L_C - L_{01})} = \frac{5}{0.25 \cdot 12.43} = 1.61 N / mm \tag{42}$$

the overall pre-stretch of, p, is calculated from (29) and K_C, is calculated from eq. (3)

$$p = (L_C - L_{01}) \cdot \frac{s_1 \cdot s_2 + s_F + 1 - s_0}{s_1 \cdot s_2} = 12.43 \cdot \frac{2.875 \cdot 1.896 + 0.25 + 1 - 0.1}{2.875 \cdot 1.896} = 15.01 mm \tag{43}$$

$$K_{2\min} = K_{2SC} = s_2 \cdot K_{1SC} = 1.896 \cdot 1.61 = 3.052\, N / mm \tag{44}$$

$$\begin{aligned}p - (L_H - L_{01}) &= (L_C - L_{01}) \cdot \frac{s_1 \cdot s_2 - s_2 \cdot s_F + s_F + 1 - s_0 \cdot (s_2 + 1)}{s_2 \cdot (s_1 + s_2)} = \\ &= 12.43 \cdot \frac{2.875 \cdot 1.896 - 1.896 \cdot 0.25 + 0.25 + 1 - 0.1 \cdot (1.896 + 1)}{1.896 \cdot (2.875 + 1.896)} = 8.16\, mm\end{aligned} \tag{45}$$

the maximum deflection of the secondary SMA spring, p-(L_H-L_{01}), is calculated from eq. (34).

The detailed design of primary SMA spring is given by combination of eq. (11) and (24) regarding the spring wire diameter, considering a spring index C=7

$$d = 4\sqrt{\frac{(C+2) K_{1SC} (L_C - L_{01})}{\pi \cdot G \cdot \gamma_{adm} \cdot (4C - 3)}} = 4\sqrt{\frac{(7+2) \cdot 1.61 \cdot 12.43}{\pi \cdot 8000 \cdot 0.02 \cdot (4 \cdot 7 - 3)}} = 0.48\, mm \tag{46}$$

the free length of the primary spring is given by the combination of eq. (12) and (28)

$$L_0 = \frac{1.15 + \pi \cdot \gamma_{adm} \cdot C^2}{\pi \cdot \gamma_{adm} \cdot C^2} \cdot \left(L_C - L_{01} \right) = \frac{1.15 + \pi \cdot 0.02 \cdot 7^2}{\pi \cdot 0.02 \cdot 7^2} \cdot 12.43 = 17.07\, mm \tag{47}$$

The same procedure is applied to calculate the design parameter of the antagonist spring.

3. Design of compensated SMA actuators

SMA based actuators develop significant forces but are usually characterized by low strokes. The stroke is primarily limited by the maximum strain that the alloy can withstand for the expected life as shown in [24], [25]. The backup element needed to recover the stroke prevents the SMA element from recovering completely its shape is a further cause of stroke loss. Furthermore, the force delivered by shape memory actuators varies linearly with the displacement while the external load is usually constant. The design of a SMA actuator ensures that the minimum actuator force is sufficient to contrast the external load [18], providing an additional cause of stroke reduction.

This section introduces a compensation system to store available power from the SMA element in high force positions and then return this power in low force positions [19]. The same principle was successfully applied for electroactive polymer actuators by introducing compliant mechanisms [27]. The compensation system adopted has a negative elastic characteristic, generating a decreasing force as the deformation increases.

3.1. Principle of elastic compensation

The force-deflection diagram in Figure 7 shows the characteristic lines of an SMA actuator in the austenitic (hot SMA, solid circles) and martensitic (cold SMA, solid squares) states. When the SMA element is backed up by a linear spring with the linear characteristic shown with hollow triangles, the net stroke under no external force is S_{spring}. Starting from the same maximum deflection and force of the cold SMA, the stroke increases to S_{weight} when the active element is backed up by a constant force (crossed horizontal line). The improvement is consequent upon the reduced stiffness (lower contrasting forces) of the backup element which allows the SMA to recover a greater share of deformation.

By learning from this positive trend, it is easily seen that an even greater stroke (S_{comp}) is achieved if the backup element displays a negative slope (hollow circles) so that the contrast force would decrease with increasing deflection. Energetically, the compensation system accumulates energy from the SMA element in the positions where the SMA force is high (right-hand side in Figure 7) and releases that energy to the actuators in the positions where the SMA force is low (left-hand side in Figure 7).

As shown in the subsequent sections, a backup element with negative slope as in Figure 7 can be achieved by exploiting one of the many spring-assisted mono or bistable mechanisms described in the technical literature. The use of an elastic compensation system requires the introduction of hard stops to prevent the SMA elements from over-straining. In the case of a single-SMA actuator (see Figure 7), a single hard stop is needed and the behaviour in the absence of power becomes monostable. In the case of a double-SMA actuator, two hard stops are required and the behaviour in the absence of power becomes bistable. Although the advantages of the compensation system in terms of force and stroke apply to both single-SMA and two-SMA actuators, the improvements are more pronounced for the two-SMA actuator. The general theory [21] demonstrates that a single-SMA actuator can generate a truly constant output force on only one direction of motion. By contrast, a two-SMA actuator can achieve a constant-force profile in both directions. Further advantages of the compensated architecture over regular SMA actuators are the existence of definite equilibrium positions when the power is shut off, the enforcement of precise mono (single-SMA) or binary (two-SMA) positioning, and the possibility of easy control strategies. The input data required for the design of a shape memory actuator are the value of the guaranteed minimum useful force in the two directions of activation (F_{ON}, F_{OFF} in the case of a single SMA element, F_{ON1}, F_{ON2} when there are two opposing SMA elements), the value of the stroke desired, S, and the type of alloy used for the active elements (i.e. s_l, s_m, s_g).

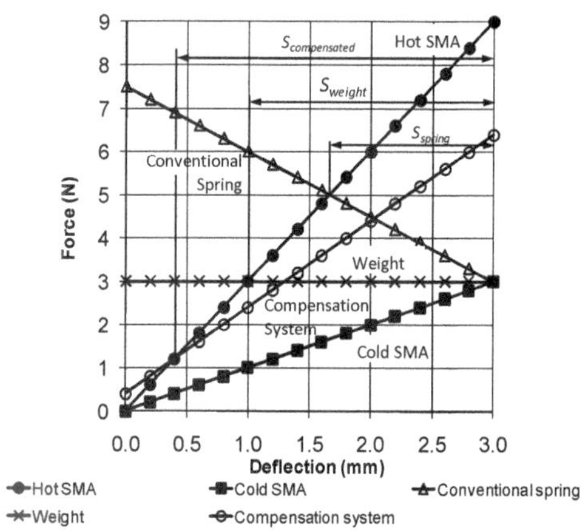

Figure 7. Force-deflection curves of a single-SMA actuator backed up by: a conventional spring, a constant force and an elastic compensation system

3.2. Material model: Definitions

In order to design compensated shape memory actuators the bilinear model for the martensitic state of SMA (Figure 1b) was used. Since ε_g (then x_g) is very small ($0.2\% < \varepsilon_g < 0.5\%$) the force-stroke characteristic of the SMA elements can be approximated as a linear trend (D'E in Figure 8), with slope K_{MB}, starting from the force F_{0m} in correspondence to zero displacement (Figure 8). The force F_{0m} is calculated as:

$$F_{0m} = \left(K_{MA} - K_{MB}\right)x_g \tag{48}$$

The force F_{SMA_ON} produced by the generic shape memory element in the activated state (austenite) is given by the line OC in Figure 8 and is:

$$F_{SMA\_ON} = K_A x \tag{49}$$

In the cold state (martensite), the force F_{SMA_OFF} is given by the line D'E in Figure 8:

$$F_{SMA\_OFF} = F_{0m} + K_{MB} x \tag{50}$$

In addition to s_1 already defined in eq. (1), to facilitate the design of the actuator, it is useful to define two other dimensionless coefficients:

$$s_m = \frac{E_{MB}}{E_{MA}} = \frac{K_{MB}}{K_{MA}} \tag{51}$$

$$s_g = \frac{\varepsilon_g}{\varepsilon_{adm}} = \frac{x_g}{x_{adm}} \tag{52}$$

The parameter s_m is the ratio between the two stiffness of the alloy at cold temperatures and it is a characteristic of the shape memory material. The parameter s_g is the ratio of the deflection, x_g, at which the change in stiffness in the martensitic state is recorded and the maximum allowable deflection x_{adm}, related to the maximum deformation, ε_{adm}, admissible for the material. To limit the size of the device, it is convenient to assume the $x_{maz} = x_{adm}$ so as to completely exploit the material.

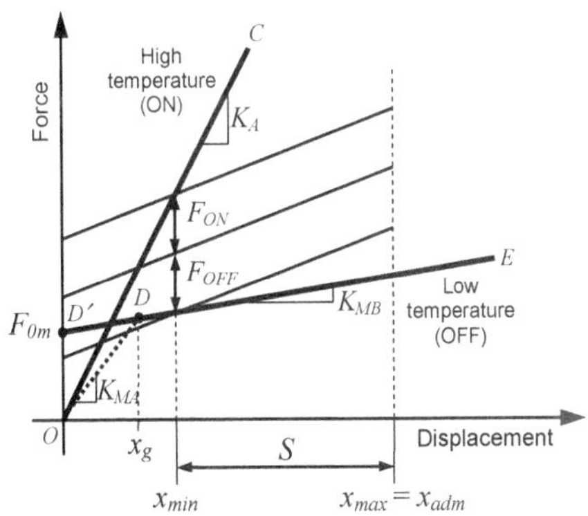

Figure 8. Force-deflection curves of a two-SMA actuator superimposed on the characteristic lines of the compensator.

3.3. Design of compensated single SMA actuators

To design compensated actuators with a single SMA element, it is convenient to introduce the ratio, γ, between the stroke and the maximum deflection of the SMA element:

$$\gamma = S / x_{adm} \tag{53}$$

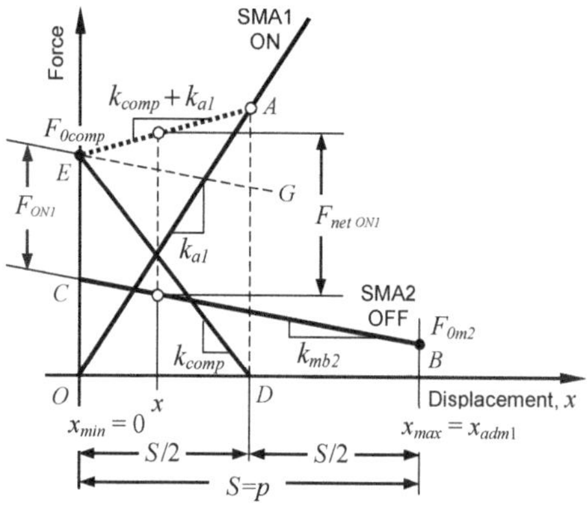

Figure 9. Force-deflection curves of a two-SMA actuator compensated by a generic negative elastic spring (line *ED*)

The axial dimension of the actuator is reduced as the ratio γ tends to one because in this case all the deformation that the material can sustain is used to produce a useful stroke. However, the value γ cannot reach the unity because the minimum deflection would be zero and the SMA element would not be able to exert any useful force.

The force delivered by the shape memory element can be calculated with (49), in the austenitic state, and with the approximated expression (50), in the martensitic state. Figure 8 shows that the high-temperature curve (OC) and the low-temperature curve (A'B) diverge. Thus, the most unfavorable position in which the design force is required is where the difference between the two curves (useful force) is minimal. In this position, the sum of the design forces in both directions must equal the distance between the ON straight line and the OFF straight line. Assuming conventionally $F_{ON} > 0$ (because it agrees with the force exerted by the SMA element) and $F_{OFF} < 0$ (because it is opposed to the force produced by the SMA) we have $(F_{ON} - F_{OFF}) = F_{SMA\_ON}(x = x_{min}) - F_{SMA\_OFF}(x = x_{min})$. Recalling (48),(49),(50) and introducing the dimensionless parameters (1),(51),(52) this condition becomes:

$$F_{ON} - F_{OFF} = \frac{k_{MA} S}{\gamma}\left[(s_1 - s_m)(1-\gamma) - s_g(1 - s_m)\right] \tag{54}$$

The stiffness needed to fulfill the minimum force in both directions and to achieve the desired stroke is obtained by solving (54) with respect to the stiffness k_{MA}:

$$k_{MA} = \frac{\gamma(F_{ON} - F_{OFF})}{S\left[(s_1 - s_m)(1-\gamma) - s_g(1 - s_m)\right]} \tag{55}$$

The next step is to set the desired characteristics of the compensation element in terms of stiffness (K_{comp}) and of force ($F_{comp}\big|_{x=x_{min}}$), delivered at position $x = x_{min} = S(1-\gamma)/\gamma$.

Target behaviour	Value of k_{comp}
Balanced force variation in ON and OFF modes	$-(K_{MB}+K_A)/2$
Constant force in OFF mode	$-K_A$
Constant force in ON mode	$-K_{MB}$

Table 2. Suggested values for the compensation stiffness to achieve specific behaviour of the actuator

From (1) and (55) the stiffness K_A of the SMA element in the austenitic state is obtained.

Though the compensation stiffness (always negative) could be set as desired in the range $-K_{MB} \div -K_A$, Table 1 shows how to choose the value of K_{comp} to obtain favourable operating characteristics: for $K_{comp} = -K_A$, the force of the activated actuator is constant and for $K_{comp} = -K_{MB}$ the force of the deactivated actuator is constant. By setting $K_{comp} = -0.5(K_A + K_{MB})$, the deviation from uniformity of both states of actuator is minimized.

The compensator force the at position $x = x_{min}$ to obtain the desired behavior must be:

$$F_{comp}\Big|_{x=x_{min}} = F_{ON} - K_A S\left(\frac{1-\gamma}{\gamma}\right) \tag{56}$$

The compensator always acts with a force increasing with x and opposite to the force delivered by the SMA element.

3.3.1. *Case study: single SMA compensated actuator*

In this section, the design procedure of a single SMA compensated actuator is carried out numerically for a generic system with the following technical specifications:

- Required stroke: 10mm
- Minimum force in the right to left displacement F_{ON}: 10N
- Minimum force in the left to right displacement F_{OFF}: -5N
- Minimization of the variation force in overall travel

The material considered is a Nitinol wire, with austenitic Young modulus E_a=75GPa, martensitic Young modulus E_{MA}=28GPa, bilinear gradient E_{MB}=5GPa, strain at the threshold of pseudo-plastic regime ε_g=0.4% maximum deformation ε_{adm}=4%.

The non dimensional groups are calculated as follows:

$$s_1 = \frac{E_A}{E_{MA}} = \frac{75}{28} = 2.68 \tag{57}$$

$$s_m = \frac{E_{MB}}{E_{MA}} = \frac{5}{28} = 0.18 \tag{58}$$

$$s_g = \frac{\varepsilon_g}{\varepsilon_{adm}} = \frac{0.004}{0.04} = 0.1 \tag{59}$$

An optimal value for the ratio γ=0.75 can be chosen as a tradeoff between axial dimension and oversizing of the active SMA elements.

From (55) it is possible to obtain the cold state stiffness of the SMA wire:

$$K_{MA} = \frac{\gamma(F_{ON} - F_{OFF})}{S\left[(s_1 - s_m)(1-\gamma) - s_g(1-s_m)\right]} = \frac{0.75\left[10-(-5)\right]}{10\left[(2.68-0.18)(1-0.75)-0.1(1-0.18)\right]} = \tag{60}$$
$$= 2.07\frac{N}{mm}$$

Considering (1) and (51) and the last expression the cold state stiffness K_A and the post elastic martensitic stiffness K_{MB} can be obtained:

$$K_A = s_1 K_{MA} = 2.68 \cdot 2.07 = 5.55 N/mm \tag{61}$$

$$K_{MB} = s_m K_{MA} = 0.18 \cdot 2.07 = 0.373 N/mm$$

From (53) is computed x_{adm} , the maximum elongation of the SMA wire:

$$x_{adm} = \frac{S}{\gamma} = \frac{10}{0.75} = 13.33mm \tag{62}$$

Taking into account the maximum deformation of the material ε_{adm} the wire length l_0 is:

$$l_0 = \frac{x_{adm}}{\varepsilon_{adm}} = \frac{13.33}{0.04} = 333.3mm \tag{63}$$

From the austenitic stiffness and the wire length the wire diameter d is immediately obtained:

$$d = \sqrt{\frac{4K_A l_0}{E_A \pi}} = \sqrt{\frac{4 \cdot 5.55 \cdot 333.3}{75000\pi}} = 0.177mm \tag{64}$$

The compensator system preload is obtained through (56).This value is negative because in its minimum stroke position $x = x_{min}$= 3.33mm, the preload is opposed to SMA active element action:

$$F_{comp}\Big|_{x=x_{min}} = F_{ON} - K_A S\left(\frac{1-\gamma}{\gamma}\right) = 10 - 5.55 \cdot 10\left(\frac{1-0.75}{0.75}\right) = -8.5N \tag{65}$$

In order to minimize the force fluctuations in both travel directions the optimum stiffness of compensation system is obtained from (Table 1) and is:

$$k_{comp} = -\frac{K_{MB}+K_A}{2} = -\frac{0.373+5.55}{2} = -2.96\frac{N}{mm} \quad (66)$$

3.4. Design of compensated actuators with two antagonistic SMA elements

The input data required for the design of a two antagonistic shape memory actuators are: the value of the minimum useful forces in the two directions of activation F_{ON1}, F_{ON2} the value of the stroke desired, S, and the active elements material characteristics (i.e. s_1, s_m and s_g). In the following equations, assuming that the axis of the actuator is horizontal, the forces are assumed positive when directed from right to left and negative when directed from left to right. The displacements are assumed positive from left to right.

For a compensated actuator with two antagonistic SMA element it can be shown [21] that it is always possible to set the ratio γ (53) to 1, so that the actuator stroke is equal to the maximum deflection imposed on the SMA elements (x_{min}=0).

If both active elements are constituted by the same alloy (same parameters s_1, s_m, s_g) is possible to demonstrate [21] that the optimum rigidity of the two elements will be equal ($K_{A1}=K_{A2}=K_A$ and $K_{MB1}=K_{MB2}=K_{MB}$).

The force generated by element 1 in the ON state is:

$$F_{SMA1\_ON} = K_A x \quad (67)$$

In the OFF state, using (3) for the martensite behavior, the force in element 1 is:

$$F_{SMA1\_OFF} = F_{0m} + K_{MB} x \quad (68)$$

For element 2, the force output depends on the difference between the prestretch p of the two SMA elements and the position of the actuator. Assuming γ=1 the prestretch p is equal to the stroke S, and the force of element 2 in the ON and OFF state is respectively:

$$F_{SMA2\_ON} = -K_A(S-x) \quad (69)$$

$$F_{SMA2\_OFF} = -F_{0m} - K_{MB}(S-x) \quad (70)$$

The force delivered from the compensator is:

$$F_{comp} = F_{0comp} + k_{comp} x \quad (71)$$

where F_{0comp} indicates the force of the compensator at position $x = 0$.

Figure 9 shows the force displacement diagram of a two-SMA actuator and helps to understand the relationship between the variables involved in the equations.

Line OA in Figure 9 represents the characteristics of the austenitic (hot) SMA1 element. Segment BC represents the martensitic (cold) response of the antagonistic SMA2 element. Line DE is the characteristic of the compensation spring, with point D corresponding to the centerpoint of the total stroke S of the actuator. Line EG represents an ideally constant external load of amplitude F_{ON1}. The situation beyond point D is obtained by extrapolating linearly all the characteristic lines shown in Figure 9. Line AE corresponds to the characteristic of the SMA1 element and the compensation spring combined. At any position x, the difference between lines AE and BC gives the net output force of the actuator ($F_{net\ ON1}$) when element SMA1 is activated. When element SMA1 is disabled and SMA2 is enabled, the chart becomes similar to Figure 9, with all the lines mirrored with respect to the line AD.

The optimal performance of the actuator in Figure 9 is achieved when lines AE is parallel to line BC, so that the net output force of the actuator ($F_{net\ ON1}$) equals the external load (F_{ON1}) at any position x. Scirè and Dragoni [21] demonstrated that the optimal actuator meets the design specifications (i.e. the given output forces F_{ON1}, F_{ON2} and the net stroke S) when the following relationships hold true:

$$K_{MA} = \frac{\left(F_{ON1} - F_{ON2}\right)}{S\left[\left(s_1 - s_m\right) + 2s_g\left(s_m - 1\right)\right]} \tag{72}$$

$$k_{comp} = -\frac{\left(F_{ON1} - F_{ON2}\right)\left(s_1 + s_m\right)}{S\left[\left(s_1 - s_m\right) + 2s_g\left(s_m - 1\right)\right]} \tag{73}$$

$$F_{0comp} = \frac{s_g\left(F_{ON1} + F_{ON2}\right)\left(s_m - 1\right) + \left(F_{ON1}s_1 - F_{ON2}s_m\right)}{\left(s_1 - s_m\right) + 2s_g\left(s_m - 1\right)} \tag{74}$$

The force generated by the actuator in the case of element 1 ON and element 2 OFF is:

$$F_{net\_ON1} = K_A x + F_{0comp} + k_{comp}x - F_{0m} - K_{MB}\left(S - x\right) \tag{75}$$

while the force generated in the case of element 2 ON and element 1 OFF is:

$$F_{net\_ON2} = F_{0m} + K_{MB}x + F_{0comp} + k_{comp}x - K_A\left(S - x\right) \tag{76}$$

From (1) and (55) the stiffness K_A of the SMA element in the austenitic state is obtained.

The actuator designed by (72)-(74) provides a constant force in both directions. In the compensated actuators with two SMAs, the force of the compensator becomes zero and changes sign at mid-stroke. In the range $0 < x < S/2$ the force of the compensator has the same direction as the force exerted by element 1, while in the range $S/2 < x < S$ it has the same direction as the force generated by element 2.

3.4.1. Case study: Double SMA compensated actuator

This section describes the design of double SMA compensated actuator under the hypothesis of using traction Nitinol springs with the following characteristics:

- Wire diameter: 0.65mm
- Mean coil diameter: 6.8mm
- Number of active coils: 25.5
- Free length: 25mm
- Austenitic stiffness: K_A: 0.0615N/mm
- Martensitic stiffness: K_{MA}: 0.0414N/mm
- Pseudo-plastic stiffness: K_{MB}: 0.0156N/mm
- Deflection at the threshold of pseudo-plastic regime x_g: 23mm

The desired output stroke is S =75mm, with the maximum possible force in both directions.

In this case the dimensional parameter γ can be taken equal to one and consequently the maximum extension of the active elements x_{adm} is equal to the desired stroke S.

The non dimensional parameters are:

$$s_1 = \frac{K_A}{K_{MA}} = \frac{0.0615}{0.0414} = 1.486 \quad (77)$$

$$s_m = \frac{K_{MB}}{K_{MA}} = \frac{0.0156}{0.0414} = 0.377 \quad (78)$$

$$s_g = \frac{x_g}{x_{adm}} = \frac{23}{75} = 0.306 \quad (79)$$

Using the above dimensionless parameters, eq. (72) gives the maximum force differential $(F_{ON1} - F_{ON2})$ that the actuator can produce in both directions as:

$$\begin{aligned} F_{ON1} - F_{ON2} &= K_{MA} S\left[\left(s_1 - s_m\right) + 2 s_g \left(s_m - 1\right)\right] = \\ &= 0.0414 \cdot 75\left[\left(1.486 - 0.377\right) + 2 \cdot 0.306\left(0.377 - 1\right)\right] = 2.26N \end{aligned} \quad (80)$$

Assuming that the actuator generates the same output force regardless of the direction of motion gives:

$$F_{ON1} = -F_{ON2} = 1.13N \quad (81)$$

From (73) the overall stiffness of the compensation system k_{comp} is calculated:

$$\begin{aligned} k_{comp} &= -\frac{\left(F_{ON1} - F_{ON2}\right)\left(s_1 + s_m\right)}{S\left[\left(s_1 - s_m\right) + 2 s_g \left(s_m - 1\right)\right]} = -\frac{2.26\left(1.486 + 0.377\right)}{75\left[\left(1.486 - 0.377\right) + 2 \cdot 0.306\left(0.377 - 1\right)\right]} = \\ &= -0.0771 N / mm \end{aligned} \quad (82)$$

Likewise, (74) gives the force F_{0comp} that the compensator must apply to the SMA elements at the position $x = 0$:

$$F_{0comp} = \frac{s_g\left(F_{ON1} + F_{ON2}\right)\left(s_m - 1\right) + \left(F_{ON1}s_1 - F_{ON2}s_m\right)}{\left(s_1 - s_m\right) + 2s_g\left(s_m - 1\right)} = \\ = \frac{0.306\left[1.13 + (-1.13)\right](0.377 - 1) + \left[1.13 \cdot 1.486 - (-1.13) \cdot 0.377\right]}{(1.486 - 0.377) + 2 \cdot 0.306(0.377 - 1)} = 2.893N \tag{83}$$

3.5. Design of the compensator system

This section describes the elastic constants of the shape memory elements and the maximum deflections that they will undergo, calculated from the design data (force and stroke required of the actuator) following the procedure of the previous section. Thanks to these parameters the shape memory elements (wires or springs) can be described using classical engineering formulas. The method also provides the properties needed by the elastic compensation to meet the required performance. Two elastic compensation systems are described in detail in this section: 1) a rocker-arm mechanism and 2) a double articulated quadrilateral.

3.5.1. Rocker-arm compensator

The compensation mechanism shown in Figure 10 is made up of a rocker-arm R hinged in G to the frame T. The (conventional) compensation spring S_c (with free length L_{0Trad} and spring rate k_{Trad}) connects the extremity F of the shortest side of the rocker-arm to point E of the frame. Point O (the output port of the actuator) is used to connect the primary elastic elements of the actuator (SMA1 and SMA2), which are also fixed at points P and Q to the frame. In the case of a single-SMA actuator, the active element (SMA1) is placed at the bottom of the device, element SMA 2 disappears and the contrasting element is made up by the conventional compensation spring, S_c.

The mechanism in Figure 10 has a position of unstable equilibrium, corresponding to the configuration where the axis EF of the compensation spring S_c passes through the hinge G of the rocker-arm. In this position, the force exerted by the compensator on the slider O is null. In the case of an actuator with a single active element, the compensation spring S_c is always placed to the right of hinge G in order to exert a contrasting force on the SMA element 1 needed to deform it in the cold state. For an actuator with two opposing SMA elements, the spring S_c is located in an unstable position (line EF passes through G) when point O is at the center of the stroke (S). In this way, the compensation mechanism helps active element SMA1 for the lower half of the stroke and the element SMA2 for the upper half of the stroke. If in Figure 10 the absolute value of the angle γ is small, i.e. if $l_0+x-(a+d) \ll c$ then the following expressions apply with good approximation:

$$F_{comp}(x) = -\frac{abF_{Trad}\beta}{c(a+b)} \tag{84}$$

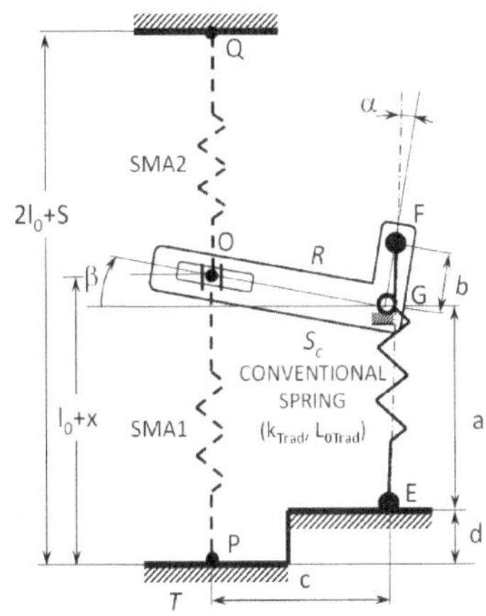

Figure 10. Oscillating rocker arm compensation mechanism

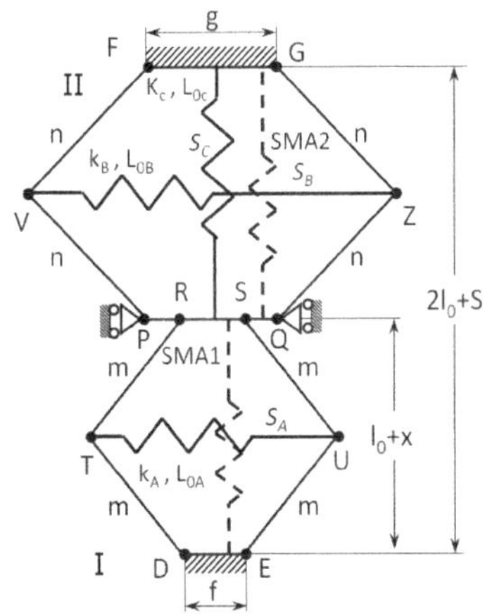

Figure 11. Double articulated quadrilateral compensation mechanism

$$k_{Comp} = \frac{\partial F_{Comp}}{\partial x} = -\frac{abF_{Trad}}{(a+b)c^2} \tag{85}$$

When lengths b,c,d have been arbitrarily decided along with the stiffness of the traditional spring k_{Trad}, from (84) (calculated for $x=x_{min}$) and (85) the expressions for a and L_{0Trad} are derived which define the correct sizing of the compensation mechanism:

$$L_{0Trad} = \frac{\left[k_{comp}\left(A-b+d-l_0\right)+F_{comp}\Big|_{x_{\min}}\right]\left(c^2\,k_{comp}^2\,\gamma-B\right)}{B\,k_{comp}} \tag{86}$$

$$a = l_0 - A - d - \frac{F_{comp}\big|_{x_{\min}}}{k_{comp}} \tag{87}$$

with:

$$A = S\left(\frac{\gamma - 1}{\gamma}\right) \tag{88}$$

$$B = b\,k_{Trad}\,\gamma\left[k_{comp}(d - l_0 + A) + F_{comp}\big|_{x_{\min}}\right] \tag{89}$$

In Equations (86) – (89) the term $F_{comp}\big|_{x_{\min}}$ is given by (56) or by (74) respectively for the single SMA or two opposing SMAs actuator. Likewise, the stiffness k_{comp}, is obtained from Table 1 or from (73) for single SMA or for the two opposing SMAs actuator respectively.

3.5.2. *Double articulated quadrilateral compensator*

Figure 11 illustrates a second elastic compensation system based on the use of two articulated quadrilaterals (I and II), which are connected together in series. The first quadrilateral (I) is made up of four equal rods DT, EU, RT and SU, hinged together in T and U and fixed to the frame with hinges D and E. In the upper part, the rods are hinged in R and S to the bar PQ, which can translate vertically. The rods DT, EU, RT and SU have length m, while the horizontal segments DE and RS have length f. The conventional spring S_A, with stiffness k_A and free length L_{0A}, is stretched between hinges T and U.

The second quadrilateral (II) is made up of four equal rods VP, ZQ, VF and ZG, connected with internal hinges in V and Z and fixed to the frame with hinges F and G. The length of the rods VP, ZQ, VF and ZG is n, while the length of the horizontal segments PQ and FG is g.

Quadrilateral II contains two conventional springs, S_B and S_C. Spring S_B, with stiffness k_B and free length L_{0B}, is stretched horizontally between hinges V and Z. Spring S_C, with stiffness k_C and free length L_{0C}, is stretched vertically between the horizontal sides FG and PQ. Member PRSQ, which is common to the two quadrilaterals, represents the output ports of the actuator. In addition to the conventional springs, S_A, S_B and S_C, the mechanism in Figure 11 contains also the primary active elements of the actuator. Active element SMA1 is hosted by the quadrilateral I and connects the base DE to the output port PRSQ. The second primary element SMA2 (if applicable) is hosted by quadrilateral II and connects the frame FG to the output port PRSQ.

For the sake of simplicity, all the springs shown in Figure 11 (both SMA and conventional) are traction springs. However, the mechanism can work just as well with all compression springs, which can be designed with the same equations presented below.

By imposing the vertical equilibrium of the member PRSQ in Figure 11 and excluding the forces exerted by the shape memory elements, it is possible to obtain the compensation force as a function of the position x of the output port.

Imposing that:

$$L_{0A} = f \qquad L_{0B} = g \qquad k_B = k_C \tag{90}$$

the compensation force becomes:

$$F_{comp}(x) = -k_A(l_0 + x) + k_C L_{0C} \tag{91}$$

i.e. it depends linearly on x as desired in an ideal compensation system. The compensation stiffness implied in (91) is:

$$k_{comp} = \frac{\partial F_{comp}}{\partial x} = -k_A \tag{92}$$

From (91) the force exerted by the compensation system at $x=x_{min}$ is:

$$F_{comp}\big|_{x=x_{min}} = -k_A(l_0 + x_{min}) + k_C L_{0C} \tag{93}$$

Conditions (92) and (93) complete the design of the compensation mechanism in Figure 11. The value k_{comp} is taken from Table 1, for the case of a single SMA, or alternatively from (73) for two SMA elements. Similarly, the value $F_{comp}\big|_{x=x_{min}}$ is taken from (56) or (74), for one or two SMA elements, respectively.

By solving (93) with respect to the product $k_C L_{0C}$, you can get the correct design of spring S_C. From condition (90), we see that the springs S_B and S_C must have the same stiffness and that the free length of spring S_B must be g. It is thus convenient to set to g the free length of spring C as well, so as to have two identical springs S_B and S_C in quadrilateral II with spring rate:

$$k_C = k_B = \left(F_{comp}\big|_{x=x_{min}} + k_A(l_0 + x_{\min}) \right) \Big/ g \tag{94}$$

4. General discussion

4.1. Performances of the uncompensated SMA actuators

The analytical framework described in Paragraph 3 considers both dissipative and conservative forces acting on the system. Closed-form relationships are developed, which are the basis of a step-by-step procedure for an optimal design of the entire actuator (both with antagonist SMA, bias elastic element or backup force). Specific formulas for dimensioning of the SMA element in the form of straight wires and helical traction springs are also presented.

The physical meaning of dimensionless design parameters s_1, s_2, s_F and s_0 deserves specific comments. Parameter s_1 is an intrinsic property of the SMA material, the higher s_1, the better the SMA behaviour. Parameter s_2 is a measure of the relative stiffness between the bias spring (be it traditional or SMA) and the primary SMA spring. Optimal values of parameter

s_2 maximize the overall travel. Low values of s_2 lead to an increase in stroke and in the actuator dimension. Parameter s_F quantifies the influence of dissipative forces on the stroke of the actuator. For low values of s_F the dissipative force is negligible in comparison with the forces expressed by the backup systems and the actuator moves freely. The upper limit of s_F is reached when the dissipative force becomes high enough to prevent the movement of the actuator. Parameter s_0, is the equivalent of s_F for the conservative force. Since the conservative force helps the backup system, high values of s_0 improve the actuator performances.

Also for the three backup systems analyzed, some specific comments are reported. Using a force as back up for the primary spring may be convenient when there is important conservative force acting on the system which helps the restoring process. Often this solution is not feasible due to dimensional constraint which prevents dead loads to be used. The binary actuator with a traditional spring backup gives the worst performances in terms of output characteristic, due to disadvantages of the elastic slope of the traditional spring. By analyzing Section 3.2 equations it is found that for a given cursor stroke, Δx the optimal value of s_2 provided by (21) minimizes the pre-stretch of the system, p, and tends to minimize the total size of the actuator. Eq.(18) shows that the pre-stretch of the system, p, acts on the stroke, Δx, as a pure gain. Figure 12a displays the normalized stroke, $\Delta x/p$, as a function of the other variables: s_0 (used as parameter of the family of surfaces), s_2 and s_F (these two variables are plotted along the axes), taking $s_1=3$ because this is a typical value for the elastic moduli ratio of commercial shape memory alloy. Eq. (18) and Figure 12a also highlight that for each combination of s_0 and s_F exists an optimal value, s_2^*, of parameter s_2 which maximizes the normalized stroke. Eq. (19) shows that for each combination of s_1 and s_F the stroke increases monotonically when parameter s_2 decreases, implying a more flexible bias spring. Both equations (18) and (19) prove that meaningful strokes ($\Delta x>0$) are only possible if parameter s_F is lower than s_{Fcr} of equation (10).

Section 3.3 shows that the binary actuator with a SMA spring backup gives the best performances in terms of output characteristic, both respect to elastic backup and bias force backup. Equation (30) shows that also in this case the pre-stretch of the system, p, acts on the stroke, Δx, as a pure gain. Figure 12b displays the normalized stroke, $\Delta x/p$, as a function of the other variables: s_0 (used as a parameter of the family of surfaces), s_2 and s_F (these two variables are plotted along the axes), taking $s_1=3$ (a typical value for elastic moduli in austenitic and martensitic phase of a commercial SMA). Equation (30) and Figure 12b also highlight that for each combination of s_1 and s_F an optimal value, s_2^*, of parameter s_2 exists that maximizes the normalized stroke. Equation (30) shows that for each combination of s_1 and s_F the stroke increases monotonically when parameter s_2 decreases, implying a more flexible SMA bias spring. Correspondingly, equation (29) shows that the pre-stretch of the system, p, tends to infinity. If (31) is used to choose s_2, the global pre-stretch, p, is minimized and the total length of the actuator tends to be reduced, but it is not minimized. To obtain a value of s_F and a value of s_2 able to minimize the total length of the actuator, expressions that depend on the embodiment of the springs are needed.

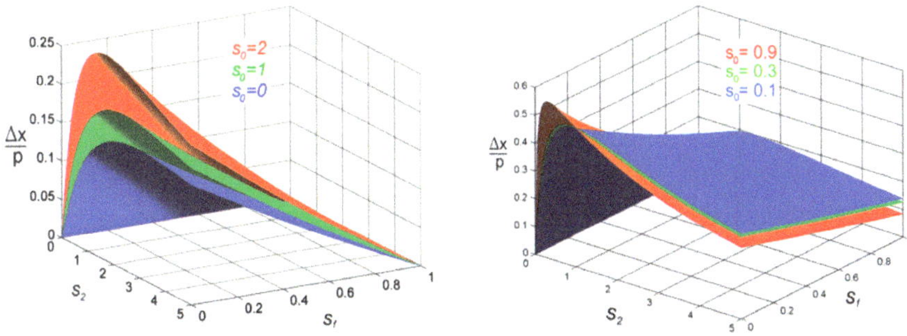

(a) (b)

Figure 12. Normalized stroke of a shape memory actuator backed up by a traditional spring (a) or by an antagonist SMA spring (b) as a function of the parameters s_2, s_F and s_0.

4.2. Performances of the compensated SMA actuators

The performances of SMA compensated actuators are improved compared to traditional SMA actuators. A comparison between a traditional and a compensated SMA actuators, being equal the active elements and their maximum deformations, shows that the compensated system has longer strokes. Conversely, being equal the output stroke, the deformations in the compensated SMA system are lower than the traditional one. In general the compensated SMA actuator is able to obtain the same performances of a traditional system with lower maximum deformation in the material, while for a given admissible deformation the active elements of a compensated SMA are more stressed because they work against both the external load and the compensation system during the first part of the stroke. The energy stored in the compensator is released in the second part of the stroke helping in keeping the output force constant.

Experimental comparative tests on a SMA actuator with or without a rocker-arm compensator [28] demonstrated that the stroke of the compensated system is always greater regardless the external load. The stroke incremented by 2.5 times with no load up to 22 times with the maximum design force applied. Furthermore, the net stroke of the compensated actuator is not dependent on the applied force. This peculiarity simplifies both the design and the selection of the actuator for a given application. Other remarkable advantages of the compensated actuator lie in the nearly-constant value of the output force and in the fastest response compared to traditional systems, even with longer output strokes. The fastest response is due to the lower difference between internal external forces in the end positions. By contrast, the response time (the time needed to achieve an incipient displacement) is considerably longer for the compensated actuator. This behaviour is due to the high stress at which the SMA springs operate before activation, which requires a greater degree of martensite-austenite transformation (hence a greater temperature) to start the motion. In general, the compensated actuator has no intrinsic positions of stable equilibrium so mechanical hard stops are needed to prevent dangerous over deformation in the SMA. One hard stop provides a monostable behaviour for the single SMA, the two hard stops needed for the double SMA produces a bistable behaviour. The characteristic of mono or bistability leads to great operative advantages because the actuator is able to: maintain a specific position even without power; achieve precise and repetitive positioning; facilitate control strategies of the overall system.

5. Conclusion

The optimum mechanical design of shape memory based actuators is performed by means of analytical description of the governing equations of the system, providing an easy to follow design procedure for engineers working with SMA devices. Since no complex thermo-mechanical models are involved, the closed form equations developed are based on two simple constitutive models: a linear or a bilinear behaviour for the martensitic phase and a linear relationship for the austenitic phase. Three backup elements are considered to recover the stroke: a constant force, a traditional elastic spring and antagonist SMA element, leaving the designer the opportunity to adopt those which fits the application best. The external load is as general as possible, because a system of both dissipative and conservative forces is taken into account. The actuator performances are improved considering an elastic compensation instead of a backup element. This negative stiffness element can enhance either the stroke or the force of the SMA actuator depending on the needs. The output force design equations both for the SMA actuators and for the compensator are given and the detailed description of the compensated SMA system grants an immediate comprehension of the whole system. In order to give an operative guideline for SMA actuators design, several numerical examples are provided. The proposed procedure is applied to real case studies and to a specific configuration of SMA active elements (wires or springs). The benefits of the described methodology are critically discussed and compared. In terms of mechanical performance the agonist antagonist SMA system gives the best performances both for compensated and uncompensated systems, while optimum values for the non dimensional parameters are provided in order to obtain compact and efficient SMA based actuators. The analytical equations developed in this Chapter allow a SMA based actuator to be correctly designed in order to match the specifications and give useful information to optimize size and mechanical response of the system with no need of complex numerical simulations.

Author details

A. Spaggiari*, G. Scirè Mammano and E. Dragoni
Department of Science and Methods for Engineering, University of Modena and Reggio Emilia, via Amendola, Campus S. Lazzaro Reggio Emilia, Italy

* Corresponding Author

6. References

[1] Otsuka K Wayman C M (2002) Shape memory Materials, Cambridge University Press, Cambrdige UK.

[2] Lagoudas DC, (2008) Shape Memory Alloys Encyclopedia of Aerospace Engineering John Wiley & Sons, New York.

[3] Boyd JG, Lagoudas DC (1996) A Thermodynamical Constitutive Model for Shape Memory Materials. Part I. Int. J. Plast 12(6): 805-842.

[4] Arghavania J, Auricchio F, Naghdabadia R and Sohrabpoura S (2009) A 3-D phenomenological constitutive model for shape memory alloys under multiaxial loadings Int. J. Plast. 26(7): 976-991.

[5] Kuribayashi, K. (1989) Millimeter-sized Joint Actuator using a Shape Memory Alloy, Sensor Actuator, 20(1-2): 57-64.

[6] Mertmann M, Vergani G (2008) Design and application of shape memory actuators, Eur. Phys. J. Special Top. 158:221–230.

[7] Kim B, Lee MG, Lee YP, Kim YI, Lee GH (2006) An earthworm-like micro robot using shape memory alloy actuator, Sensors and Actuators A: 125(2): 429-437,

[8] Reynaerts D. Van Brussel H (1998) Design aspects of shape memory actuators Mechatronics, 8: 635-656.

[9] Lu A, Grant D, Hayward V (1997) Design and comparison of high strain shape memory alloy actuators" Proc.IEEE Robotics and Automation 1(20): 260-265.

[10] Bergamasco M, Dario P Salsedo F (1990) Shape Memory Alloy Microactuators Sensors Actuators A 21(1-3): 253-257.

[11] Nespoli A, Besseghini S, Pittaccio S, Villa E, Viscuso S (2010) The high potential of shape memory alloys in developing miniature mechanical devices: A review on shape memory alloy mini-actuators. Sensors and Actuators A: Physical 158(1): 149-160.

[12] Jansen S, Breidert J Welp EG (2004) Positioning actuator based on shape memory wires, ACTUATOR Proc. of 9th International Conference on New Actuators.

[13] Strittmatter J, Gümpel P (2004) Shape memory actuator for hydraulic valve ACTUATOR Proc. of 9th International Conference on New Actuators.

[14] Bellini A, Colli M, Dragoni E (2009) Mechatronic Design of a Shape Memory Alloy Actuator for Automotive Tumble Flaps: A Case Study, IEEE Trans on Industrial Electronics, 56(7):2644-2656.

[15] Haga Y et al. (2005) Dynamic Braille display using SMA coil actuator and magnetic latch, Sens. Actuators A, 119: 316-322.

[16] Elwaleed AK, et al., (2008) A new method for actuating parallel manipulators, Sens. Actuators A 147:593–599.

[17] Migliavacca F. et al. (2004) Stainless and shape memory alloy coronary stents: a computational study on the interaction with the vascular wall, Biomechan Model Mechanobiol 2:205-217.

[18] Spinella I, Dragoni E (2009) Design equations for binary shape memory actuators under dissipative forces. Pro Insn Mech Engrs C: J. Mech. Eng. Sci 223(C3): 531–543.

[19] Spinella I, Scirè Mammano G, Dragoni E (2009) Conceptual design and simulation of a compact shape memory actuator for rotary motion. J. Mater. Eng. Perform. 18(5-6): 638–648.

[20] Spaggiari A, Spinella I, Dragoni E (2012) Design equations for binary shape memory actuators under arbitrary external forces, Journal of Intelligent Material Systems and Structures (In press) DOI: 10.1177/1045389X12444491

[21] Scirè Mammano G, Dragoni E (2011) Increasing stroke and output force of linear shape memory actuators by elastic compensation. Mechatronics 21(3): 570-580.

[22] Spaggiari A, Spinella I, Dragoni E (2011) Design of a Telescopic Linear Actuator Based on Hollow Shape Memory Springs, J. Mater. Eng. Perform. 20: 489-496.

[23] Scirè Mammano G, Dragoni E (2011) Modelling of Wire-on-Drum Shape Memory Actuators for Linear and Rotary Motion". J. Intell. Mater. Syst. Struct. 22(11): 1129-40

[24] Lagoudas D C, Miller D A, Rong L, Kumar P K (2009) Thermomechanical fatigue of shape memory alloys. Smart Materials and Structures 18:085021.

[25] Scirè Mammano G, Dragoni E (2012) Functional fatigue of Ni–Ti shape memory wires under various loading conditions. Int. J. fatigue (In press) http://dx.doi.org/10.1016/j.ijfatigue.2012.03.004

[26] Budynas R, Nisbett R, (2006) Shigley's Mechanical Engineering Design, McGrow Hill, New York.

[27] Berselli G, Vertechy R, Vassura G, Parenti Castelli V (2009) Design of a single-acting constant-force actuator based on dielectric elastomers. J Mech Robot 1 (3): 031007.1:031007.7.

[28] Scirè Mammano G, Dragoni E (2011) Design and Testing of an Enhanced Shape Memory Actuator Elastically Compensated by a Bi-Stable Rocker-Arm. Proc.ASME SMASIS Conference 18 – 21, Scottsdale, Arizona, USA, ISBN: 978-0-7918-4415-1.

Chapter 3

SMA-Based Muscle-Like Actuation in Biologically Inspired Robots: A State of the Art Review

William Coral, Claudio Rossi, Julian Colorado, Daniel Lemus and Antonio Barrientos

Additional information is available at the end of the chapter

1. Introduction

New actuation technology in functional or "smart" materials has opened new horizons in robotics actuation systems. Materials such as piezo-electric fiber composites, electro-active polymers and shape memory alloys (SMA) are being investigated as promising alternatives to standard servomotor technology [52]. This paper focuses on the use of SMAs for building muscle-like actuators. SMAs are extremely cheap, easily available commercially and have the advantage of working at low voltages.

The use of SMA provides a very interesting alternative to the mechanisms used by conventional actuators. SMAs allow to drastically reduce the size, weight and complexity of robotic systems. In fact, their large force-weight ratio, large life cycles, negligible volume, sensing capability and noise-free operation make possible the use of this technology for building a new class of actuation devices. Nonetheless, high power consumption and low bandwidth limit this technology for certain kind of applications. This presents a challenge that must be addressed from both materials and control perspectives in order to overcome these drawbacks. Here, the latter is tackled. It has been demonstrated that suitable control strategies and proper mechanical arrangements can dramatically improve on SMA performance, mostly in terms of actuation speed and limit cycles.

Due to their limitations, SMAs have not raised the attention of the robotics technology for several years. However, recent studies have demonstrated that by (i) finding suitable niches of application, (ii) dedicated mechatronics design, and (iii) ad-hoc control strategies, SMAs can effectively be used as an alternative actuation technology in a wide spectrum of applications and robotic systems. Indeed, as it will be introduced in this chapter, careful control design that takes into account the particular characteristics of the material coupled with proper mechanic design, play a significant role for an efficient use of SMAs. Even so, it is clear that SMAs (and smart materials in general) cannot, at this stage, be thought as a universal substitute for classical servomotor technology. However, niches of applications can be found that greatly benefit from this technology. Bio-inspired artificial systems are one such niche.

Although SMAs are mostly used as actuators, they also have sensing capabilities. Despite most of the SMA physical parameters are strongly related in a nonlinear hysteresis fashion, the electrical resistance varies linearly with the strain of the alloy. Because strain is kinematically related to the motion of the actuator (either linear motion or rotational), the electrical resistance and the motion produced by the actuator are both linearly related. This linear relationship between resistance variation and motion is achieved because the martensite fraction is kinematically coupled to the motion, and the martensite fraction is what drives the resistance changes. This issue is an advantage for developing closed-loop position controllers that regulate the SMA actuation. In fact, most of the applications involving position linear control of SMAs, feedback electrical resistance measurements to estimate the motion generated by the actuator. This avoids the inclusion of external position sensors for closing the control loop.

SMAs are used in a variety of applications [46],[40],[56],[29],[27],[80]. Their special properties have aroused great expectations in various technologies and industries; it can be used to generate a movement or storing energy. In addition, its scope covers many sectors ranging from the use in deployable satellite antennas for different sensors to machinery, to materials for the construction of suspension bridges or anti-seismic devices. In general, all applications somehow depend on the effect of action-reaction of the material and the conditions under which particular application takes place, which make the SMAs a functional material.

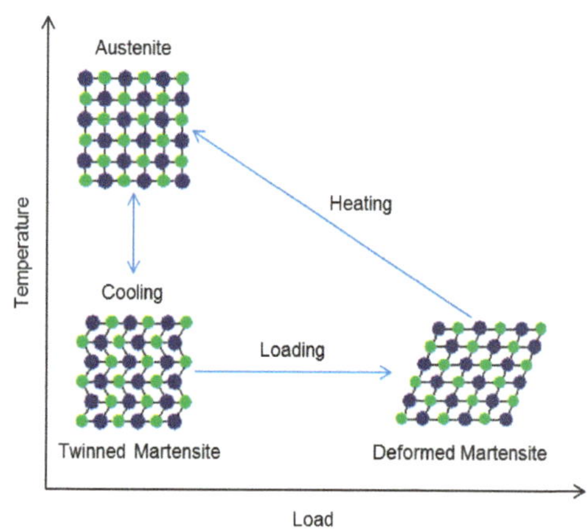

Figure 1. Microscopic viewpoint of the Shape Memory Effect

For instance, they are being used in many non-invasive surgery devices [45],[21],[62],[23],[43] and biomedicine, taking advantage of their large strains and their capability to recover the shape when the load is removed. This property allows applications in devices such as stents, tubular prosthetic devices, because it restores the ability of flow of any bodily duct affected by a narrowing.

In classical robotic systems, linear actuation systems have been proposed using SMAs. The focus of this chapter is on bio-inspired robotics. SMA-based actuators provide a suitable technology as muscle-like actuation mechanisms, which resemble the mechanics of muscles in biological systems. For this reason in the last years a number of bio-inspired robots have been designed adopting SMA technology. In this paper, we review the main prototypes, organizing them according to the mean (water, air, ground), and on main morphological characteristics (full body actuation or appendices only).

2. Shape memory alloys background

2.1. Principle of operation

Shape Memory Alloys are metallic materials with the ability to "remember" a determined shape, even after a severe deformation produced by a thermal stimulus. In the case of metallic alloys, the shape memory effect consists on a transition that occurs between two solid phases, one of low temperature or martensitic and other of high temperature or austenitic. The material is deformed in the martensitic phase and retrieves, reversibly, its original dimensions by heating above a critical transition temperature. The terms martensite and austenite originally referred only to the steel phases, however these terms have been extended referring not only to the material but also the kind of transformation. Thereby, the martensite steel involves a change of volume and shape, while the SMA has basically a change of length.

In general, NiTi (Nickel-Titanium) SMAs are the most common alloys used. This is basically because these materials are intrinsically susceptible of use both as sensors and actuators, which makes them suitable for integration in smart structures. NiTi SMAs work based on the shape memory effect, which essentially takes place by the influence of temperature change of the material; i.e. the temperatures at which the martensitic and austenite phase transformations begin and end. Figure 1 depicts how these changes occur at the microscopic level of the material. The phase transition occurs when the material is heated or cooled. In general, there is a certain temperature range for the transition, which is mainly defined by the manufacturer.

SMAs normally exhibit one-way shape memory effect, also called memory effect in a simple manner. The alloy deforms upon heating but cooling does not change the shape unless it is stressed again. The percentage of deformation of NiTi alloys (% of strain) is about five percent, a range considerably higher if one considers that the deformation of common steel allows only an average of two percent. Currently, SMAs that exhibit two-way shape memory effect are also manufactured. In this case, the alloy expands by heating above the range of transition temperature and spontaneously contract when cooled again below this temperature [47]. To produce the double shape memory effect, the material is subjected to heat treatment, also called training. This training-phase forces the material to remember both heating and cooling states.

From the microscopic viewpoint (Figure 1), all the physical properties of the alloy vary depending on the phase, i.e. from cooling to heating and vice versa. Some of these properties refer to corrosion resistance, elasticity, damping capacity, strain, stress, electrical resistance, and temperature. Therefore, shape memory alloys behave in a thermo-mechanical way, with all these variables strongly coupled within a nonlinear hysteresis fashion.

Table 1 shows the commercial characteristics of SMAs depending on the diameter of the wires (NiTiNol®). From the table it can be noticed their high electrical power consumption. In robotics applications, power consumption is a critical issue due to the level autonomy of the robotic system is fully dependent on the capacity of the onboard batteries.

Diameter Size *inches(mm)*	Resistance	Pull Force *pounds* (*grams*)	Approximate Current for 1 Second Contraction (mA)	Cooling Time $158°F, 70°C$ "LT" Wire (seconds)	Cooling Time $194°F, 90°C$ "HT" Wire (seconds)
0.001 (0.025)	36.2 (1425)	0.02 (8.9)	45	0.18	0.15
0.0015 (0.038)	22.6 (890)	0.04 (20)	55	0.24	0.2
0.002 (0.050)	12.7 (500)	0.08 (36)	85	0.4	0.3
0.003 (0.076)	5.9 (232)	0.18 (80)	150	0.8	0.7
0.004 (0.10)	3.2 (126)	0.31 (143)	200	1.1	0.9
0.005 (0.13)	1.9 (75)	0.49 (223)	320	1.6	1.4
0.006 (0.15)	1.4 (55)	0.71 (321)	410	2	1.7
0.008 (0.20)	0.74 (29)	1.26 (570)	660	3.2	2.7
0.010 (0.25)	0.47 (18.5)	1.96 (891)	1050	5.4	4.5
0.012 (0.31)	0.31 (12.2)	2.83 (1280)	1500	8.1	6.8
0.015 (0.38)	0.21 (8.3)	4.42 (2250)	2250	10.5	8.8
0.020 (0.51)	0.11 (4.3)	7.85 (3560)	4000	16.8	14

Table 1. Characteristics of NiTinol®SMA wires [6].

2.2. Improving the performance of SMA actuators

One of the main limitations in SMA actuation speed is due to high latency that the the cooling time of the wire implies. Despite increasing the input heating power can reduce the heating time, large cooling times limit the operation frequency of the actuator. On average, NiTi wires with a diameter of $127\mu m$ typically requires an electrical current input about $320mA$ to contract in about $1s$ (nominal heating time) and relax in approximately $1.4s$ (nominal cooling time). In this case both contraction and recovery times would set a nominal actuation frequency about $0.416Hz$, quite slow for many applications requirements.

Research to overcome this limitation has been oriented towards developing cooling systems for SMAs, aimed at decreasing the nominal cooling time involved during the recovery process. In this direction, temperature control methods have been proposed in [41]. Cooling systems based on Peltier cells [18] or active cooling [67], have been commonly used. However, nowadays bio-inspired robotic systems tend to be small and light, therefore other methodologies for enhancing SMA actuation speed must be addressed. For several years different strategies have been proposed to implement rapid control in the SMA wires [12],[19],[70],[72],[11],[71],[75]. A system consisting of rapid heating of the SMA was proposed by [12] aimed at increasing the overall actuation frequency by means of overloading the operation of SMAs. The term overloading refers to increasing the amount of input heating power to be delivered to the SMA wires. In [74] experiments carried out using a two degree-of-freedom Pantograph robot actuated by an antagonistic pair of SMA wires acting as linear actuators have shown how the nominal actuation frequency was increased from $0.416Hz$ to $1Hz$.

Overloading should be monitored in order to avoid overheating problems that may cause physical damage of the shape memory effect. In [71], further research in this direction allowed for the introduction of a force control architecture with the proper mechanisms for safe overload the operation of SMA actuators. In the prototypes described in Section 4, we have used a control architecture similar to the one described in [71], which makes use of proper mechanisms to overload the operation of SMAs. However, these mechanisms have been adapted to work within a position control scheme, avoiding the need of including external force sensors. Section 4.2 will detail on this issue.

Besides rapid heating techniques to overload SMA operation, further investigations have been also carried out to verify whether SMAs can respond to high frequencies. In [64] and [73] experiments have demonstrated that NiTi SMA wires with a diameter of $0.1mm$ can respond up to frequencies of $2KHz$. This high-frequency response corresponds to small-signal heating currents inputs with frequencies of that magnitude. These results allow for the development of small-signal high-bandwidth controllers capable of improving SMA performance, but more important, eliminating the limit cycles of operation of SMAs. In other approaches, $20 - 30Hz$ limit cycles have been observed, whereas in [19], [50] at approximately $100 - 200Hz$. In this regard, the use of high-bandwidth force sensors might be suitable for developing a SMA force feedback control system.

2.3. Modeling and control

The physical behavior of SMAs is more complex than many common materials: the stress-strain relationship is nonlinear, hysteresis is presented, large reversible strains are exhibited, and it is temperature dependent. This thermo-mechanical relationship can be described by formulating phenomenological models. Tanaka in [69] was one of the pioneers to study a stress-induced martensite phase transformation, proposing an unified one-dimensional phenomenological model that makes use of three state variables to describe this process: temperature, strain, and martensite fraction. His main contribution was to demonstrate that the rate of stress is a function of strain, temperature and martensite fraction rates. Later, Elahinia [7], [8] proposed an enhanced phenomenological model compared to other works [40], [31], [69] and also addressed the nonlinear control problem. This model was able to better describe the behavior of SMAs in cases where the temperature and stress states changed simultaneously. Their model was verified against experimental data regarding a SMA-actuated robotic arm [10].

Phenomenological models may provide some insights of SMA thermo-mechanical behavior that facilitate the development of control procedures. To control purposes, parameters' tuning is highly dependent of a modeling stage, but definitively phenomenological models are not the best choice for control design, especially if the goal is related to improving actuation speed. In this direction system identification is a promising alternative. As noted by [40], [19], [11], [71], [20], identified linear models for SMA can be developed. It has been demonstrated that the AC response of NiTi SMA wires behave as a first order low-pass filter.

Section 4 of this paper details two different approaches for modeling SMAs; one based on identifying how electrical resistance change as a function of the input current [59], and the other based on identifying how the output torque produced by an antagonistic pair of SMA actuators change as as a function of the applied power [5]. Furthermore, [5] details how to take advantage of phenomenological models for simulating overheating problems when a SMA wires are overloaded. Attempting to perform this analysis on the real SMA actuators might cause several damage to the structure. Phenomenological models are really useful for determining the upper limits of applied input heating currents.

The control methods presented in [59] and [5] have been conceived for controlling a pair of antagonist SMA actuators. The antagonistic configuration is useful for having SMA actuators where each direction of motion can be controlled. In [59], the antagonistic pair of actuators must bend the structure of the fish robot, whereas in [5] the antagonistic SMA actuators are connected to a joint for providing the rotational motion. Other approaches in [11], [30], [41], [28] have demonstrated the advantages of using an antagonistic arrangement in terms of controllability. When the active actuator is being heated while the passive (antagonistic) is cooling, hysteresis effects are reduced due to the external stress that the active actuator applies on the inactive one above the austenite finish temperature.

3. Bio-inspired robots with SMA muscle-like actuation

The use of SMAs as artificial muscles allows for more realistic bio-inspired actuation presented in nature [26]. SMA wires acting as muscle fibers can respond upon electrical signals, taking advantage of the large pull force and its excellent strength-weight tradeoff. Currently, the use of SMAs in biomimetic robotic systems [3],[13],[9] can be found in ground, water and air robots, in many sizes, including those micro-robots or microstructures [25],[78],[53],[35],[63],[82],[37],[37]. In the following sections, we describe the most representative bio-inspired robots and structures that integrate SMAs as muscle-like actuation mechanisms.

3.1. Water

Biologically inspired robots that operate in water can be found in two categories. Firstly, robots that use SMAs for actuating appendices (fins), and secondly, robots that use SMAs to actuate the robot's body. In the latter body actuation is used for undulatory motion (fish-like robots). Some animals can move by bending their body in such a way to produce a backward-propagating propulsive wave. The movement obtained by bending a continuous structure is much more natural than others where joints are presented. In Section 4.2 our bending structure prototype is presented.

3.1.1. A micro-robot fish with embedded SMA wire actuated by flexible biomimetic fin

In [78] it has been proposed a micro robot fish that uses a flexible biomimetic fin propeller with embedded SMA wires to mimic the musculature and flexible bending of squid fin. The propulsion consists of an active component (the biomimetic fin) and a passive component (the caudal fin). The biomimetic fin-based propulsion mechanism is an actuator that combines the SMA wire and an elastic substrate.

This micro-robot fish introduces a new concept in the world of biomimetic robotics due to its ability to swim noiseless. This means the robot avoid the use of any traditional components like gears, bearings and joints, only using the SMAs as actuators that produce the propulsion. The robot is able to achieve a swimming speed of $112mm/s$ when the SMAs actuate at $2.1Hz$ (contracting upon electrical heating), and a minimum turning radius of $136mm$, which makes the robot the fastest micro robot-fish compared to other prototypes that use IPMCs like actuators [14],[54],[33]. The authors have measured the robot performance based on the Strouhal number [15], [24], which typically varies between 0.25 to 0.35 for the biological counterparts. Their robot has a Strouhal number of 0.58 (at maximum swimming speed). This upper value highlights the optimal movement of the robot, however, high amounts of input power have been required to actuate the SMAs.

3.1.2. *Towards a biologically inspired small-scale water jumping robot*

In [63], the locomotion description of a water-jumping robot that mimic the ability of the water striders and the fishing spider to jump on the water surface. This biomimetic robot achieves a vertical jumping motion by pushing the water surface. The motion is triggered with a latch driven by the SMA actuator.

As a result of the research, quantification of Re = 260 (Reynolds number is the ratio of inertial over viscous forces), Bo = 0.0054 (Bond number is the ratio of the buoyancy to the surface tension) and We = 4.7 (Weber number is the ratio of the inertia to the surface tension) and the B_a (Baudoin number is the ratio of the body weight to the surface tension) suggest that the physics of jumping in this robot is similar to those of the fishing spider. The Bond, Weber number and Baudoin numbers are explained by [15], [24]. In terms of actuation, the SMA allows the robot to be extremely light (mass of $0.51g$), which it is essential to ensure the buoyancy on water. The maximum jumping height is $26mm$, 26% of the height reached when jumping on ground ($53.1mm$). This prototype is the first concept of jumping robot that integrates SMAs within a structure with an overall mass of $1g$. The robot requires $2W$ of power consumption in order to generate a force of $1.35mN$.

3.1.3. *A micro biomimetic manta ray robot fish actuated by SMA*

In [79] a manta ray robot fish actuated by SMA wires is designed. Figure 2 shows the prototype of the robot. Two pectoral fins arranged in triangular-shaped made of latex with a thickness of 0.2mm form the fin surface.

This micro manta ray was the first prototype that uses SMAs to generate thrust. This robot is capable to swim forward and turn. The sweep back angle of the pectoral fins is 20°. A maximum swimming speed of $57mm/s$ was achieved and the maximum amplitude of the motion was $40mm$. All the biomimetic fins are open-loop controlled.

3.1.4. *Controlling a lamprey-based robot with an electronic nervous system*

In [81] a sea Lamprey has been developed. The robot consists of a cylindrical electronics bay propelled by an undulatory body axis. SMA actuators generate propagating flexion waves in five undulatory segments of a polyurethane strip. The lamprey robot Figure 3(a) consists on a cylindrical hull that houses the electronics and battery pack. In this application, the authors use a neuronal network that allows the robot to be controlled in real time. This neuronal

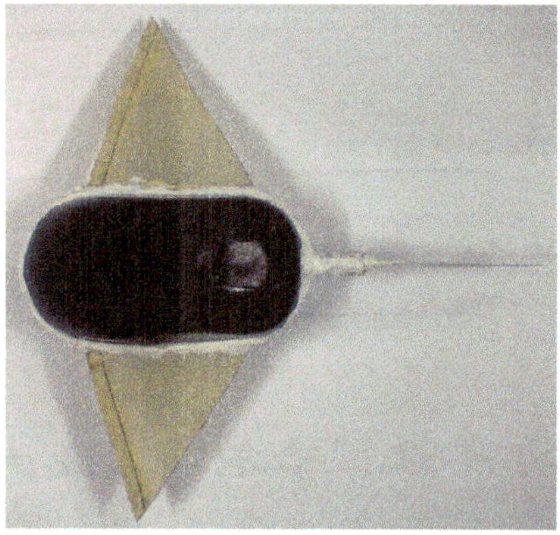

Figure 2. Micro biomimetic manta ray robot fish [79].

network also drives the control of the SMAs. The results have shown the system can reject disturbances thanks to the robustness of the nonlinear controller [55]. Each SMA wire drains 1.5A of electrical current when activated.

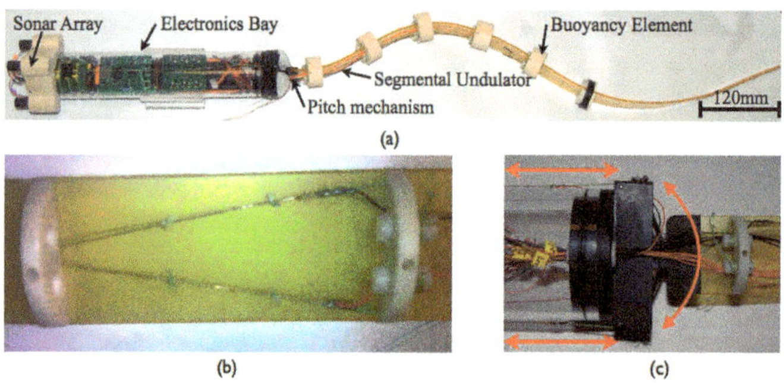

Figure 3. (a) Lamprey Robot with sonar array, (b) Lateral view of tail segment showing nitinol actuator, Teflon vertebra and tensioning nuts and (c) Lateral view of pitch mechanism [81].

3.1.5. A biomimetic robotic jellyfish (Robojelly) actuated by shape memory alloy composite actuators

The newest and more advanced aquatic robot that uses SMA actuators is a jellyfish robot designed by [68]. The hydrogen-fuel-powered robot called "Robojelly" mimics the propulsion, morphology, kinematics and physical appearance of a medusa (jellyfish); the Aurelia aurita species. The bio-inspired actuators are made of silicone, SMA wires and spring steel.

The development of Robojelly has introduced a systematic method for the design and fabrication of SMA-based actuators called BISMAC (bio-inspired shape memory alloy composite). This method allows for bending the structure of the robot by means of SMA contraction [77]. Thanks to the BISMAC SMA arrangement, this robot was capable to mimic the physics and swimming characteristics of jellyfish in terms of A. aurita's bell geometry, passive relaxation mechanism, neutral buoyancy, frequency of motion, and deformation-to-flap motion profiles. The structure can be bended by the SMAs actuators (deformation), and then a flap motion of the bell-segment structures provide the propulsion. The Robojelly was able to produce enough thrust to propel itself and achieve a proficiency of $0.19s-1$ which is comparable to the natural medusa at $0.25s-1$. The robot consumes an average of 16.74W over its 14th cycle of actuation. This robot confirms the fact that most aquatic biomimetic robots use SMA wires combined with other materials to create SMA-based actuators. This characteristic shows the flexibility of the SMA to work in combination with other materials.

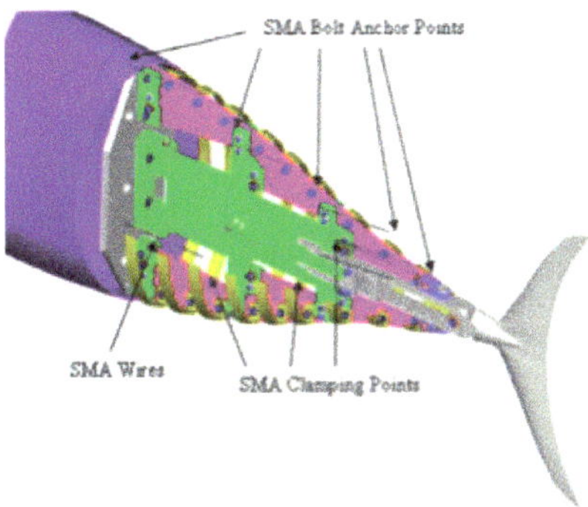

Figure 4. A CAD representation of the tail of Suleman's Tuna [65].

3.1.6. Design and testing of a biomimetic tuna using shape memory alloy induced propulsion

In [65], an SMA actuated tail inspired by a Bluefin tuna is proposed. Figure 4 shows the tail cutaway. This fish-like robot has length of $1m$ capable of the biological carangiform swimming mode. The maximum tail beat frequency was $0.5Hz$ due to the limitations of the SMA. Even at this low frequency, the power requirements were significant. The minimum and maximum power consumptions were calculated to be $292.8W$ and $333.6W$ respectively.

3.2. Air

In aerial bio-inspired robots most of the applications are appendices. Here, we can identify two main categories: insects and birds. To the best of the authors' knowledge only one robotic flying insect has been developed, apart form the jumping robot described earlier. This can be explained by the flapping frequency needed, far form the SMA's capabilities, and also by their power requirements. For these reasons insect-like flying robots mostly adopt piezo-electric actuators. In fact, the flying insect prototype described below uses SMA to fold and unfold the wings, and not for the primary flapping motion. Despite SMA actuation speed does not allow the actuation of flapping wings, it could allow for other kind of wing actuation, such as morphing-wings.

3.2.1. Recent progress in developing a beetle-mimicking flapping-wing system

In [51] a beetle-like insect robot inspired by the Allomyrina Dichotomapresents is presented. This robot features a morphing-wing airfoil capable of folding and unfolding the hind wing using SMA wires. A single small size DC motor drives the flapping mechanism. Figure 5 shows the prototype and the unfolding of the artificial flapping/morphing wing device.

Similar folding ratio of the robot's wings has been observed in comparison with the biological counterpart, accounting for 1.7 of value. On average, wing unfolding was completed within about 3*s* and the wing folded in about 4*s*.

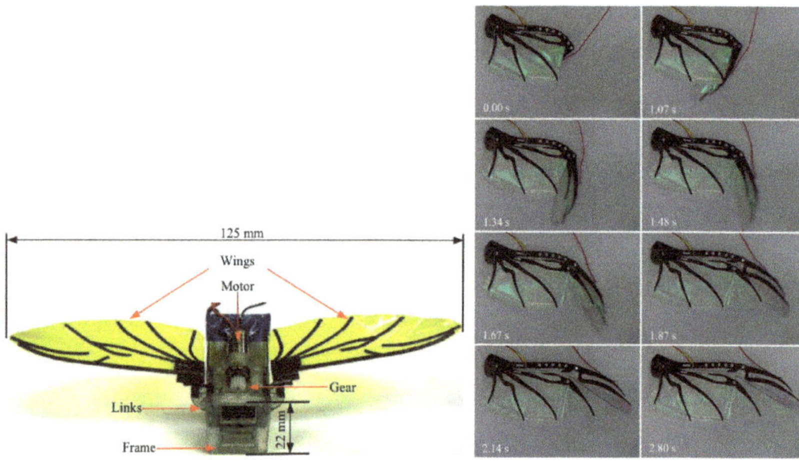

Figure 5. Prototype of the robotic beetle and detail of the unfolding of the artificial wing [51].

3.2.2. BATMAV-a biologically inspired micro-air vehicle for flapping flight: artificial-muscle based actuation

The BATMAV is a biologically inspired bat-like Micro-Aerial Vehicle (MAV) with flexible and foldable wings capable of flapping flight [2]. The robot features bat-inspired wings with a large number of flexible joints that allow mimicking the kinematics of a real bat flyer. Figure 6 details the overall structure of the robot, and the main connections of the SMA-like muscles.

BATMAV is the first robot that uses the SMA wires to play a dual role: first, as muscle-like actuators that provide the flapping and morphing wingbeat motions of the robot, and second, as super-elastic flexible hinges that join the wing's bone structure. Most of the experiments in [?] were carried out with a two-degree of freedom wing capable of flapping at 3*Hz*. Despite the fact that their robot is able to achieve accurate bio-inspired trajectories, the results presented lack experimental evidence of aerodynamics measurements that might demonstrate the viability of their proposed design.

3.3. Ground

Ground bio-inspired robots have been divided in two categories: the ones that uses actuated appendices (i.e, legged robots) and those that actuate the whole body, i.e. crawling robots such as snakes and worms.

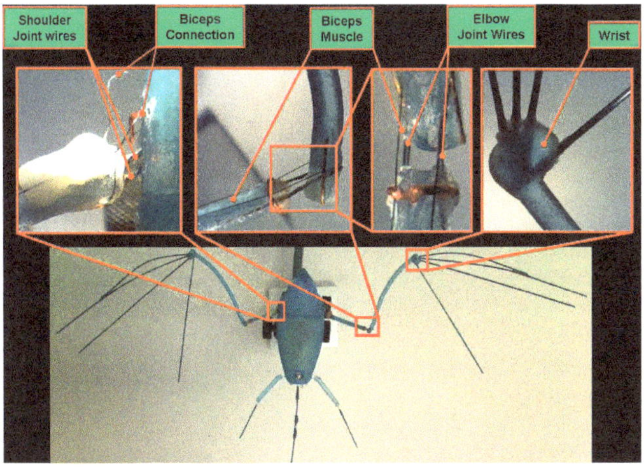

Figure 6. BATMAV. Dual Role of Shape Memory Alloy wires: as actuation muscles, and super elastic joints. (Picture from `https://sites.google.com/site/gheorghebunget/research/batmav`, with permission).

3.3.1. Sensor fusion in a SMA-based hexapod bio-mimetic robot

In [44] SMABOT is presented, a hexapod biomimetic robot with two SMA actuators that allow for the motion of the two degree-of-freedom robot. Each SMA actuator produces $300 gram - force$ of pull force. Figure 7 shows the SMABOT IV. SMABOT IV incorporates two-dimensional inertial navigation system for position control. The average speed when moving with tripod gait is $30 cm/min$. Its maximum power consumption is about $25W$ (the mass is $290g$).

Figure 7. SMABOT IV, a SMA based hexapod robot with the IMU module, compass sensor and step touch sensors [44].

3.3.2. Omegabot: Crawling robot inspired by Ascotis Selenaria

In [38] a robot inspired by the inchworm Ascotis Selenaria is presented. The robot, called Omegabot, is named after the omega (Ω) shape of the crawling motion of the inchworm. Figure 8 shows the Omegabot platform. Previous work about this robot can be also found in [37].

Experimental results report the first step for establishing an inchworm-like robot that can crawl on various terrains where conventional robots cannot move. The Omegabot uses a SMA coil actuator that requires a current of $200mA$ for activation. The frequency of motion is about $1Hz$, limited by the response time of the SMA wires. The inchworm robot is manually controlled by an IR remote operation, and it achieves a maximum linear velocity of $5mm/s$. The robot travels a distance of $5mm$ per stroke.

Figure 8. Omegabot, a biomimetic inchworm robot, grasps the branch of a wood, raises its head, and turns right. Bottom right: Proleg of Omegabot [38].

3.3.3. An earthworm-like micro robot using shape memory alloy actuator

In [34] a bio-mimetic micro earthworm-like robot with wireless control is proposed. The actuation mechanism consists on a SMA spring that contract and extend the earthworm muscle respectively. The proposed mechanism is simple but effective when traveling in narrow and rough environments, such as human digestive organs, bended long pipeline and so on. Also, this micro robot incorporates both control and power supply onboard. The theoretical speed of the micro robot is approximately $3.4mm/cycle$, where the total time per cycle is $8s$ (the contraction time of the SMA is $2s$, whereas the recovery time is $6s$). The fabricated micro robot can move with the velocity of $10mm/min$ during 8 minutes. The stroke per cycle is $2.0mm$.

3.3.4. Other peristaltic motion concepts

A concept similar to the one depicted in Figure 9 has been proposed in [48]. A SMA-spring has been used for changing the axial length of worm's modules, and consequently changing their width. The robot has a length and diameter of $1cm$. SMA springs of 50 and $100\mu m$ wires have been tested. The resulting prototype, made of 4 modules, was able to achieve a speed up to $0.22mm/sec$, with a power consumption of $600mA$. The previous prototype was composed of a number of identical segments attached in series, each of which alternately contract axially and expand radially. To conclude this section two prototypes are worth mentioning that use SMAs structures for peristaltic motion.

The proposed structure in these cases is a tubular mesh made of SMA wires that convert radial contraction into longitudinal lengthening. Such a structure has been proposed independently in [61] (see also [36]) and in the SoftWorm project [4].

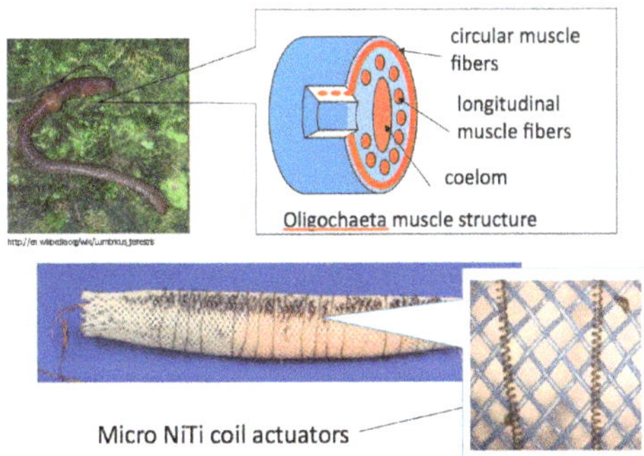

Figure 9. Arrangement of Antagonistic circular muscles in Oligochaeta [48] (Image of the real Oligochaeta from http://en.wikipedia.org/wiki/Lumbricus_terrestris).

3.4. Other SMA-based actuation systems

In this section we present two works that do not address the development a full robot, but rather studying and developing appendices to be added to future full robotic systems.

3.4.1. Research on Development of a Flexible Pectoral Fin Using Shape Memory Alloys

In [83], experimental research on pectoral fin structure is presented. The design of the pectoral fin actuator is based on SMAs wires composed by a couple of plates with the opposite functions. Figure 10 shows the biomimetic pectoral fin.

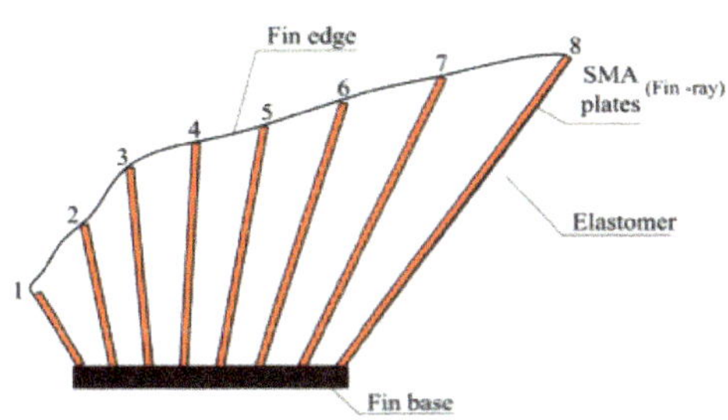

Figure 10. Biomimetic pectoral fin driven by eight couples of SMA plates [83].

This pectoral fin was the first and today continues being the only designed with only SMA wires. In fact, most research on fish-like robots is focused on studying propulsion (how to generate thrust), while maneuvers is largely unexplored.

3.4.2. Development of a dexterous tentacle-like manipulator using SMA-actuated hydrostats

Novel design principles and technologies for a new generation of high dexterity soft-bodied robots inspired by the morphology and behavior of the octopus are being developed in the framework of the OCTOPUS-IP project[1].

The imitation of the internal muscular structure of octopuses' tentacles is being studies and imitated. Longitudinal cables and transverse SMA imitate the arrangement of muscle fibers, controlling contractions as soft actuators within the robot arm [42]. Moreover, this manipulator is surrounded by a sensitive skin, with contact sensors embedded into silicone rubber, equipped with passive suckers that allow the grasping of objects. SMA actuators are used to change the section of the tentacle in several locations, inducing its bending.

Figure 11. The SMA-based tentacle (See `http://www.octopusproject.eu/`).

3.4.3. Development of a Shape-Memory-Alloy actuated biomimetic hydrofoil

The development and testing of a biomimetic active hydrofoil using Shape Memory Alloy (SMA) actuators is presented in [16]. This work describes the development and testing of a six-segment demonstration foil and the control schemes used.

4. Review on recent advances: iTuna and BaTboT

In this section, we report our most recent results on two SMA-actuated bio-inspired robots. The first, called iTuna, ia an underwater robot that according to our classification falls into the "full-actuated-boy" category. The second is an aerial robot, which implements the concept of morphing wings by means of SMA-based muscles.

[1] http://www.octopusproject.eu/

4.1. iTuna: a bending structure swimming robotic fish

The iTuna [59] is a swimming fish-like robot that apart from the external appearance, imitates some key features of the internal morphology of fishes.

This mechatronic concept takes inspiration from the arrangement of the red or slow-twitch muscles (see inset in Figure 12). In live fishes, such muscles are used for bending a flexible but nearly incompressible axis. Such axis is either composed of a (visco) elastic beam (notochord) or a series of vertebrae connected through intervertebral discs. The main structure of the iTuna robot fish is inspired by the former solution, and is composed by a continuous flexible backbone. The backbone is composed of polycarbonate of $1mm$ thickness actuated by SMA muscles acting as red muscles.

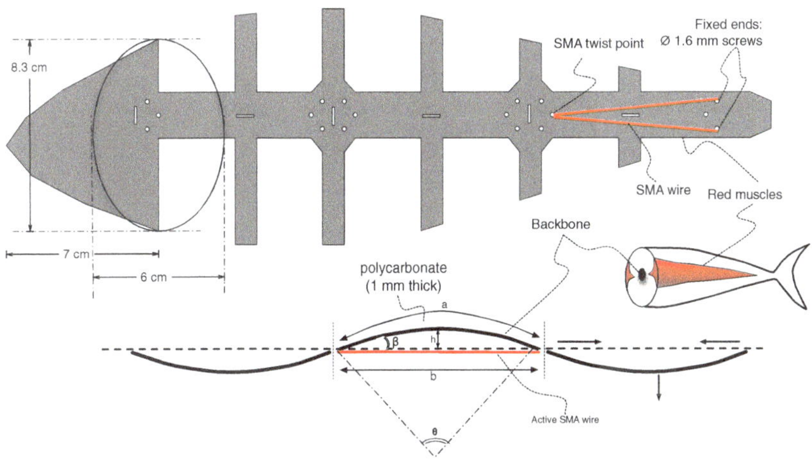

Figure 12. Main structure of the iTuna robot fish. a=8.5 cm. Under nominal operatrion, $b \cong 96\%$ a = 8.16 cm, h=1.02 cm, b=28° [59].

Six SMA-based actuators whose length is 1/3 of the body length are positioned in pairs, parallel to the body in such a way to produce an antagonistic movement on three body segments of $8.5cm$ length. This antagonistic configuration of SMA wires has some advantages in terms of increasing the range of controllable actuation, since both directions of motion (contraction and elongation) can be actively controlled. Figure 12 shows the location of the SMA wires within the skeleton structure of the prototype.

A V-shaped configuration of the wires, where each artificial muscle is composed of a single V-shaped SMA wire, twisted around the tension screw, allows to double the pull force without a significant increase of power consumption. NiTi SMA wires with a diameter size of $150\mu m$ have been adopted. These have a pull force of $230grams - force$ at consumption of $250mA$ at room temperature, and a nominal contraction time of 1 second.

Under nominal operation such SMAs can bend the body segments up to 28 degrees (angle β of Figure 12), even if SMA wires only contract approximately 4% of their length. By increasing the input electrical current and including a suitable control that handles an overloaded SMA operation, contraction time of $0.5s$ was achieved, and strain could be increased up to 6%, corresponding to a bending of approximately 36° (Fig. 13).

4.1.1. SMA control in the iTuna

After identification, a low-level PID controller has been designed to address two main limitations of SMAs: slack in the fibers, and limited actuation speed. Slack issues appear when SMA wires develop a two-way memory effect during operation [11]. Limitation in actuation speed occurs due to the large switching time between cooling and heating phases. To address such problems, a pre-heating mechanism has been developed that works in conjunction to the antagonistic arrangement. The pre-heating avoids the temperature on both wires drops below the 10% of the maximum applied electrical current, preventing the inactive alloy from complete cooling. On the other hand, the antagonistic arrangement provides an external stress to the cooling wire (provided both by the elastic backbone and by the active antagonistic wire). Working with an already-warm wire allows for a faster stretch and slack issues are avoided. Note that the PID controller is based on the experimental observation that the hysteresis on the electrical resistance curve was smaller than the hysteresis on the temperature curve. Resistance measurements are used as a feedback signal for closed-loop control (see [59] and [60] form more details).

The control developed allows overloading the SMA with up to $350mA$ peak current (note that power signals are sinusoidal, hence overloading only lasts a brief period of time). Overloading has allowed for achieving a $1Hz$ oscillation time (i.e. 0.5 seconds contraction and cooling times) and a bending angle of 36 degrees of each body segment.

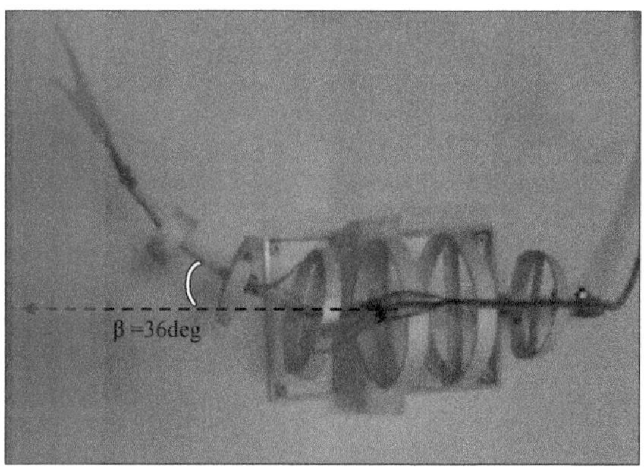

Figure 13. Bending under SMA overloading [59].

4.1.2. Control architecture

A key feature of SMAs is the possibility to develop closed loop control systems without the need of external sensor hardware. The feed back signal is provided by the detection of inner electrical resistance, that allows an indirect measurement of the temperature.

The main components are described in the following. A micro controller implements the PID algorithm. The PID controller receives the input reference position (set point) and the feedback of SMA's voltage and current that allows calculating the heating current to drive the SMA actuator. The digital output of the PID controller is converted to a reference current in two steps. First, it is converted into an analog signal using a 2-wire serial 8-Bit DAC (Digital to Analog Converter) with Rail-to-Rail outputs. Then, a Voltage Controlled Current Source (VCCS) transforms the DC voltage in a constant current that feeds the SMA. This stage has a power consumption of less than $10mA$.

The DAC output ranges from 0 to 5 Volts with a resolution of $0.02V$. The measured voltage (VSMA) and current (ISMA) on the SMA fiber are fed-back to the micro controller in order to close the control loop. The hardware used (16F690-PIC) had a 12-bits A/D converter with a resolution of $0.537mV$ (considering the maximum voltage measured at the SMA $VSMA = 0.55V$.

On the other hand, taking into consideration the maximum current through the wire ($500mA$), SMA resistance variations about 1.074mΩ can be measured. Therefore, since the maximum variation in the SMA length is $0.34cm$, and the maximum variation of the resistance is 1.6Ω, the theoretical position error of the system based on the SMA length is 0.067%. i.e.,$0.12mm$.

4.2. BaTboT: a biologically-inspired bat-like aerial robot

BaTboT is a bio-inspired bat robot that uses Shape Memory Alloys (SMAs) as artificial muscles for powering the morphing motion of the wings. The morphing motion is related to the capacity of the robot to modulate its wings by contracting and extending the membrane in sync with the flapping motion. It is precisely this characteristic what makes biological bats more agile to maneuver than any other flying creature within the same Reynolds number range ($10^3 - 10^4$) [66], [22]. In addition, biological studies in [32], [58] have revealed that real bats are able to maneuver because of the inertial changes produced by the wings' modulations. Attempting to mimic this functionality using an artificial counterpart -BaTboT- mainly presents a twofold challenge: i) biomechanical design of the wings, and ii) proper control/actuation to module BaTboT's wings.

Prior work in [5] presented experimental results regarding both challenges. The investigation carried out in [5] does not only describes the biomechanical design of BaTboT's wings, but also focuses on evaluating the implications of using SMAs as artificial muscles to power the change of wing's morphology. Figure 14 shows the design-flow process to evaluate key issues of SMA performance and their implications to the application at hand.

The use of SMAs as artificial muscles has been concretely evaluated in terms of two issues:

- Functionality: SMA Power-to-Force.
- SMA Performance: actuation speed and fatigue.

4.2.1. *Functionality*

The SMA actuators shown in Fig. 14 (step-1) are the commercial Migamotor NanoMuscle model RS-70-CE [49]. Each NanoMuscle consists of several short strips of SMA NiTi wire with a thickness of $150\mu m$ attached to opposite ends of six metal strips stacked in parallel. Each SMA segment pulls the next strip about $0.67mm$ relative to the previous strip, and the relative movements sum to make a stroke of $4mm$. As depicted, two Migamotors muscles have been arranged into an antagonistic configuration working as artificial biceps and triceps that provide the rotation motion of the wing elbow's joint θ_3. The range of motion of the joint

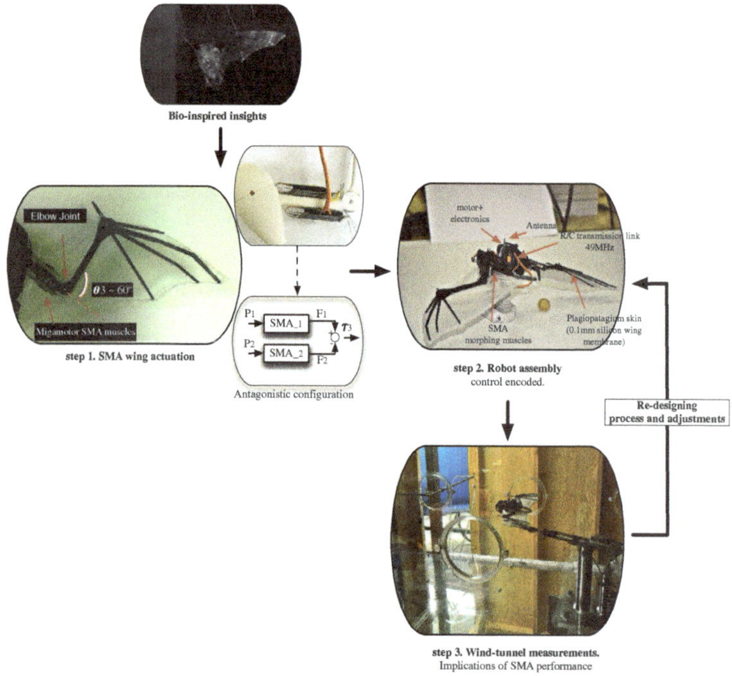

Figure 14. Flow-process for SMA evaluation in the BaTboT prototype [5].

is about 60°. The joint motion is achieved when each SMA actuator contracts upon heating, thus generating a pull-force (F_{sma}). Because both actuators are connected to the joint in an antagonistic fashion, the pull-force F_{sma} generates a joint torque, denoted as τ_3. Therefore, each actuator requires an input heating power P, to produce and output torque τ_3. The input heating power (P) is proportional to the input electrical current (I): $P = I^2R$, being R the nominal electrical resistance of the NiTi wires, $R = 8.5\Omega$.

In [5] simulations and experiments have been carried out aimed at quantifying the Power-to-Force tradeoff of the SMA muscles working under two operation modes: nominal and overloaded. To the application at hand, nominal-mode implies an input heating current (I) between $175mA$ and $350mA$, whereas overloaded-mode, between ~$400mA$ and $600mA$. Overloading allows for increasing the heating-time of the SMA, therefore increasing the contraction speed and the overall actuation frequency. However, overloading must be monitored to avoid overheating issues that may cause physical damage of the shape memory effect. In [5], SMA limitations were found under simulation and validated under experimentation.

Fig.15 shows the Power-to-Force tradeoffs: i) simulation using a SMA phenomenological model, ii) SMA response using an identified model, and iii) experimental measurements of output torque. Herein, our goal is not focused on describing the models, but on comparing the correlations between models (i)-(ii) and experimental data (iii). Further details about the models can be found in [5].

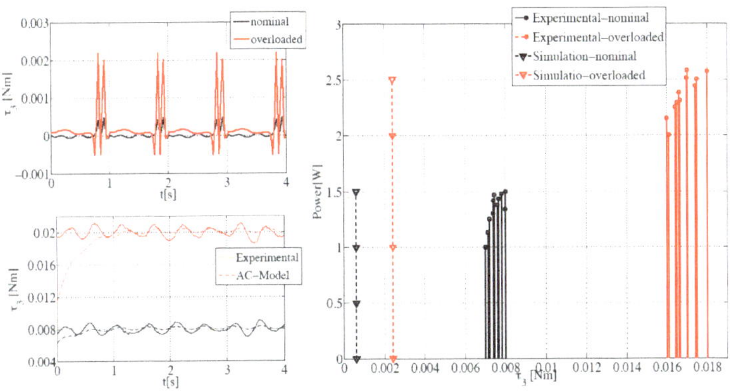

Figure 15. SMA Power-to-Force response. The input power is of the form: $a + b\sin(2\pi ft)$. a) Simulation of elbow's torque response under nominal and overloaded SMA operation, b) AC small-signal response of SMAs at f=~2Hz, c) maximum peak-values of output torque as a function of the input power

In order to simulate how the SMA muscles response upon electrical heating, a phenomenological model was used. This model is composed by thermo-mechanical equations that describe how the temperature, strain, and stress of the SMA change during its hysteresis loop, going from austenite to martensite and vice versa. Likewise, the SMA phenomenological actuation model is coupled with the dynamics equations that govern the motion of the robot's wing (refer to [5] for further details).

Figure 15a shows simulation results of joint torques achieved by applying nominal and overloaded values of input power of the form: $a + b\sin(2\pi ft)$, where a is the mean input power P, the term b is the small-magnitude of the sinusoidal component, and f is the commanded frequency for SMA contraction. Simulation data determines that by applying a nominal input power of $P = 1.04W$, the SMAs are able to generate a torque around the elbow joint of $\tau_3 = 0.0007Nm$, whereas by applying an overloaded input power of $P = 2.57W$, the output torque can be increased up to $\tau_3 = 0.0022Nm$. This implies that by increasing the input power about 2.5 times, the output torque can be increased about three times.

In order to verify the accuracy of simulation results, Fig.15c shows experimental measurements of output torques. In this trial, several profiles of input power were applied, and the respective forces were measured. Note that under nominal-mode, the output torque varies between 0.007 - 0.008Nm, whereas under overloading-mode, from 0.016 - 0.018Nm. Likewise, at the left side of the plot, the maximum peaks of nominal and overloaded torques corresponding to the simulation in Fig. 15 a, are also shown. One can note there is almost an order of magnitude of difference between simulation and experimental data. The error is produced because the phenomenological model does not take into account the joint friction, and even most important, the anisotropic loading of the silicone-based membrane. However, despite the error in magnitude, note that by doubling the input heating power, the measured

	Parameter	Theoretical[1]	Simulation[2]	Experimental[3]	AC-model[4]
Nominal	Input power P [W]	0.26	1.04	0.87-1.5	1.36
	Input current I [mA]	175	350	320-420	400
	SMA Pull-force F [N]	0.012	0.007	0.07 - 0.08	0.08
	Output torque tau_3 [Nm]	0.0012	0.0007	0.007 - 0.008	0.008
Overloaded	Input power P [W]	–	2.57	2 - 2.57	3.06
	Input current I [mA]	–	550	485 - 550	600
	SMA Pull-force F [N]	–	0.022	0.16 - 0.18	0.2
	Output torque tau_3 [Nm]	–	0.0022	0.016 - 0.018	0.02

Table 2. SMA-muscle Power-to-Force performance. [1] Nominal values can be found in Migamotors website [49]. The manufacturer does not provide overloading values. [2] Simulation data corresponds to Fig. 15 a. Please refer to Fig.9 in [5] for details about the SMA phenomenological model. [3] Experimental data corresponds to Fig. 15 c. [3] Identified AC Power-to-force model response. Data corresponding to Fig. 15 b.

output torque increases about 2.5 times. This scaling factor is quite similar compared to the simulation prediction.

The use of a phenomenological SMA model can provide a useful insight into the implications of overheating, by increasing the input power beyond the limits of SMA stress, and thus defining the upper threshold for overloading when using the real system. Unfortunately, the phenomenological model has not proven to be accurate for modeling the response of the SMAs when applying small-signal power inputs. Therefore, the next step is to identify the Power-to-Force AC behavior of the SMA actuators.

As observed by [11], [71] the AC response of NiTi SMA wires behaves as a first order low-pass filter. In [40], [19], [20] similar first-order behavior in their SMA actuator response had also been observed. Therefore in [5], we have followed a similar frequency response analysis procedure for finding a transfer function model that matches the measured gain and phase data of the frequency range of interest. Fig. 16 shows the results.

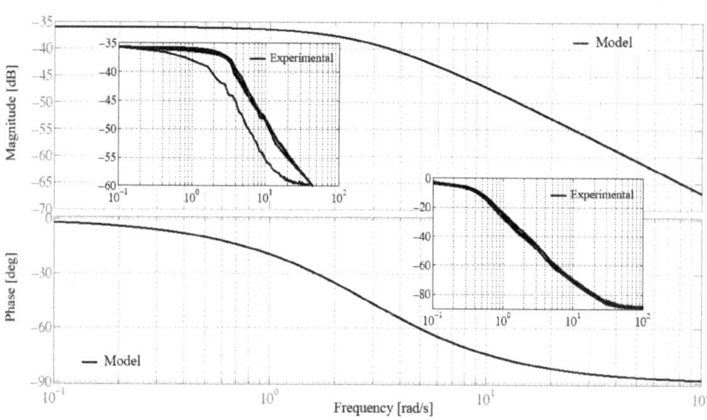

Figure 16. (Experiment VS model) Bode magnitude and phase plots for NiTi 150μm SMA Migamotor actuator. Input heating power of the form: $a + b\sin(2\pi ft)$.

The suitable transfer function that fits the experimental data from Fig. 16 is: $\tau_3(s) = 0.016(0.35s+1)^{-1}P(s)$. The response of the AC model is shown in Fig. 15 b. The AC model is compared against the experimental response of the SMA actuator to the AC small-signal. Note that by applying an input power of ~$1.36W$ (I = $400mA$), the output torque τ_3 stabilizes around ~$0.008Nm$. This response corresponds to the real nominal mode of the SMA actuator. Increasing the input power up to ~$3.06W$ (overloaded mode, $I = 600mA$), the registered output torque stabilizes around ~$0.02Nm$. In both experiments the anisotropic loading of the $0.1mm$ silicone-based membrane has been taken into account.

Note how well the AC model correlates against the real behavior of the SMA muscles. A detailed comparison of the obtained data is summarized in Table 2. In conclusion, to properly evaluate the functionality of using SMA muscles based on the criteria of the application at hand:

- First of all, identify the required force the SMA muscle must provide to drive the motion of the robot. This quantification requires a dynamic model of the robot.
- Use a phenomenological model for defining the limits of SMA actuation. The best models in the literature can be found in [69],[1],[7],[8],[10].
- Classify the response of the SMA in terms of nominal and overloaded input power. Verify that under overloaded operation, SMAs are safe from overheating, and yet achieving the desired output torque - to - input power tradeoff.
- Using the real system, apply an AC input power of the form $a + b\sin(2\pi ft)$, being a the mean power (nominal or overloaded),b the small-signal amplitude of the sinusoidal component, and f the desired actuation frequency. Subsequently, measure the small changes of pulling-force generated by the SMA when contracting. Note that forces can be mapped to torques (if using a rotational joint).
- Follow the frequency response analysis procedure described in [11], [71]. Note the AC behavior of a NiTi SMA wire resemble a first order low-pass filter. Find the transfer function that matches the data. The Power-to-Force model of the SMA will be of the form: $F(s) = k(\tau s+1)^{-1}P(s)$. Use the identified model for control tuning. Note that a simple PID control technique is suitable for driving SMA actuation. Depending on the available sensors, force control or position control could be applied. The following section provides a brief insight about the advantages and implications of using both control schemes: force or position feedback. For detailed insights about SMA force control, refer to [71], whereas for SMA position control, refer to [5], and [59].

4.2.2. *SMA Performance: actuation speed and fatigue*

As concluded in [29], [57], [11], [17], [76], PID technique is sufficient for allowing accurate and faster SMA actuation. The key issue for achieving an outstanding SMA performance does not necessarily depend on the control technique, but in complementary mechanisms that monitor the lower and upper limits of input power.

In [11], and [71], these mechanisms are introduced: i) anti-slack, and ii) anti-overload. The former deals with the two-way shape memory effect [39], improving accuracy and speed, whereas the latter limits the amount of input heating power to prevent physical damage when SMAs are overloaded. As a result, the force controller presented in [71] was capable of tracking fast and accurate force references when compared with other works reported in [20], [84].

The improvement in accuracy and speed are due to two factors: i) by avoiding wire entanglement, and ii) by ensuring the passive wire of the antagonistic configuration does not cool completely. Wire entanglement can be produced when the SMA wires extended upon cooling. The passive SMA wire can develop a few millimeters of slack as it cools, which consequently affects the accuracy of the control. This slack-phenomenon is only presented in the antagonistic configuration, due to the two-way shape memory effect produced by each actuator.

To avoid the aforementioned issues, the anti-slack mechanism defines a minimum threshold of input heating power that ensures the inactive wire does not cool completely. The improvement in actuation speed is due to the fact that the already-warm SMA wire can begin to pull as soon as the heating current is raised, whereas a cold wire would first need to be raised to its operating temperature. It has been observed from experimental results in [71] and [5] that a suitable minimum threshold of input heating power is about 10% of the power applied.

On the other hand, the anti-overload mechanism is in charge of ensuring that the maximum input power does not increase above an upper limit. This approach avoids overheating the SMAs in case the controller delivers a large amount of power to the wires. As mentioned before, this upper limit can be found using a phenomenological model, or by performing real measurements of SMA temperature and stress on the wires.

Both anti-slack and anti-overload mechanisms are key for improving on SMA performance under a force control architecture. The advantages of using a force-control scheme are twofold: i) high-bandwidth response, and ii) SMA fatigue avoidance. High-bandwidth response requires the use of force sensors capable of providing the force feedback. It has also been demonstrated in [71], that by using high-bandwidth force feedback, limit cycles of SMA operation are eliminated.

Nevertheless, for some systems, the use of force sensors could be a hardware limitation. In [5], it has been demonstrated that both anti-slack and anti-overload mechanisms can be implemented in a position control scheme. The position feedback can be achieved by measuring the electrical resistance of the SMA wires, which is a linear function of the strain. The key disadvantage of using a position scheme that forces the SMA to behave in overloaded operation mode relies on fatigue. As experimentally observed in [5], overloaded operation mode could be maintained only for about five minutes of SMA continuous operation before decreasing performance to nominal mode. For the application at hand, overloaded mode implied an actuation frequency of $2.5Hz$, while nominal mode, an actuation frequency of $1.3Hz$. In this case fatigue issues caused a decrease in actuation speed performance about 56%. Further investigations shall be devoted to quantifying the lifetime of SMAs when subjected to higher stresses and larger heating currents within a position control scheme.

5. Conclusions

SMA technology allows the development of a wide variety of robotics designs, exploiting their advantages in terms of weight, volume and sensing capabilities. Their particular features open a wide range of possibilities hardly achievable with classical technology: hydrofoils, morphing shapes, and hydrostats are just some of the new concepts that have been made possible to develop thanks to SMAs. The prototypes described in this chapter have been classified first according to the operation environment (water, air and ground) and second according to their application in the robotics system.

Researchers have successfully overcome SMA disadvantages in terms of power consumption, actuation speed and low strain, by:

- Finding suitable niches of application
- Designing dedicated mechatronics
- Developing ad-hoc control strategies

Biomimetics has emerged as a very promising field together with bio-medical applications. Even if SMAs cannot substitute classical servomotor and hydraulic technology in general, niches can be found where they can effectively compete with, and even outperform standard actuation technology. Clearly, merely replacing servomotors with SMA-based actuators in classical linear-actuation mechanic setups would make little sense. But dedicated mechatronic design, such as embedding SMA fibers in silicone obtaining morphing materials, bending of continuous structures and hydrostats make the best out of this technology, regardless the relatively small strain achievable from the fibers. Finally, cleverly designed control strategies, that exploit the knowledge of the physics of the material and of its behavior over time, coupled with dedicated mechanic setups can dramatically reduce response time.

Our experience with SMA-based actuators has revealed a relatively little considered drawback, which relates to fatigue. In fact, extensive testing with our prototypes has demonstrated that even if SMA performance can be improved by overloading, the effects of overloading disappear quite rapidly. We believe that further investigations shall be devoted to study the behavior of the alloys when exposed to high stresses and large heating currents over a large period of time.

In conclusion, we can say that Shape Memory Alloys have demonstrated great capabilities, as it has been demonstrated by this review of the state of the art. Most importantly, they have the potential to be used in future robotic systems: the more researchers will apply and study them, the more their limitations are smoothed and their employment becomes more effective. After all, they are being extensively investigated only since a short time in comparison with electric motors and hydraulic actuators, that can boast more than a century of history.

Author details

William Coral, Claudio Rossi, Julian Colorado, Daniel Lemus and Antonio Barrientos
Centre for Automation and Robotics (CAR) of the Technical University of Madrid, Spain
CSIC (National Council for Scientific research)

Acknowledgements

This work was supported by the Robotics and Cybernetics Group at the *Centre for Automation and Robotics (CAR) UPM-CSIC*, and funded under the project *ROBOCITY 2030* sponsored by the Community of Madrid (S-0505/DPI/000235) and The project *ROTOS (MultiRobot Systems for Protection of large outdoor infrastructures)* sponsored by the Ministry of Science And Innovation of Spain (DPI2010-17998)

6. References

[1] Brinson, L. C. [1993]. One-Dimensional Constitutive Behavior of Shape Memory Alloys: Thermomechanical Derivation with Non-Constant Material Functions and Redefined Martensite Internal Variable, *Journal of Intelligent Material Systems and Structures* 4(2): 229–242.
URL: *http://jim.sagepub.com/cgi/doi/10.1177/1045389X9300400213*

[2] Bunget, G. & Seelecke, S. [2010]. BATMAV: a 2-DOF bio-inspired flapping flight platform, *Proceedings of SPIE* 7643: 76433B.

[3] Caldwell, D. G. & Taylor., P. [1988]. Artificial muscles as robotic actuators, *IFAC Robot Control*, Karlsruhe, pp. 401–406.

[4] *Case Western Reserve University. Continuous Wave Peristaltic Locomotion, In: Center For Biologically Inspired Robotics Research, 2012* [2012].
URL: *http://biorobots.cwru.edu/projects/softworm/*

[5] Colorado, J., Barrientos, A., Rossi, C. & Breuer, K. [2012]. Biomechanics of smart wings in a bat robot: morphing-wings using SMA actuators, *Bioinspiration & Biomimetics* 7(3).
URL: *http://iopscience.iop.org/1748-3190/7/3/036006*

[6] Dynalloy [2012, Available from: http://www.dynalloy.com/TechDataWire.php]. Dynalloy, Inc. FLEXINOL®Actuator Wire In: Technical and Design Data.
URL: *http://www.dynalloy.com/TechDataWire.php*

[7] Elahinia, M. H. [2004]. *Effect of System Dynamics on Shape Memory Alloy Behavior and Control*, PhD thesis, Virginia Polytechnic Institute and State University.

[8] Elahinia, M. H. & Ahmadian, M. [2005]. An enhanced SMA phenomenological model: I. The shortcomings of the existing models, *Smart Materials and Structures* 14(6): 1297–1308.
URL: *http://stacks.iop.org/0964-1726/14/i=6/a=022?key=crossref.356645df188b7494fe51e5de94b4f281*

[9] Eren, Y., Mavroidis, C. & Nikitczuk, J. [2002]. B-Spline Based Adaptive Control of Shape Memory Alloy Actuated Robotic Systems, *Dynamic Systems and Control*, Vol. 2002, ASME, pp. 471–478.
URL: *http://dx.doi.org/10.1115/IMECE2002-33425*

[10] Esfahani, E. T. & Elahinia, M. H. [2007]. Stable Walking Pattern for an SMA-Actuated Biped, 12(5): 534–541.

[11] Featherstone, R. [2008]. An Architecture for Fast and Accurate Control of Shape Memory Alloy Actuators, *The International Journal of Robotics Research* 27(5): 595–611.
URL: *http://ijr.sagepub.com/cgi/doi/10.1177/0278364908090951*

[12] Featherstone, R. & Teh, Y. H. [2004]. Improving the Speed of Shape Memory Alloy Actuators by Faster Electrical Heating, *9th International Symposium on Experimental Robotics* pp. 1–10.

[13] Fujita, H. [1989]. Studies of micro actuators in Japan, *International Conference on Robotics and Automation*, IEEE Comput. Soc. Press, pp. 1559–1564.
URL: *http://ieeexplore.ieee.org/lpdocs/epic03/wrapper.htm?arnumber=100200*

[14] Fukuda, T. & Asaka, K. [2003]. A new type of fish-like underwater microrobot, *IEEE/ASME Transactions on Mechatronics* 8(1): 136–141.
URL: *http://ieeexplore.ieee.org/lpdocs/epic03/wrapper.htm?arnumber=1187383*

[15] G. K. Batchelor, U. o. C. [2000]. *An Introduction to Fluid Dynamics*, cambridge edn, Cambridge University Press, Cambridge.
URL: *http://www.cambridge.org/gb/knowledge/isbn/item1161783/?site_locale=en_GB*

[16] Rediniotis, O. K., Wilson, L. N., Lagoudas, D. C. & Dimitris, C. [2002]. Development of a Shape-Memory-Alloy Actuated Biomimetic Hydrofoil, *Intelligent Materials Systems and Structures* 13(1): 35–49.

[17] Gorbet, R. B. & Russell, R. A. [1995]. A novel differential shape memory alloy actuator for position control, *Robotica* 13(4): 423–430.
URL: *http://www.journals.cambridge.org/abstract_S0263574700018853*

[18] Granito, M. [2011]. *S.M.A. Shape Memory Alloy Cooling System by Peltier Cells: A Cooling System for Shape Memory Alloy Based on the Use of Peltier Cells*, LAP Lambert Acad. Publ.
URL: *http://books.google.es/books?id=IFvSygAACAAJ*

[19] Grant, D. [1999]. *Accurate and rapid control of shape memory alloy actuators*, PhD thesis.

[20] Grant, D. & Hayward, V. [1997]. Variable structure control of shape memory alloy actuators, *Control Systems Magazine IEEE* 17(3): 80–88.
URL: *http://ieeexplore.ieee.org/xpls/abs_all.jsp?arnumber=588180*

[21] Hashimoto, M., Tabata, T. & Yuki, T. [1999]. Development of electrically heated SMA active forceps for laparoscopic surgery, *IEEE International Conference on Robotics and Automation Cat No99CH36288C* 3(May): 2372–2377.
URL: *http://ieeexplore.ieee.org/lpdocs/epic03/wrapper.htm?arnumber=770460*

[22] Hedenström, A., Johansson, L. C. & Spedding, G. R. [2009]. Bird or bat: comparing airframe design and flight performance., *Bioinspiration biomimetics* 4(1): 015001.
URL: *http://www.ncbi.nlm.nih.gov/pubmed/19258691*

[23] Ho, M., McMillan, A. B., Simard, J. M., Gullapalli, R. & Desai, J. P. [2011]. Toward a Meso-Scale SMA-Actuated MRI-Compatible Neurosurgical Robot: Robotics, IEEE Transactions on, *Robotics IEEE Transactions on* PP(99): 1–10.

[24] Homsy, G., Aref, H., Breuer, K.S., Hochgreb, S., Powell, K., Munson, B. & Robertson, C. [2008]. *Multimedia Fluid Mechanics DVD-ROM*, second edi edn, Cambridge University Press.

[25] Honma, D., Miwa, Y. & Iguchi. [1985]. Micro Robots and Micro Mechanisms Using Shape Memory Alloy to Robotic Actuators, *Robotic Systems* 2(1): 3–25.

[26] Hunter, I. W. & Lafontaine, S. [1992]. A comparison of muscle with artificial actuators, *SolidState Sensor and Actuator Workshop 5th Technical Digest IEEE*, pp. 178–185.
URL: *http://ieeexplore.ieee.org/xpls/abs_all.jsp?arnumber=228297*

[27] Hunter, I. W., Lafontaine, S., Hollerbach, J. M. & Hunter, P. J. [1991]. Fast reversible NiTi fibers for use in microrobotics, *Micro Electro Mechanical Systems*, Ieee, pp. 166–170. URL: *http://ieeexplore.ieee.org/lpdocs/epic03/wrapper.htm?arnumber=114789*

[28] Ihálcz, I. M. [2001]. Fundamental characteristics and design method for nickel-titanium shape memory alloy, *Periodica Polytechnica Mechanical Engineering* 45(1): 75–86. URL: *http://www.pp.bme.hu/me/2001_1/pdf/me2001_1_10.pdf*

[29] Ikuta, K. [1990]. Micro/miniature shape memory alloy actuator, *International Conference on Robotics and Automation*, Vol. 3, IEEE Comput. Soc. Press, pp. 2156–2161.

[30] Ikuta, K., Tsukamoto, M. & Hirose, S. [1988]. Shape memory alloy servo actuator system with electric resistance feedback and application for active endoscope, *International Conference on Robotics and Automation*, Vol. 1, IEEE, pp. 427–430. URL: *http://ieeexplore.ieee.org/xpls/abs_all.jsp?arnumber=12085*

[31] Ikuta, K., Tsukamoto, M. & Hirose, S. [1991]. Mathematical model and experimental verification of shape memory alloy for designing micro actuator, *IEEE Micro Electro Mechanical Systems*, pp. 103–108.

[32] Iriarte-Díaz, J., Riskin, D. K., Willis, D. J., Breuer, K. S. & Swartz, S. M. [2011]. Whole-body kinematics of a fruit bat reveal the influence of wing inertia on body accelerations., *Journal of Experimental Biology* 214(Pt 9): 1546–1553. URL: *http://www.ncbi.nlm.nih.gov/pubmed/21490262*

[33] Kim, B., Kim, D.-h., Jung, J. & Park, J.-o. [2005]. A biomimetic undulatory tadpole robot using ionic polymer-metal composite actuators, *Smart Materials and Structures* 14(6): 1579–1585. URL: *http://stacks.iop.org/0964-1726/14/i=6/a=051?key=crossref.f252f55e78ad58af50f72847f0cd41a3*

[34] Kim, B., Lee, M. G., Lee, Y. P., Kim, Y. & Lee, G. [2006]. An earthworm-like micro robot using shape memory alloy actuator, *Sensors and Actuators A: Physical* 125(2): 429–437. URL: *http://linkinghub.elsevier.com/retrieve/pii/S0924424705003067*

[35] Kim, B., Park, S. & Park, J.-O. [2009]. Microrobots for a capsule endoscope, *2009 IEEE/ASME International Conference on Advanced Intelligent Mechatronics* pp. 729–734. URL: *http://ieeexplore.ieee.org/lpdocs/epic03/wrapper.htm?arnumber=5229926*

[36] Kim, S., Hawkes, E., Choy, K. & Joldaz, M. [2009]. Micro artificial muscle fiber using NiTi spring for soft robotics, *Robots and Systems* pp. 2228–2234. URL: *http://ieeexplore.ieee.org/xpls/abs_all.jsp?arnumber=5354178*

[37] Koh, J. [2009]. Omegabot: Biomimetic inchworm robot using sma coil actuator and smart composite microstructures (scm), *Robotics and Biomimetics (ROBIO), 2009* pp. 1154–1159. URL: *http://ieeexplore.ieee.org/lpdocs/epic03/wrapper.htm?arnumber=5420752* *http:// ieeexplore. ieee.org/xpls/abs_all.jsp?arnumber=5420752*

[38] Koh, J.-s. & Cho, K.-j. [2010]. Omegabot: Crawling robot inspired by Ascotis Selenaria, *International Conference on Robotics and Automation*, IEEE, pp. 109–114. URL: *http://ieeexplore.ieee.org/lpdocs/epic03/wrapper.htm?arnumber=5509425*

[39] Kohl, M. [2004]. *Shape Memory Microactuators*, Springer. URL: *http://www.amazon.com/Shape-Memory-Microactuators-Manfred-Kohl/dp/3540206353*

[40] Kuribayashi, K. [1986]. A New Actuator of a Joint Mechanism Using TiNi Alloy Wire, *The International Journal of Robotics Research* 4(4): 47–58.

URL: *http://ijr.sagepub.com /cgi/doi/10.1177/027836498600400404*
http://ijr.sagepub.com/cgi/content/abstract/4/4/47

[41] Kuribayashi, K. [1991]. Improvement of the Response of an SMA Actuator Using a Temperature Sensor, *The International Journal of Robotics Research* 10(1): 13–20. URL: *http://ijr.sagepub.com/cgi/doi/10.1177/027836499101000102*

[42] Cecilia Laschi, Matteo Cianchetti, Barbara Mazzolai, Laura Margheri, Maurizio Follador & Paolo Dario [2012]. Soft Robot Arm Inspired by the Octopus, *Advanced Robotics* 26: 709-727.

[43] Li, W. D., Guo, W., Li, M. T. & Zhu, Y. H. [2006]. Design and Test of a Capsule Type Endoscope Robot with Novel Locomation Principle, *2006 9th International Conference on Control Automation Robotics and Vision* 20(02): 1–6. URL: *http://ieeexplore.ieee.org/lpdocs/epic03/wrapper.htm?arnumber=4150137*

[44] Liu, S.-H. & Yen, J.-Y. [2008]. Sensor fusion in a SMA-based hexapod bio-mimetic robot, *Workshop on Advanced robotics and Its Social Impacts*, IEEE, pp. 1–6. URL: *http://ieeexplore.ieee.org/lpdocs/epic03/wrapper.htm?arnumber=4653611*

[45] Liu, X., Luo, H.-Y., Liu, S.-P. & Wang, D.-F. [2011]. Pilot study of SMA-based expansion device for transanal endoscopic microsurgery, *2011 International Conference on Machine Learning and Cybernetics*, IEEE, pp. 1420–1424. URL: *http://ieeexplore.ieee.org/lpdocs/epic03/wrapper.htm?arnumber=6016875*

[46] M. Hashimoto, Takeda, M., Sagawa, H., Chiba, I. & Sato, K. [1985]. Application of shape memory alloy to robotic actuators, *Journal of Robotic Systems* 2(1): 3–25.

[47] McNaney, J. [2003]. An experimental study of the superelastic effect in a shape-memory Nitinol alloy under biaxial loading, *Mechanics of Materials* 35(10): 969–986. URL: *http://linkinghub.elsevier.com/retrieve/pii/S0167663602003101*

[48] Menciassi, A., Gorini, S., Pernorio, G. & Dario, P. [2004]. A SMA Actuated Artificial Earthworm, *IEEE International Conference on Robotics and Automation 2004 Proceedings ICRA 04 2004* 4: 3282–3287 Vol.4. URL: *http://ieeexplore.ieee.org/lpdocs/epic03/wrapper.htm?arnumber=1308760*

[49] Migamotors [n.d.]. Migamotors Company, 2012, Avaliable from: http://www. migamotors.com/. URL: *http://www.migamotors.com/*

[50] M.W.M. van der Wijst [1998]. *Shape control of structures and materials with shape memory alloys*, PhD thesis, Universiteit Eindhoven. URL: *http://en.scientificcommons.org/17599133*

[51] Nguyen, Q., Park, H., Byun, D. & Goo, N. [2010]. Recent progress in developing a beetle-mimicking flapping-wing system, *World Automation Congress (WAC)*, pp. 1–6.

[52] Pons, J. L. [2005]. *Emerging Actuator Technologies: A Micromechatronic Approach*, Wiley.

[53] Pornsin-sirirak, T., Tai, Y., Nassef, H. & Ho, C. [2001]. Titanium-alloy MEMS wing technology for a micro aerial vehicle application, *Sensors and Actuators A: Physical* 89(1-2): 95–103. URL: *http://linkinghub.elsevier.com/retrieve/pii/S0924424700005276*

[54] Punning, A., Anton, M., Kruusmaa, M. & Aabloo, A. [2004]. A Biologically Inspired Ray-like Underwater Robot with Electroactive Polymer Pectoral Fins, *IEEE Confrence "Mechatronics and Robotics*.

[55] Rabinovich, M., Varona, P., Selverston, A. & Abarbanel, H. [2006]. Dynamical principles in neuroscience, *Reviews of Modern Physics* 78(4): 1213–1265.
URL: *http://link.aps.org/doi/10.1103/RevModPhys.78.1213*

[56] Raynaerts, D. & Brussel, H. V. [1991]. Development of a SMA high performance robotic actuator, *Fifth International Conference on Advanced Robotics Robots in Unstructured Environments*, Ieee, pp. 61–66.
URL: *http://ieeexplore.ieee.org/lpdocs/epic03/wrapper.htm?arnumber=240475*

[57] Reynaerts, D. & Brussel, H. V. [1998]. Design aspects of shape memory actuators, *Mechatronics* 8(6): 635–656.
URL: *http://linkinghub.elsevier.com/retrieve/pii/S0957415898000233*

[58] Riskin, D. K., Iriarte-Díaz, J., Middleton, K. M., Breuer, K. S. & Swartz, S. M. [2010]. The effect of body size on the wing movements of pteropodid bats, with insights into thrust and lift production., *Journal of Experimental Biology* 213(Pt 23): 4110–4122.
URL: *http://www.ncbi.nlm.nih.gov/pubmed/21075953*

[59] Rossi, C., Colorado, J., Coral, W. & Barrientos, A. [2011]. Bending continuous structures with SMAs : a novel robotic fish design, *Bioinspiration biomimetics* 045005(4): 045005.
URL: *http://iopscience.iop.org/1748-3190/6/4/045005*

[60] Rossi, C., Coral, W. & Barrientos., A. [2010]. SMA Control for Bio-mimetic Fish Locomotion, *International Conference on Informatics in Control, Automation and Robotics (ICINCO)*, Madeira.

[61] Seok, S., Onal, C. D., Wood, R., Rus, D. & Kim, S. [2010]. Peristaltic locomotion with antagonistic actuators in soft robotics, *2010 IEEE International Conference on Robotics and Automation* 0060: 1228–1233.
URL: *http://ieeexplore.ieee.org/lpdocs/epic03/wrapper.htm?arnumber=5509542*

[62] Shi, Z., Liu, D., Ma, C. & Zhao, D. [2011]. Accurate controlled Shape Memory Alloy Actuator for Minimally Invasive Surgery, *Self* pp. 817–822.

[63] Shin, B., Kim, H.-Y. & Cho, K.-J. [2008]. Towards a biologically inspired small-scale water jumping robot, *2008 2nd IEEE RAS & EMBS International Conference on Biomedical Robotics and Biomechatronics*, IEEE, pp. 127–131.
URL: *http://ieeexplore.ieee.org/lpdocs/epic03/wrapper.htm?arnumber=4762896*

[64] Shin, D. D., Mohanchandra, K. P. & Carman, G. P. [2004]. High frequency actuation of thin film NiTi, *Sensors And Actuators* 111(2-3): 166–171.
URL: *http://linkinghub.elsevier.com/retrieve/pii/S0924424703004771*

[65] Suleman, A. & Crawford, C. [2008]. Design and testing of a biomimetic tuna using shape memory alloy induced propulsion, *Computers and Structures* (86): 491?499.

[66] Swartz, S. M., Bishop, K. L. & Ismael-Aguirre, M. F. [2005]. Dynamic complexity of wing form in bats: implications for flight performance, *Functional and evolutionary ecology of bats* pp. 110–130.

[67] Tadesse, Y., Thayer, N. & Priya, S. [2009]. Tailoring the Response Time of Shape Memory Alloy Wires through Active Cooling and Pre-stress, *Journal of Intelligent Material Systems and Structures* 21(1): 19–40.
URL: *http://jim.sagepub.com/cgi/doi/10.1177/1045389X09352814*

[68] Tadesse, Y., Villanueva, A., Haines, C., Novitski, D., Baughman, R. & Priya, S. [2012]. Hydrogen-fuel-powered bell segments of biomimetic jellyfish, *Smart Materials*

and Structures 21(4): 045013. URL: *http://stacks.iop.org/0964-1726/21/i=4/a=045013?key=crossref.fb72cd34852f3dfd4f1efbce ff7c630b*

[69] Tanaka, K. [1986]. A thermomechanical sketch of shape memory effect: one-dimensional tensile behavior, *Res Mechanica* 2(3): 59–72.
URL: *http://www.sciencedirect.com/science/article/pii/0749641986900161*

[70] Teh, Y. H. [2003]. *A Control System For Achieving Rapid Controlled Motions From Shape Memory Alloy (SMA) Actuator Wires*, PhD thesis, Final Year Project, Australian National University.

[71] Teh, Y. H. [2008]. *Fast , Accurate Force and Position Control of Shape Memory Alloy Actuators*, PhD thesis, The Australian National University.

[72] Teh, Y. H. & Featherstone, R. [2004a]. A new control system for fast motion control of SMA actuator wires, *The 1st International Symposium on* .
URL: *http://axiom.anu.edu.au/ roy/SMA/smart04paper.pdf*

[73] Teh, Y. H. & Featherstone, R. [2004b]. Experiments on the Audio Frequency Response of Shape Memory Alloy Actuators, *Information Sciences* .

[74] Teh, Y. H. & Featherstone, R. [2004c]. Experiments on the Performance of a 2-DOF Pantograph Robot Actuated by Shape Memory Alloy Wires, *Information Sciences* pp. 1–8.

[75] Teh, Y. H. & Featherstone, R. [2007]. Accurate Force Control and Motion Disturbance Rejection for Shape Memory Alloy Actuators, *Information Sciences* (April): 10–14.
URL: *http://ieeexplore.ieee.org/lpdocs/epic03/wrapper.htm?arnumber=4209783*

[76] Troisfontaine, N., Bidaud, P. & Dario, P. [1997]. Control experiments on two SMA based micro-actuators, *The Fifth International Symposium on Experimental Robotics V Barcelona Spain* pp. 490–500.

[77] Villanueva, A., Smith, C. & Priya, S. [2011]. A biomimetic robotic jellyfish (Robojelly) actuated by shape memory alloy composite actuators., *Bioinspiration & biomimetics* 6(3): 036004.
URL: *http://www.ncbi.nlm.nih.gov/pubmed/21852714*
http://stacks.iop.org/1748-3190/6/i=3/a =036004

[78] Wang, Z., Hang, G., Li, J., Wang, Y. & Xiao, K. [2008]. A micro-robot fish with embedded SMA wire actuated flexible biomimetic fin, *Sensors and Actuators A: Physical* 144(2): 354–360.
URL: *http://linkinghub.elsevier.com/retrieve/pii/S0924424708001283*

[79] Wang, Z., Wang, Y., Li, J. & Hang, G. [2009]. A micro biomimetic manta ray robot fish actuated by SMA, *2009 IEEE International Conference on Robotics and Biomimetics (ROBIO)*, IEEE, pp. 1809–1813.
URL: *http://ieeexplore.ieee.org/lpdocs/epic03/wrapper.htm?arnumber=5420423*

[80] Waram, T. [1993]. *Actuator Design Using Shape Memory Alloys*, T.C. Waram.
URL: *http://books.google.es/books?id=fjZBAgAACAAJ*

[81] Westphal, A., Rulkov, N. F., Ayers, J., Brady, D. & Hunt, M. [2011]. Controlling a lamprey-based robot with an electronic nervous system, *Smart Structures and Systems* 8(1): 39–52.

[82] Yang, Y., Ye, X. & Guo, S. [2007]. A New Type of Jellyfish-Like Microrobot, *2007 IEEE International Conference on Integration Technology*, number 2006, IEEE, pp. 673–678. URL: *http: //ieeexplore.ieee.org/lpdocs/epic03/wrapper.htm?arnumber=4290404*
http://ieeexplore.ieee.org/xpls/abs_all.jsp?arnumber=4290404

[83] Zhang, Y., He, J. & Yang, J. [2006]. Initial research on development of a flexible pectoral fin using Shape Memory Alloy, *Automation, Proceedings of* pp. 255–260.
URL: *http://ieeexplore.ieee.org/xpls/abs_all.jsp?arnumber=4026090*

[84] Zhong, Z. W. & Yeong, C. K. [2006]. Development of a gripper using SMA wire, *Sensors and Actuators A: Physical* 126(2): 375–381.
URL: *http://linkinghub.elsevier.com/retrieve/pii/S0924424705005935*

Chapter 4

Shape Memory Actuators for Medical Rehabilitation and Neuroscience

Simone Pittaccio and Stefano Viscuso

Additional information is available at the end of the chapter

1. Introduction

Actuators based on shape memory alloys (SMA) proved to be particularly advantageous with respect to other actuation technologies when they are embedded in applications requiring strict compliance to a set of compatibility (e.g. mechanical, biological, weight, …) and environmental constraints. Most noteworthy are the uses in miniaturised components, lightweight systems, sensing-actuating systems (e.g. actuators interacting with changing environmental variables), low-noise or low-impact appliances (e.g. actuators with reduced interaction with the environment) and self-sensing controllable devices. With particular regard to these preferred applications, SMA can also play a role in solving specific actuation problems in the fields of Medical Rehabilitation and Neuroscience. The main characteristics expected of SMA actuators for these fields are light weight and portability; self-adjustment to evolving needs; applicability of actuation in shielded environments (with bioimaging and diagnostic devices: magnetic resonance imaging (MRI, fMRI), magnetoencephalography (MEG), and the like).

Taking the move from these key ideas, this Chapter will discuss some innovative uses and some implemented examples of biomedical devices based on the Shape Memory Effect (SME), explaining their advantages and limitations. In particular, it will address the following main topics: analysis of the relevant literature and background; the derivation of technical constraints and required material properties given the expected biomechanical and or clinical cases; the mathematical design of sample actuators and their implementations; ideas for their clinical application and control. It is important to notice that these same concepts can be easily transferred to other important industrial engineering fields.

1.1. Biomedical background

Neuromuscular Rehabilitation is the science and practice of supporting the recovery of lost or impaired motor function, particularly due to damage of the Central Nervous System

(CNS). The established approach to treatment is physiotherapeutic, pharmacological and surgical, the choice of a specific line of action depending on the exact diagnosis, the general state of the patient, the severity of the damage, the time distance from the acute event, the evolution of sequelae, the presence of co-morbidities, etc.

In the general case, a patient undergoing neuromuscular rehabilitation was struck by a primary event (e.g. a stroke, a traumatic brain injury, an infection, hypoxia, a tumour,...) producing severe neuronal loss. According to the affected area of the CNS different scenarios may result, including a state of paresis (essentially, tetra- or hemi-plegia), and possibly other types of impairment, such as cognitive, sensory and so on. The recovery process from these initial insults often does not proceed independently of a series of interconnected secondary phenomena, which, if left uncontrolled, can undermine all hopes of an optimal final outcome. These phenomena include muscular contracture, spasticity, CNS remodelling and learned non-use.

With no ambition to be exhaustive, it can be pointed out here that the main mechanisms by which the human body, and in particular the neuromuscular system, react to the primary event are mediated by two functional disruptions of the physiological status, namely, *immobility* and *disuse*. Immobility is the absence of movement, especially of a limb or body joint; disuse is the lack of muscular activation and consequent mechanical loading of the skeleto-muscular structure. It is evident that both conditions are tightly connected to and often caused by the state of paresis. *Contracture* (shortening of the muscles) can occur due to prolonged (days or weeks) permanence of the body segment in a given position. This can be the case e.g. for a flexed elbow when patients sitting on wheelchairs keep their hands in their laps; or for an ankle joint that is maintained in an overly extended posture by gravity during continuous rest in bed. Muscular contracture produces, as a major consequence, a forceful and rigid ill-positioning of the adjacent joint, reduction of the available articular range of motion, and malformation. It can lead, in time, through reduced mobility, to structured calcifications of joints and retraction. Alongside contractures, the development of *spasticity* is one of the most common and severe secondary events in the sub-acute phase of paresis. Spasticity is an abnormal and speed-dependent stiffening of the muscles due to hyperactive stretch reflex. This exaggerated response to muscle lengthening exacerbates the impossibility of movement and subjects the patient to discomfort and pain as soon as the limb is mobilised. In turn, contrasting immobility becomes an even harder challenge. The picture resulting from the instauration of contracture and spasticity is given the name of *spastic paresis*, which is a painful, disfiguring and severely disabling condition, and constitutes a serious obstacle to patients' recovery [1-2]. In fact, the increased rigidity of muscles and joints further hinders any voluntary or involuntary motion or use of the limb, thus creating a vicious circle towards greater disruption of neuromuscular functionality.

Besides producing peripheral modifications, the prolonged immobility and disuse of the limb appears to affect negatively the plastic reorganisation and remapping of brain structures, by which live neurons take over the role of lost ones. The manner of such remapping is bound to have a paramount influence on ultimate functional healing, and this is one more motivation for trying and avoiding those conditions in the first place. Their removal could shun the worsening of muscle stiffness and hypertone, while a sustained proprioceptive (sensorial) stimulation through movement may support appropriate cortical reorganisation. However, a complete picture of brain remodelling after a CNS insult and the effect of rehabilitation in modulating such a reorganisation is still not available. Studies of Neuroscience investigating in particular rehabilitation processes are certainly of great interest for physicians, neurologists, researchers and physiotherapists, as they can shed some light on those phenomena.

Recent trends in neuromuscular rehabilitation rely on pharmacological treatments against mild and severe spasticity, such as denervating neurotoxins (BOTOX) or myorelaxation agents (Baclofen) [3]; in most severe cases of contracture the approach can resort to surgical interventions such as tendon lengthening or transfer. The most widely used technique for all severities and types of spastic paresis is however physical therapy, which consists in static repositioning by means of splints and casting, controlled and gradual mobilisation of the limbs, muscular stretching and active exercising. It has been widely acknowledged that active work-out is a particularly important part of the rehabilitation program [4-5]. However, if autonomous control by the patient is absent or insufficient, passive mobilisation is utilised to try and maintain the viscoelastic properties of muscles and periarticular tissues. Furthermore, it is reasonable to think that minimising immobility and disuse, even by means of passive motion, can be advantageous for the prevention of the other adverse sequelae of CNS damage or to contrast their worsening.

Robotic tools are becoming very popular, as they make it possible to extend the duration of therapeutic sessions without the need for an individual involvement of human therapists, and reduce costs. In fact, though time-extensive physiotherapy is known to influence positively the rehabilitation outcome, organisational issues may limit the practical application of this principle, especially for passive mobilisation, in which case the physiotherapist has to move each affected joint one by one. In most situations, robotic tools provide repetitive and repeatable passive mobilisation of the whole limb [6], or joint by joint [7]. Moreover, several rehabilitation robots can assist active work-out, by completing patient's efforts to move [8] with the aim of improving muscle recruitment, by perturbing limb trajectories to enhance active control, or by contrasting movement [9] to strengthen muscles. Some drawbacks can be identified concerning the current use of robots in rehabilitation. First, as many of these machines are large, massive and costly, in a typical clinical environment the number of robots is limited and their use at home is ruled out. Second, most part of them require minimal skills by the patients, e.g. trunk control, equilibrium or cognitive abilities: that excludes very early or severe subjects, although they could benefit from extended sessions of passive mobilisation as well.

Robotic devices have been used for the mobilisation of body parts also in the field of Neuroscience [10-15]. Machines to manage the motion of the effectors (muscles) can in fact be used to support the measurement and visualisation of the interconnections between the activities of peripheral segments and the brain structures involved in their control. In this type of applications tight compatibility with Bioimaging and Biosignal technologies is mandatory [13].

1.2. Overview of shape memory alloys

Robotic devices for Rehabilitation, like all robotic devices, are based on one or more actuators to transfer motion and displace a load. The load, in the case of Rehabilitation, will often consist of body limbs and joints, comprehensive of their mechanical properties, inertia and gravity. As was reported in Section 1.1, there is still a want of appropriate robots, meeting the needs of frailer patients and organisational requirements of the clinical structure. Much investigation in this respect is devoted to the manufacture of light-weight, portable and/or wearable devices. To the ears of the materials scientist or technologist these properties and the capability for actuation are typical features of a special class of metallic alloys, namely Shape Memory Alloys (SMA).

SMA are a compositionally heterogeneous class of materials, comprising a number of intermetallic compounds made up of transition and post-transition metals of different groups. It is beyond the scope of this Chapter to relate about all the different available compositions and their properties, which can be found in reference texts and scientific literature [16-19]. The most relevant compound of this class, employed in most technological applications to date is nearly-equiatomic NiTi, which is also supplied commercially in many semi-finished forms. The general name of NiTi (or TiNi) or Nitinol refers to several grades of the alloy found in the quasi-stoichiometric range from 49%at Ni to 51%at Ni. There are also some important ternary compositions based on NiTi, such as NiTiCu, NiTiHf, NiTiPt, NiTiNb, NiTiCr, with optimised characteristics for dedicated uses.

The Shape Memory Effect (SME) is an athermal reversible martensitic transformation producing macroscopic strain recovery upon heating above a certain characteristic temperature, generally referred to as A_f in the specialised literature. In the case of binary NiTi, the stable low-temperature phase (below the M_f temperature) is a B19′ martensite, while the high-temperature structure is a B2 parent phase. M_f and A_f are separated by a temperature difference of around 20°C in the solution-treated state. The effect of cold working is to strengthen and embrittle the material, and suppress the transformation. A controlled work-hardening however produces beneficial effects on the mechanical properties. In this state the material forward and backward transformations occur across spread-out temperature ranges and the hysteresis is also increased. Ageing treatments are necessary to optimise mechanical and functional properties and to adjust the characteristic temperatures. Typically, NiTi is aged at 350-650°C and water quenched, and this process also shape-sets the material in the shape "to be remembered". Treatment temperature and duration have to be honed to the application and size of the specimen. The mechanical properties, as described by a stress-strain tensile loading curve, may vary but, in the most representative cases, show an initial linear-elastic range, followed by a long flat plateau and a final increase. Most of the macroscopic shape change happens along the plateau region (up to 6% long) and corresponds to the microstructural phenomenon of martensite detwinning.

There are two phenomenological varieties of the unloading curve, according to the temperature of the test. If the test is carried out below M_f, unloading occurs along a line and strain decreases only minimally, so no macroscopic shape recovery is often observable. Full recovery is obtained by successive heating above the A_f temperature. This is called the SME proper. However, if the test is carried out above A_f the unloading curve is different and shows a long recovery plateau, at lower stresses than the loading one, and a final linear decrease towards zero-strains. The curve, in this case, is hysteretic and the shape recovery is attained without any need to heat the material. When the test is at room temperature T_r ,and A_f lies below T_r, SME is given the special name of Pseudoelastic Effect (PE). In all cases where the test temperature falls between M_f and A_f the behaviour is intermediate. Cold-work, ageing and precipitation of second phases can have very significant effects on all aspects of the mechanical curves, and in particular, on the height and separation of the plateaux, and cycling stability. So has the precise test temperature: first of all, in the case of SME loading occurs through the deformation of preformed martensite, resulting in lower stress values of the plateau and lower linear elastic modulus than in the case of PE, where austenite is initially loaded until stress starts inducing the formation of detwinned martensite; furthermore, through the Clausius-Clapeyron effect, stress is proportional to temperature and thus plateau stress levels are increased by an increase in test temperature, in particular in the case of PE. Finally, the compositions in the range 49%at-50.5%at Ni (Ti-rich) tend to show SME proper, while the Ni-rich ones (50.6%at-51%at Ni) have a pseudoelastic behaviour at room temperature. As a general trend, the higher the Ni content, the lower are the transformation temperatures.

1.3. State of the art of SMA in rehabilitation and neuroscience

Many efforts were made in the last 20 years for developing SMA-rehabilitation tools. Figure 1 shows the evolution of the number of papers, patents and conference contributions over this period. It is evident that in the last five years several groups focused on studying the matter, issuing some two thirds of the overall production. All papers dealing with implanted devices were not included in the search, for a number of reasons: because the typical fields of use of implantable devices are hardly connected to Neuromuscular Rehabilitation; because the design of parts to be utilised inside the human body is based on assumptions and limitations rather different from those distinctive of external actuators; because the domain of implantable devices in NiTi is well covered by extensive and comprehensive reviews and is now an established application. A choice was made to include only papers addressing directly the issue of Neuromuscular Rehabilitation or, indirectly, suggesting manners of applying static or dynamic external forces to reposition or move body parts. Particular attention was paid to applications including SMA *actuators*, i.e. devices exploiting the SME to transform heat into mechanical work.

1.3.1. Repositioning

Repositioning is the set of procedures aiming at contrasting ill-postures and malformations caused by contractures and spasticity. Static orthoses are often used to impart these treatments and may be as rigid as castings or partially compliant. Many authors imagined the use of SMA in orthotics as passive components that apply static corrective forces [20-30]. In these applications SMA are mostly employed for their pseudoelastic properties, even though some designs are also based on SME. A series of papers [20-25] showed applications of pseudoelastic NiTi for spastic limb repositioning (elbow, ankle). The authors suggest that pseudoelasticity can be an interesting solution to the disuse and immobility problems during the orthotic repositioning therapies, in that it safeguards residual motor capabilities (voluntary or reflex) and decreases contact pressure by yielding under muscular jerks. They also showed possible advantages of this type of devices for correcting equinus gait [25]. In [26], pseudoelastic NiTi was utilised to remodel deformed auricles. Though not strictly correlated to the main subject of Neuromuscular Rehabilitation, this paper presents an approach in which the anatomic and biomechanical constraints are integrated in the design procedure. In [27], PE is employed to try and correct flat foot malformation and provide increased stability.

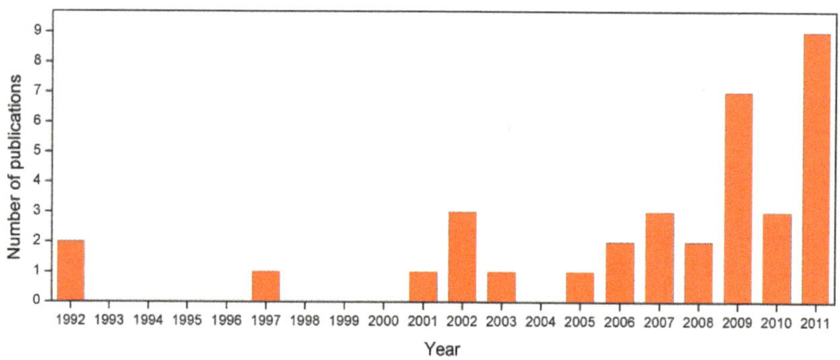

Figure 1. Histogram of the number of publications about SMA-based rehabilitation devices in the last 20 years.

On the other hand, [28] used the Shape Memory Effect to stretch gradually spastic wrist and fingers of paretic patients. The authors of this particular report also carried out measurements of joint resistance to movement, and designed the NiTi elements on the basis of these results. Thanks to the biomechanical design, the joints could be positioned at the neutral angle when the SMA was activated. The main characteristic of SMA exploited for this application was its ability to change shape, which eased putting on the device on malformed limbs. In an international patent [29], SMA are cited as a possible means to produce adjustable degrees of knee joint distraction and assuage contact overload between femur and tibia in unicompartmental osteoarthritis. Another international patent [30] discloses the use of SMA elements as possible actuators in a dynamically adjustable shoe to adapt to congenital or acquired deformities of the foot.

1.3.2. Functional exercise

Passive physiotherapy consists in repetitions of movements, muscular stretching and proprioceptive stimulation imparted by the hands of a therapist or by robots. Some authors

[31-41] investigated the possibility of employing SMA to make portable or wearable devices to act as functional exercising robots.

Some papers focused on finger movement, in particular the review [31] reported an example of rehabilitation glove fully actuated by SMA wires showing the two-way Shape Memory Effect. Another attempt to move fingers is reported in [32,33]. This paper describes the characterisation of bending wire actuators on a dummy finger and a temperature-controlled switch, and clearly evidences the issue of the force-speed trade-off. In [34] another device for passive mobilisation of the flaccid fingers is presented. This paper dealt with biomechanical constraints and design issues connected to the use of SMA springs as actuators, also proposing several modelling equations. Interestingly, both [33,34] employed the composition NiTiCu, which is known to have a narrower thermal hysteresis and lower detwinning stress than binary NiTi.

Other papers [35-41] concentrated on the lower limb. In [35], a concept for a paediatric boot is described. In the intention of the authors, a control system would activate two NiTi wires producing slow movement of the ankle and provide the possibility for home passive mobilisation. Two different implementations of a wearable ankle mobiliser were published by another group [36-41]. In the first paper [36] a description of the design procedure is presented, including mechanical and power dimensioning of SMA linear actuators based upon biomechanical considerations. Pre-clinical trials are also reported in the text. The following publications [40-41] dealt with an amagnetic rotary actuator and its use for the construction of a biosignal-controlled ankle exerciser that could be used in two different modes: fully passive and haptic-assistive, according to the evolving state of the patient during recovery. Trials on healthy subjects are described in [41].

A different use of SMA is proposed in [42], where a NiTi-wire was utilised to produce fingertip tactile stimulation with the purpose of providing a means for sensory feedback in haptic rehabilitation. Several implementations are presented and also qualitative tests on healthy volunteers are reported.

1.3.3. Muscle toning

Some robotic tools provide resisting forces that contrast active exercising by the patients. This could help increase muscle strength and control abilities. A few publications propose SMA-based devices for manifestly therapeutic uses and for generic muscle strength training. In the international patent [43], a set of wearable pseudoelastic garments are proposed to favour muscle toning during daily life activities. Though this application was not described as a therapeutic tool, it shows similarities with techniques that are used in Neuromuscular Rehabilitiation. Two papers by another group [32,33] suggest the use of SMA actuators for applying resisting forces to active movement. Unfortunately, the authors did not discuss either the level of force required or the consequent dimensioning of the actuator.

1.3.4. *Assistive robotics*

Other proposed applications of SMA in Rehabilitation comprise a number of studies that present active tools to support [44-47] or take over [48-51] impaired motor functions.

An M.Sc. thesis and a following paper by the same authors [44,45] report about the design and control of a downscaled bio-inspired mechanical model of an ankle joint wearing a SMA-actuated orthosis to support the walking capabilities of plegic subjects. The thesis discusses the biomechanical constraints and both publications clearly explain the dimensioning of the device. The authors also discuss the relationship between the size and number of SMA elements and the actuation properties. Another group issued a paper [46] and a patent [47] describing the use of SMA actuators to manufacture soft wearable orthoses for the knee joint. The garments were intended to provide sufficient torque to support the impaired gait of neurological patients, but the authors eventually found that the cycling speed was somewhat slower than appropriate.

An international patent [48] describes an orthotic device for restoring the grasping function in paretic subjects. The SMA actuation is intended to substitute entirely the lost function by allowing index and middle finger tips to reach jointly the tip of the thumb. A more complex multi-joint device with the same general purpose was developed in a Ph.D. thesis [49], where motors (possibly SMA-based ones) are however not mounted on the patient's arm, but lie at a distance, being connected to the limb by Bowden cables. This manuscript also includes biomechanical measurements and a control system. Another patent [50] imagines orthotic devices activated by SMA elements aiming at accurately simulating the mobile capabilities of human limbs. An unusual application is provided by [51], where SMA wires are employed as actuators to regain lost facial mobility and the ability to smile. The activation of the wires, aligned along the main affected muscle groups, is triggered by bioelectric signals acquired on the contralateral healthy side of the face.

1.3.5. *Neuroscience*

The scope of limb-moving actuators in the field of Neuroscience is mainly to provide proprioceptive, kinaesthetic or tactile stimulation in order to investigate neural responses to those stimuli. Increasing interest has being expressed during the last few years [10-15] to the effect of finding specialised tools that can be compatible with the instrumentation used to conduct Neuroscience experiments, e.g. magnetic resonance imaging (MRI, fMRI), magnetoencephalography (MEG), electroencephalography (EEG), near-infrared spettroscopy (NIRS).

SMA actuators have found little application in this field as yet: only two examples of SMA-based devices for use in Neuroscience were identified in the literature [37-41]. These publications describe a linear and a rotary actuator designed to comply with electromagnetic noise constraints, and be compatible in particular with MEG and fMRI. Most other devices were implemented using alternative technologies such as pneumatic and piezoelectric. The main reported advantage of SMA actuators is again portability and wearability.

1.3.6. *Final remarks*

Most authors recognised as a major advantage of using SMA in orthotics the light weight and compactness of the actuators. These factors, in fact, suggest that these materials can be used to build portable and/or wearable devices, and oriented the concepts of many designs towards compact robots, light static or dynamic splints, or even soft wearable garments, often for use in the home environment. Going from concept to actual implementation of the devices, the most challenging issue in this field appears to be the dimensioning of the actuators to address at the same time long-strokes and fairly-high-force requirements. In particular, human joints, bones and muscles move or deform across broad ranges of angular motion and several centimetres in length, and possess non-negligible masses. Understanding this is a first fundamental step of an engineering approach to the problem of actuation for Rehabilitation. Some papers did not find an optimal solution to the design question, because there is a serious trade-off between size and power variables to be overcome. In particular, high forces require large cross-sections of SMA: big diameters or many units working in parallel, or both. Any one of these choices, however, brings along some drawbacks: large diameters generally imply larger currents (for Joule's heating) and longer heating and especially cooling times; thin actuators are faster, but building large bundles or arrays of those require cross-insulation of the parallel fibres, larger volumes or surfaces to store them on board and the right strategy of electric connection (in series to minimise the current but requiring high voltages, or *vice versa* for the parallel configuration). It must be kept in mind, as an extra constraint, that patient's safety issues would tend to favour solutions with both low currents and voltages. Furthermore, when precise timing of the movements, matching some physiological rhythm (such as gait), or speed control are parts of the desired functionality of the device, actuation frequency is also an important factor, imposing tighter limitations on the acceptable surface-volume ratio and thermal exchange properties (viz. heating and cooling times) of the SMA actuators. The most successful approaches appeared to be characterised by a precise preliminary analysis of the biomechanics of the joint to be moved, and an optimal and simultaneous modelling solution of the overall functioning and sizing of the SMA actuator: this method made it possible to balance advantages and drawbacks against all standing trade-offs.

In terms of control systems, some designs implemented fine closed-loop architectures to set, maintain or change speed and position of the effector. These systems attracted the attention mostly of designers of dynamic tools, such as those to support the gait or the motion of the fingers. Simpler approaches were used to obtain automatic switching on and off of the actuators, based on some measurable variable (position, temperature, time,...). A few papers employed biosignals (particularly muscular bioelectric activity measured from the skin surface) to trigger actuation. We find this idea particularly interesting, dealing with Neuromuscular Rehabilitation, because it may bring the residual abilities of the patients into the design picture.

Actuators for Neuroscience share a number of characteristics with those described for the field of Rehabilitation. Besides the constraints depending on biomechanics and SMA properties, special requirements must be fulfilled in order for these devices to be able to be used in electromagnetically shielded environments. In particular, if the Joule's Effect is chosen as a means to heat SMA elements and produce phase transformation, care will have to be taken in controlling any unwanted stray magnetic fields generated by the current flow.

2. Uses of SMA actuators in medical rehabilitation

This section will provide guidelines for designing SMA actuators aiming at moving human joints. General statements and suggestions will be accompanied by the discussion of a case study taken from the authors' experience, i.e. an ankle mobiliser conceived as a lightweight and portable tool for providing neurologic patients (stroke, traumatic brain injury, paresis) with motor support and sensorial stimulation not only during the very early phase of rehabilitation when they are still unable to exercise, but also during the subsequent phase of motor relearning, thanks to a specially-designed biofeedback-based control system.

2.1. Biomechanical, bioengineering and clinical specifications

The initial step of a biomechanical design is identifying the anatomical compartment, the pathology of interest and aim of the medical device. This is however just the basic set of information to tackle the problem and provides a very rough description of the actual boundary conditions affecting the following steps. Further human-related design constraints that ought to be considered in developing rehabilitation devices span from the mechanical properties of the joint, to the typical characteristics of target patients, and as far as considering the psychological impact.

2.1.1. Tissues and joints

Any biomedical device, either implantable or external, should take into account the peculiar characteristics of living tissues. Focusing on the musculoskeletal system, a great variety of tissues can be described: cortical bone, cancellous bone, muscular fibres, tendon, ligament, cartilage, synovial fluid, fibrous tissues… All these tissues show non-linear, viscoelastic and hysteretic properties. Some of them, such as the muscular tissues, also display active mechanical behaviour. Moreover, tissue characteristics may vary significantly with age, sex, ethnicity, fitness, pathology or malformation. The properties of biological tissues work together to produce the observed features of joints and limbs, which therefore also display non-linear, viscoelastic and patient-dependent behaviour.

Human joints may differ significantly from one another in terms of dimensions, degrees of freedom, neutral position, range of motion, associated muscular power and weight of the adjacent limb segments, neighbouring structures, etc. A "general purpose" actuator is hard to imagine, but dedicated devices can be tailored to target specific joints and pathologies. Furthermore, once the application has been selected, a statistical analysis of the target population could be carried out and possibly *ad-hoc* assessment of joint characteristics, biometric parameters, classic associated malformations or impairments and comorbidities, so that an appropriate average set of biomechanical parameters is acquired for a "typical user".

Modelling of the joint behaviour is often an essential step in the design process. The desired application can direct modelling choices as to whether a detailed or simplified description of the joint should be included. For example, it may be a purpose of the new device to activate only one degree of freedom (d.o.f.) of a di- or tri-axial joint; in this case, the extra d.o.f's can be excluded from the joint model, but it must be realised that the neglected mobility can produce unwanted stresses or sliding, or pain when the actuator is in use, and it is up to the designer to decide if not controlling those events is acceptable, or else they ought to be controlled by some other means (e.g by fixing and preventing movements along the extra d.o.f.'s). Joint characteristics cannot be considered as unchangeable data, but have to be connected to the current design case: for instance, in judging what a suitable range of motion (ROM) for the actuator should be, the designer must consider the joint type, but also the application (what is the required movement for the rehabilitation exercise?), the pathology (is it likely to change the ROM? Does the actuator have to work to restore the physiological ROM?), the relationships with posture (e.g. certain positions of adjacent joints can affect the available ROM of the target one), the influence of other d.o.f.'s, etc. In relation to this, it can also be decided if joint mechanical properties can be linearised within the ROM of interest.

2.1.2. *Expected loads*

Considering the force or torque generated by the actuator, it is important to keep in mind that joints have passive and active torque-angle characteristics that can be activated by the patient voluntarily or unconsciously.

A passive response is elicited from the muscles and periarticular structures when the joint is moved from its neutral angle, i.e. a resting position where forces producing joint flexion are balanced by those producing joint extension (including or not including gravity, according to limb position in space). As mentioned above, this response is generally nonlinear, viscoelastic and hysteretic, even in physiologic conditions. It can be evaluated by means of a motor-dynamometer-encoder system, in such a manner that the joint can be mobilised passively at controlled speed across a set range of motion, and resisting torque and position can be acquired simultaneously throughout the movement. By changing movement speed the viscoelastic characteristics can be evidenced, while conducting tests both in the flexion and extension direction (or inversion and eversion; or adduction and abduction; etc.) can reveal the hysteresis. The obtained torque-angle curves can be used as load curves in the design of the actuator. The absolute values of passive joint torques strongly depend not only on pathology (viz. contracture, retraction, fibrosis, etc.), but also on age, sex and muscular trophism, and of course change from joint to joint in the body. For this reason we can only give here some examples to clarify the order of magnitude of these torques. Maximum physiological passive torques can be as low as 50-60Ncm for finger joints [52], 100-300Ncm for the elbow [53] and up to 500-1000Ncm for the ankle joint [53-54], also depending on the angle of the knee. In healthy conditions, joint passive stiffness is not prevalently affected by movement speed. The same may not be true in some pathological cases. Muscle contracture and joint retraction can cause the reported values of joint passive stiffness to increase several times and can even make joints completely rigid to any practical purpose.

Apart from a passive response, joints can resist imposed movement also through the activation of the associated muscle groups. Muscular power can be delivered voluntarily (active motion) or by reflex activation. Involuntary active motion can also be produced by neurological conditions such as cloni, dystonia, chorea, etc. During limb and joint passive mobilisation muscle stretching can occur: a dedicated physiologic reflex produces instantaneous muscular contraction when the muscle fibres are elongated over a certain limit. This so called *stretch reflex* is much exasperated in spastic syndromes and may produce an involuntary, uncontrolled increase of resisting loads. Spasticity strongly depends on the speed of stretching; preconditioning the limb by a few cycles of slow manual stretching can help reduce momentarily the intensity of stretch reflex. Very severe spasticity, however, may correspond to the absolute impossibility of mobilising a joint.

Furthermore it must be remembered that the limbs adjacent to a joint possess a mass, and are therefore subjected to gravity and inertial forces connected to accelerations produced even in different parts of the body. One important case is gait. Where and when any or all of these components of the resisting or facilitating loads for the actuation must be taken into account has to be decided for each new application separately, in particular considering the typical position held by the target patient while the actuator is functioning, whether the patient will be prone to uncontrollable movements (jerks, dystonia, convulsions, etc.), or, on the contrary, whether some physiological movements will be typically impeded or suppressed (paresis, coma, castings, etc).

2.1.3. Interfaces

The forces and strokes generated by the actuator must be suitably transmitted to the patient's limb, so connective or fastening elements have to be carefully considered in relation to local anatomical constraints, such as the size of the limbs or body parts and the available surrounding space. In particular, when designing a wearable device, additional dimensional and weight constraints must be taken into account. Generally, the device should be attached to the limb segments that articulate at the target joint, in order to promote their reciprocal movement. Designing such connections should take into account the dimensions of limb segments (that strongly depend on age, sex and pathology) and the available space (determined primarily by joint position and orientation relative to the rest of the body, sex and malformation, etc.). For example, a device to move the shoulder should be designed taking into account the sex of the patient, especially if a frontal strap is required; in the case of a knee device, on the other hand, the lateral aspect of the joint could provide more free space for attachment than the medial one, so that interference with the other leg can be avoided. Another important factor linked to the transmission of forces and movement is the device-skin interface. This must be carefully designed so that no excessive pressure, friction or relative motion occurs during activation of the actuator. Some of these problem can arise due to misalignment of the actuator rotation axis with respect to the corresponding anatomic joint axis. In case of wearable applications, also the weight and stability of the device on the limb should be taken into account.

2.1.4. Other issues

Safety is a mandatory constraint when designing medical devices in general. For the particular case of rehabilitation actuators based on SMA, most risks could arise from electric, mechanical or thermal hazards. IEC 60601-1 is a widely accepted standard for medical electric equipment, and many countries tend to require compliance with IEC 60601-1 before commercialization of an electrical medical device is permitted [55]. Even if the medical device is intended for use only in the scientific laboratory, reference to this set of prescriptions may suggest precautions and guide designers and engineers in the development of safe devices. In connection with this latter standard, international standard ISO 14971 specifies a process for identifying the hazards associated with the use of medical devices and carrying out a suitable risk analysis [56]. Consequently, good practice requires to try and minimise all the identified risks. Moreover, when the medical device is approaching clinical validation, another norm to be considered is ISO 14155, which addresses the issues of designing, conducting, recording and reporting of clinical investigations carried out on human subjects [57]. All of these guidelines are valid also for SMA-based devices and may help consider further implications of new designs for the human body.

Dealing with patients, there are still other questions to be considered. Even if the device is functional and efficient and the actuator fulfils all the technical requirements, they will not be employed in the clinical setting if they are not acceptable to the patient. At the basis of *acceptability* there is a set of physical and psychological factors. Of course, the medical device should not produce nuisance or pain during short or prolonged use. Those problems, in the particular case of rehabilitation appliances and actuators, may depend on badly-designed interfaces with the skin or on movements imparted too fast. In fact, as already stated, contracted and spastic muscles are very responsive to stretch, especially when it is applied rapidly. During mobilisation, the increased muscle tone due to spastic reflexes may result in much discomfort for the patient. The use of SMA actuators in general has the advantage of providing low movement rates; however, severe spasticity may be hard to treat even with slow active devices. Other acceptability issues are often connected to psychological aspects. In particular, wearable devices should not embarrass the patient with noisy, bulky or cumbersome structures. For children, shape and colour can also be important features. SMA actuators are likely to be silent and miniaturised, and thus suitable in this respect, but all other components may not (i.e. power supply). Other impediments for patients employing rehabilitation devices actively or for assistive therapies may arise from their possible co-existing cognitive impairment: this consideration supports the use of very clear instructions and user interfaces, especially using non-verbal pictorial communication, lights, vibrations or sounds.

2.2. Actuator design

2.2.1. Strokes and loads

Rehabilitation devices in most cases are intended to generate a mutual rotation of limb segments around an anatomical joint. By employing rotary actuators, it is possible to provide directly angular strokes and torques. On the other hand, linear actuators can be coupled with suitable lever arms and achieve the same results indirectly. The first step in designing an SMA-based rehabilitation actuator is to define, by careful analysis of the biomechanical constraints and clinical requirements, the appropriate neutral position, range of motion and resisting loads. The neutral position and range of motion will help determine a correct linear or angular stroke for the actuator; the resisting loads will set the linear force (and lever arm) or torque requisites. Of course, the two quantities will influence each other, for example different choices of lever arm may affect the linear stroke.

Once the required stroke and load have been determined, the geometrical parameters of SMA element can be worked out. Different types of actuators need different formulas for determining the parameters: for example, wire actuators have only two parameters, length and cross-section that directly depend on stroke and load respectively, while springs (helical, torsion or flat) parameters comprise coil dimensions, wire diameter and number of turns, which all depend on both stroke and load. Moreover material properties influence actuator parameters, especially in springs where anisotropy may play an important role [34]. Attention should be paid in particular to the critical stresses needed to detwin martensite both in the heated (above A_f) and cooled (below M_f) states. This consideration can guide the choices about material composition and thermomechanical treatment, because those factor have great impact on the height of the plateaux and the mechanical hysteresis. In most actuation applications, in fact, it must be guaranteed that the load to be displaced is low enough that stress in the material is lower than the critical stress for the heated state; on the other hand, sufficient loading should be provided to detwin martensite and ensure actuator deformation and position reset during cooling for most cyclic applications. Whatever the SMA geometry, it can be stated in general terms that cross-section depends mainly on the maximum desired work output, through the mean level of stress acting on it. The question of choosing the level of stress is quite complex, and also involves the expected fatigue life of the actuator and the strains recovered upon activation. For example, it is reported that commercial NiTi-wire actuators can remain stable for more than 10^5 cycles at 200MPa and 4% of strain in traction [58], but other optimal compromises can be found, provided that stress does not exceeds 400MPa and strains 5%. The use of parallel actuators can help reduce required work from each actuator, i.e. given a stress level, cross-section can be reduced. Dimensioning cross-section should also take into account cooling rates, as will be discussed in the next section.

2.2.2. *Power*

Heating and cooling are key points concerning SMA actuation. The general equation that drives these processes is

$$P_{in}(t) = c_p m \frac{dT}{dt} + m\Delta H \left|\frac{d\xi(T)}{dt}\right| + \frac{dW(\theta(t),t)}{dt} + P_{diss}(t). \quad (1)$$

Considering the heating (actuation) phase, $P_{in}(t)$ is the heating power, $c_p m \frac{dT}{dt}$ is the power needed to increase SMA element temperature (c_p is the constant-pressure specific heat and m

is the mass), $m\Delta H\left|\frac{d\xi(T)}{dt}\right|$ is the power involved in phase transformation (ξ is the fraction of austenite and increases in time as transformation goes on, ΔH is the transformation enthalpy), $W(\theta(t),t)$ is the mechanical output (useful work, inertia and frictional losses) which varies along the movement trajectory $\theta(t)$, and $P_{diss}(t)$ is the power dissipated by thermal convection, thermal conduction and (less important in this case) thermal irradiation. Considering the cooling phase, and imagining that a bias force acts to restore the starting actuator configuration (by deforming the SMA element), $P_{in}(t)$ is again a heating power that could be used to help control the cooling rate, $c_p m\frac{dT}{dt}$ is the power to be dissipated in order to decrease SMA element temperature, $m\Delta H\left|\frac{d\xi(T)}{dt}\right|$ is the power involved in phase transformation (enthalpy in this case is negative), $W(\theta(t),t)$ accounts for the mechanical energy stored in detwinned martensite plus inertial and loss terms, and $P_{diss}(t)$ is the power dissipated. The contribution of $P_{diss}(t)$ is present both during the heating and cooling phases of an actuation cycle, but it may differ significantly if any active cooling system is included. Equation 1 gives a simplified description of the process, but permits to focus on some key points. First, SMA mass is an important factor when designing power consumption. Second, as all the diverse forms of dispersed energy depend on the amount of exposed surface, surface/mass ratio is the driving contribution to determine cooling time. Therefore, thin actuators (maximising the surface/mass ratio) should be preferred to bulky ones if cyclic, especially high-frequency actuation is desired. Having in mind which use is intended for the device provides a guidance for the choice of heating source and consequently gives some hints for the actuator design. The best strategy for providing controllable, cyclic actuation is heating by Joule's effect, but, especially for non-cyclic and one-shot activation, also direct thermal transfer may be suitable. Focusing on Joule's effect, the energy provided depends on the injected current and the electric resistance of the SMA element according to Equation 2:

$$P_{in}(t) = P_{Joule}(t) = I(t)^2 \cdot R_{SMA}(\xi) \tag{2}$$

where the electric resistance $R_{SMA}(\xi)$ can be expressed as

$$R_{SMA}(\xi) = \rho(\xi) \cdot \frac{L(t)}{S} \tag{3}$$

with $\rho(\xi)$ representing the SMA resistivity, S the cross-sectional area of the SMA element, L(t) the length over which the current flows, that obviously changes as actuation goes on. Given a length of SMA wire (which depends directly on the stroke required from the actuator), the thinner the wire, the higher is the electrical resistance. Moreover, mass-dependent contributions to Equation 1 are also minimised, and heating efficiency increased.

SMA actuators are subjected also to the Clausius-Clapeyron relation, that can be expressed as:

$$\theta^* = \theta + \frac{\sigma}{C} \tag{4}$$

where $\boldsymbol{\theta}$ is one of the transformation temperatures, which is effectively increased to $\boldsymbol{\theta}^*$ when a stress σ is applied to the material. This is a very important issue when purchasing shape memory materials or dimensioning SMA actuators powering: the transformation temperatures as provided by DSC analysis will be shifted when the material is loaded. Typical values of *C* for NiTi range from 5 to 9 MPa/°C, i.e. an actuator designed to bear 200MPa would have transformation temperatures increased by as much as 40°C. However, actuation temperatures cannot be increased indefinitely: in fact, above a certain temperature (M_d), part of the deformation cannot be recovered and material undergoes plastic slip.

2.2.3. *Choice of the SMA element*

SMA actuators can be provided in a number of shapes, which may differ significantly from one another in terms of available stroke and force. The first distinction is based on the deformation mode imposed onto the material. Wires are subjected mainly to traction, which produces a uniform loading of the material cross-section: this fact makes it possible to produce high force actuators, but, on the other hand, the available stroke is limited by the fact that linear deformation can be as high as 8% only, 4% in most cyclic operating modes. Other simple SMA actuators are linear springs (tension or compression springs), whose principal mode of deformation is torsion. Springs of this kind are helical coils of SMA wire arranged around a central axis: this allows for large elongations along the central axis depending on spring ratio and the number of turns. On the contrary, force is limited by the fact that torsional loading is not uniform in the material cross-section. Some authors proposed also bending actuators, made of SMA ribbons [27] or SMA wires [33]. For the same purpose, also torsion springs and flat springs can be employed, which share the same principal mode of deformation, i.e. flexion. The advantage of this actuation is that angular stroke is directly available, but the non-uniform cross-section loading limits the available torque.

The problem of moving human joints generally involves large angular strokes coupled with fairly high levels of force or torque. The simple configurations just described most of the time provide solutions only for very long wires or thick cross-section springs, ribbons or bars, which are often deemed impractical for a number of reasons. Long wires have to be housed in a suitable manner, to limit length into a compact three-dimensional structure. A way to do so is coiling the SMA wire along a series of pulleys, whose diameter should be sufficiently large with respect to the wire diameter, to avoid strain concentrations. On the other hand, large cross-section actuators are impractical because they need high electric power to reach the transformation temperature. Moreover, cooling down bulky wires or ribbons may require too long a time for most cyclic applications. A possible solution for increasing force output in springs and ribbons is arranging many actuators in parallel or in bundles. Provided space around the joint is sufficient, attention should be posed to electrical insulation and electrical connections, in order to provide suitable solution to tension-current requirements for actuation, keeping in mind also the safety issues.

2.3. Actuator control

The problem of controlling actuation is very important for many applications in Rehabilitation, as patient's needs and responses may vary during the evolution of therapy or even during the same session of exercise. Control strategies apply mainly to assistive robots, but may also be employed in passive mobilisation devices. As the focus of this Chapter is not on control systems, only a brief overview of possible applications to SMA rehabilitation devices will be discussed.

2.3.1. *Control of passive mobilisation cycles*

Two major strategies exist to control cyclic heating and cooling of SMA actuators: in *open-loop control* parameters are predetermined by a set routine, whereas *closed-loop control* employs a feedback signal of any nature to adjust heating and cooling parameters (timing, current or voltage, active cooling systems, etc). The main drawback of open-loop control is that any perturbation to the trajectory would not be compensated, as the current profile is predetermined. This could happen, for example, if patient exerts any unexpected active contraction or in the case of unpredicted changes in environmental conditions. However, there are applications in which open-loop control is still feasible, for example repetitive passive motion of flaccid limbs. Closed-loop controls need monitoring of a feedback variable, which could be the SMA temperature, the joint angle, actuation speed, SMA force or any other measurable physical quantity. Apart from rare examples of feedback loops based on monitoring of SMA resistance [59], closed-loop controls rely on dedicated sensors that are better included from the initial steps of the device design.

Another distinction can be made on different types of control of movement trajectory. In many applications an ON/OFF actuation is suitable: the focus is only on the initial and final positions, while the detailed trajectory is controlled only mechanically and the speed only in average terms. In this case, open-loop control may be practical, and heating can be achieved with a very simple current profile, e.g. a step or a ramp. Experimental tests are a viable solution to adjust the heating and cooling parameters. In particular, attention should be paid to the experimental conditions, including the orientation of the actuator in the gravitational field, as thermal convection is strongly dependent on that.

On the other hand, when a precise trajectory of movement over time is required for the application, alternatively open-loop or closed-loop control strategies can be applied. In the open-loop approach, Equation 1 should be solved by imposing the desired mechanical output $W(\theta(t),t)$ and by calculating the time-varying current profile to be injected. However, the most reliable way to provide trajectory control is closed-loop control, in which the current input is continuously adjusted to match the prescribed position, thus counterbalancing perturbations.

2.3.2. *Assistive control*

As rehabilitation implies that patient's capabilities evolve during therapy, the characteristics of therapeutic tools could be devised to change accordingly. A way to implement this is to trigger SMA actuators only when the patient makes some effort to move the paretic limb: thus, a measure of the "effort" should be chosen to implement such a biofeedback-based closed-loop control. In some cases, mechanical variables could be employed: for example, the SMA actuator completes movement only if the patient generates at least a certain level of voluntary muscular force or flexes the joint over a minimal extent. On the other hand, possible biosignals for closed-loop control can be the surface electromyographic signals from the muscles that control the movement to be rehabilitated. The level of the biofeedback variable above which actuators are switched on may depend on the patient, pathology and progression of therapy, and therefore should be adjusted to match patient's capabilities with rehabilitation goals.

Besides the biofeedback signal, another key point for assistive rehabilitation is how to encourage the patient to participate actively in the rehabilitative session. Visual or verbal or acoustic feedback are the most common way to prompt, instruct and reward the patients during the exercise, but when selecting the communication format attention should be paid to their possible co-existing cognitive impairment.

2.3.3. Gait triggering

A very peculiar example of precise trajectory control is gait, during which any wearable active device should be able to activate according to the walking rhythm of the patient. A closed-loop control is desirable, and can be implemented with mechanical sensors (which detect angles at different joints or contact forces when the foot contacts the ground) and/or electromyographic monitoring of the leg muscles. Gait is a quite demanding application for SMA in terms of frequency of activation, as even slow walk takes place at over 1Hz. This influences dramatically actuator design, i.e. diameter and number of SMA elements. For gait applications, movement trajectory must be controlled fully within the step and from step to step: in other words, both heating and cooling phases should have adjustable parameters so that actuation can adapt to changing stride or external perturbations. The possibility of modulating input current profiles has already been discussed and may be sufficient for the control of heating; cooling, on the other hand, shall not be left to natural convection in this case, but an active system for accelerating and tuning heat transfer will often have to be adopted.

2.4. Case study: Passive/assistive SMA device for ankle dorsiflexion

2.4.1. Design aims and clinical constraints

Ankle is one of the body joints that often suffer from the sequelae of stroke, i.e. immobility, disuse, contracture and spasticity. After stroke, a rehabilitation program should be carried out as soon as the patient's conditions are sufficiently good, but in the acute phase the patient is unable to exercise actively. In these circumstances, moving the limb passively can help maintaining joint flexibility and normal muscle tone, while also contrasting contracture, spasticity and sensory deprivation. A rehabilitation device for promoting repetitive dorsiflexion of the ankle in the first few weeks after the acute event could help improve clinical outcomes. The fact of limiting mobilisation to just one degree of freedom (sagittal plane of the joint) is appropriate because dorsi-plantarflexion is the principal motion involved in gait. It is regarded as a matter of importance that the device be able to manage patient's changing conditions during recovery. As needs evolve during patient's improvements, this therapeutic device should be able to guide and sustain gradual recovery by providing commensurate aid. This includes exploiting even initial attempts at voluntary motion and turn those into effective workout. To this end appropriate control will have to be included.

The device is intended for patients in the first weeks after a stroke, i.e. we should expect adult subjects having a flaccid ankle (no active control of the muscles, mild or no hypertone) and possibly increased ankle stiffness. Figure 2 shows the passive viscoelastic ankle characteristic of a healthy subject at rest (no spasticity, no contracture, no voluntary movement) and a typical chronic flaccid patient with mild contracture. The ankle joint can move in the sagittal plane in the approximate range -30°/+20° (negative being towards plantarflexion) even though the range of -15° to +10° is the functional range of utmost interest in gait. These ranges become reduced with evolving contracture and spasticity, and in particular the positive degrees are progressively lost. In the range of interest (-15°/+10°) the characteristics of the presented healthy and paretic joints differ mostly for dorsiflexed angles (greater than around +2°, in this case), where increased stiffness in the paretic ankle can be observed. The typical acute flaccid patient will have intermediate characteristics between the healthy one and the chronic.

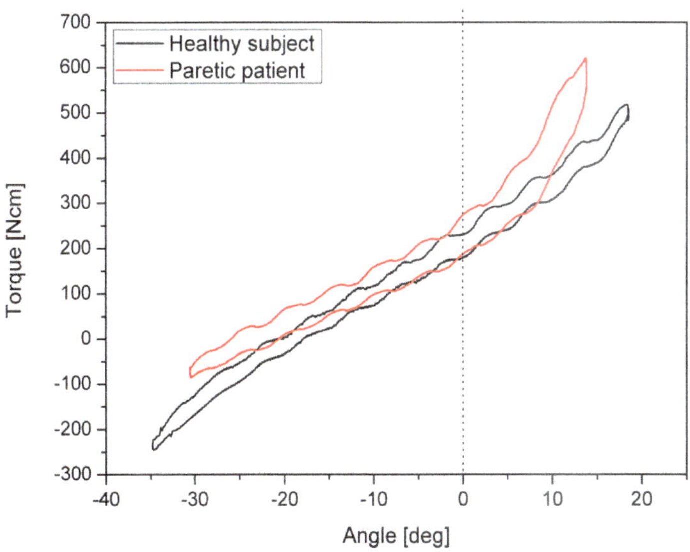

Figure 2. Measured passive characteristic of the ankle in a healthy subject vs. a paretic patient with mild contracture and ROM limitation. The main difference appears for positive dorsiflexion angles

Given this information, the biomechanical constraints on the design can be summarised as follows:

1. the resisting forces which have to be considered in actuator dimensioning are the viscoelastic passive characteristic of the joint and the foot weight. No active contribution from the muscles are expected.

2. Choosing a reduced range for the device, biased towards dorsiflexion, has the advantage that the calf muscles are kept under a relative stretch (physiological neutral angle is around -25°); furthermore it is appropriate to avoid forcing the joint excessively towards the extreme dorsiflexion position (conservative approach to limit insurgence of spastic reflexes). A target ROM for mobilisation was therefore was chosen as -5°/+10°. Of course, the actual working range will depend on the passive characteristic of the specific ankle, but is not expected to be much different from the target one.
3. For a target population represented by this type of subjects it is reasonable to assume that passive resisting torque does not exceed 400Ncm in the range of interest.

Considering the general conditions of the patients in the first weeks after a stroke, it would be wise to design a device that can be employed in the bed. However, the knee joint should be sustained in a flexed position, so that the bi-articular *gastrocnemius* muscle is not pre-stretched and full ankle range of motion is available. As a way of compromise, an angle of 10° for the leg rest can be assumed, with the foot positioned lower than the knee. It should be taken into account that, in such a configuration, the contribution of foot weight to the resisting force is limited. Figure 3 shows the total resisting torque, comprising different foot weights, and the same viscoelastic resistance from a typical patient. The expected maximal resisting torque can be approximately calculated in 400Ncm. As the curve hysteresis is not large for this joint (cf. Figure 2), the same values used for dorsiflexion can be also utilised when considering the movement towards plantarflexion and the loading level associated to martensite detwinning.

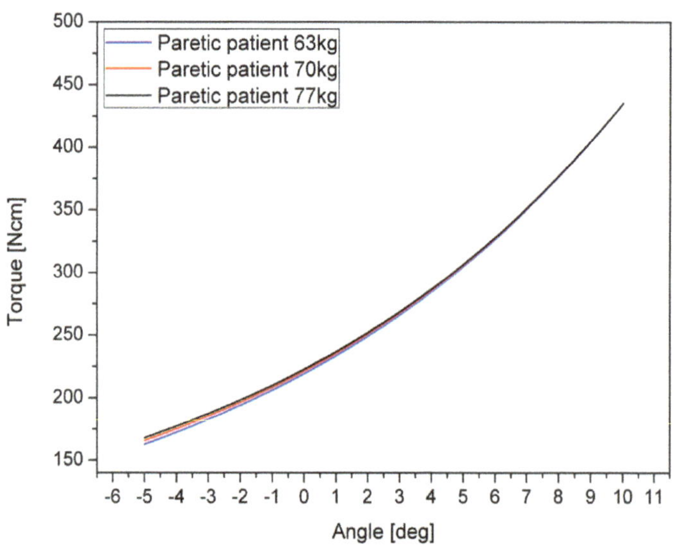

Figure 3. Influence of foot weight on the total resisting torque at the ankle. Differences in foot weight (proportional to the body weight) account for little change due to the lying position of the limb.

As the device is not intended for walking, there are no direct constraints on cycle duration, provided that a suitable number of cycles can be delivered to the patient's ankle in a therapeutic session (at least 100 cycles/hour), and movement speed is not too elevated, lest spastic responses are aroused. The actuator should provide repeatable movement and maintain the same characteristics for a number of sessions, as it may be impractical to adjust or change the SMA actuator too often. For this reason, maximal strain should be limited to 3% and stress to 300MPa.

2.4.2. General concept of the device

Figure 4 shows a simple embodiment of the concept. Two thermoplastic shells lined with soft foam are modelled on a prototype human lower limb of average size. These shells are hinged together and strapped by Velcro® bands respectively on the frontal aspect of the shin and the foot, in such a manner that the ankle and the hinge are perfectly aligned along the same axis (this minimises unwanted friction and sliding between the orthosis and the skin). The choice of planar hinges was also made to control (fixate) ad/abduction and in/eversion, although this solution makes the device unsuitable for patients that have already developed severe malformations out of the sagittal plane (unusual in the acute phase). With this structural configuration, two linear actuators are fixed on the front of the leg shell and transfer actuation pull to the foot shell through inextensible threads.

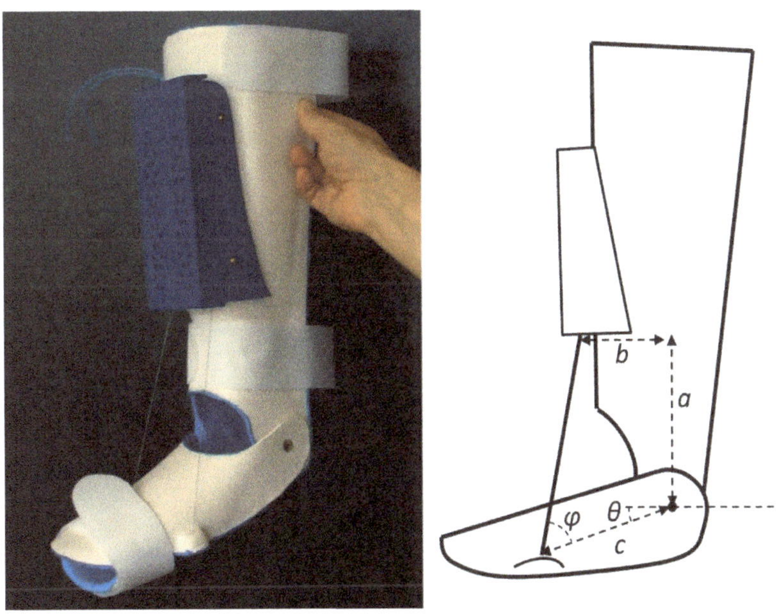

Figure 4. Implementation of an exerciser for the ankle joint with linear SMA actuators. The device is designed to provide cyclic joint flexion in the rehabilitation of neurological patients in the acute phase. On the left, schematic representation of the main design dimensions to calculate the lever arm.

The linear stroke and force output required of the SMA actuators depend, through the lever arm, on the fixation points for the actuators (on the shin), and the distal ends of the inextensible threads (foot shell). Actuators produce dorsiflexion, while plantarflexion and actuators position reset is left to gravity and viscoelastic resistances.

Power supply is to be provided by a dc-generator, which is not on board the wearable device, as the intended use is not for walking. This choice helps limiting weight and keep the device stable on the patient's leg during use.

An open-loop control strategy is adopted for the use as passive exercise device, whereas a simple closed-loop control for assistive rehabilitation employs the electromyographic activity from *tibialis anterior* as control signal, picked up by surface electrodes. As for the power supply, it was decided that all the components needed for control are not mounted on board the wearable device, even though integration would be possible by developing *ad-hoc* electronics.

2.4.3. Actuator design and characterisation

With the selected set of geometrical parameters (Figure 4: a=16cm, b=8.5cm, c=15cm), a linear stroke of 2.5cm is required to produce an angular movement across the range -5°/+10°. Considering a linear deformation of 3%, the required amount of wire for linear actuation would be 83cm. However, in our case it would be impractical to have a free-standing wire of such a length as a linear actuator. For this reason, the NiTi wire should be confined into a more compact actuator. In dimensioning the length of wire now we must take into account also the local deformations that inevitably will be imposed on the SMA wire when coiling it. The actuator was designed as an insulated cartridge wherein the necessary reference length of NiTi wire is led back-and-forth between two arrays of ten pulleys. An end of the wire is connected to an inextensible thread that transmits the force to the foot, while the other one is fixed to the housing. A pseudoelastic spring is connected to the moving end of the NiTi wire in order to keep it just taut, and its pull is negligible during actuation.

Considering the force requirement, each actuator should bear half of the resisting torque T_r showed in Figure 3, and the amount of force varies at varying ankle angles θ (defined between the foot direction and the horizontal, Figure 4):

$$F(\theta) = \frac{T_r(\theta)}{c \cdot \sin\varphi} \tag{5}$$

where φ is the angle between the inextensible thread and the foot and depends on θ according to:

$$\varphi(\theta) = \frac{\pi}{2} - \theta - \arccos\left[\frac{c \cdot \sin\theta + a}{\sqrt{(c \cdot \cos\theta - b)^2 + (c \cdot \sin\theta + a)^2}}\right] \tag{6}$$

With a maximum torque of 200Ncm, Equations 5 and 6 give a maximum load of 13N on each actuator. By selecting a 250μm-diameter commercial NiTi wire, stabilised for actuation, the maximum stress on the wire cross-section is 266MPa. In order to complete plantarflexion, full martensite detwinning should be achieved. It is certainly possible to find stabilised wires for which martensite detwinning occurs with less than 100MPa, corresponding to around 5N pulling on a 250μm-diameter wire. With the expected resisting torque in Figure 3 and the geometrical parameters, force requirement for martensite detwinning is always satisfied in the range -5°/+10° and thus cyclic actuation will take place in that range. Considering a diameter of the pulleys of 12mm, the maximum amount of localised strain is 2%. In order to respect the 3% limit for the maximal deformation, only 1% of strain is available for actuation and can be employed for calculating the length of wire to be coiled within each actuator, that is to say, 250cm.

Having limited strains to 3% and stress to 300MPa, a cycling life of 30k-100k cycles is expected [58].

2.4.4. *Power dimensioning*

In order to evaluate the parameters for Joule's effect heating, various experimental tests were carried out on samples of NiTi wire. First, the transformation temperatures were investigated by means of differential scanning calorimetry (on DSC 220 SSC/5200 - Seiko Instruments, Tokyo, Japan), showing A_f = 351K and M_f = 274K. However, calorimetry gives information about the material with no loads applied, whereas Equation 4 suggests that transformation temperature depends on the load applied. For this reasons, tensile tests on the material at different temperatures were conducted using an MTS 2/M thermo-mechanical test machine (MTS Systems, Eden Prairie, MN, USA) equipped with a 2kN load cell. The material was deformed up to an engineering strain of 5%, at 365K, 380K, 390K and 400K. The pseudoelastic plateau values varied as a function of temperature with a ratio of 8.402 MPa/K, which is exactly the constant C in Equation 4. With the estimated stress level of 266MPa, full transformation and strain recovery can be achieved at 383K.

Activation tests on the wire were conducted injecting different currents (0.65A, 0.7A, 0.75A, 0.8A) in the wire at a constant strain of 3% for a set period of 13s, in order to evaluate what current value is most appropriate to provide the working load of 13N, minimizing current expenditure and heating time. As expected, the higher the current the faster is actuation. Patient safety considerations should be taken into account, as well. A compromise solution can be accepted with 0.7A, which allows for reaching 13N within 6s. With the selected current, voltage is limited to 35V, considering an average resistance of 49.5 Ω (45-54Ω during transformation from fully austenitic to fully martensitic). Notice that the rehabilitation device mounts two actuators, which need a 0.7A current each. The electrical configuration of the two actuators could be a series or a parallel circuit: in the first case, the dc-generator should be able to provide 0.7A at 70V, in the latter case 1.4A at 35V. According to IEC 60601-1 specifications, dc tension preferably does not exceed 60V, i.e. in our case the two linear actuators are better connected in parallel.

Heating parameters can be extracted by testing a free-standing NiTi wire, and by applying 35V to the ends of the cartridge actuator suspending a 13N weight. It was demonstrated that nominal linear stroke is reached in around 6s (mean dorsiflexion speed 2.5°/s). On the other hand, cooling times depend strictly on the geometrical arrangement of the wire in space. A compact actuator may have a considerably slower cooling and position reset time, as tests on the actuator confirms. Basically, if the vertical free-standing wire cools down in approximately 10s, cooling and actuator position reset takes place in 30s. The full cycle thus lasts 36s, which makes it possible to deliver 100 cycles/hour to the patient's ankle, as required by the design specifications.

2.4.5. Control strategy

An open-loop control is adopted for using the device as a passive exerciser. A computer routine was written in LabView (National Instruments, Austin, TX, USA) to control an electronic relay (NI9481 - National Instruments, Austin, TX, USA) that closes the electric circuit between the dc-generator and the actuators.

For the assistive therapy, a closed-loop architecture is preferred, with surface electromyographic signal (sEMG) from *tibialis anterior* as control variable representing the patient's attempt to move his ankle. Three Ag/AgCl electrodes (positive, negative and reference) were used to pick up the signal. Analogical waveforms were acquired, pre-amplified and band-pass filtered (18-478Hz) before being digitalised (sampling at 1000Hz) using an NI9205 (National Instruments, Austin TX, USA) connected to an ordinary laptop computer running a LabView routine. Pre-amplifying and filtering stages were assembled using conventional 8-pin PDIP components, and included also a feed-back loop towards the body (akin to the driven right leg stage used in electrocardiographers). Further signal manipulation is built into the control software, including rectification and low-pass filtering (3Hz), so that the obtained waveform is sufficiently smooth that it can be employed as a measure of instantaneous muscular activation.

Through the user interface of the control routine two patient-specific sEMG threshold values can be selected. The lower one is set to the minimal required level of exercise (which can lie even lower than the muscular motor threshold, in some cases); the upper one is set to an appropriate activation representing the ultimate (or a higher) therapeutic goal, i.e. an effective motion.

At the start of the assistive exercise session, a visual cue is presented to the patient to dorsiflex the ankle. Then the system is set on hold waiting for the sEMG from *tibialis anterior* to cross one of the threshold values. If the lower threshold is reached, then the system waits a few milliseconds for the upper threshold also to be reached. If this latter event does not occur before the time-out, then power is supplied to the actuators and the motor task is completed for the patient. If, on the contrary, the upper threshold is reached, then a visual feedback is provided to the patient that the higher goal was hit, while the device does not intervene to support the movement.

2.4.6. Experimental tests on the device

The assembled device was tested with a static load of 40N fixed on the foot shell at a distance of 10cm from the hinge axis, while measuring the angular displacement by means of electrogoniometer SIM-HES-EG 042 (Signo Motus, Messina, Italy). The device was tested for 4000 cycles, with the actuators powered with a 6s step of 1.4A (0.7A per cartridge) and 30s allowed for cooling. The results of these tests (Figure 5) demonstrated that stroke is quite stable across cycling. Plantarflexion occurs more rapidly after cycling because of the instauration of two-way shape memory. Following this result, by employing cycled wire or wire stabilised for two-way shape memory, cycle duration could be reduced from 36s to 30s without changing the stroke, if desired (reaching 120 cicles/hour).

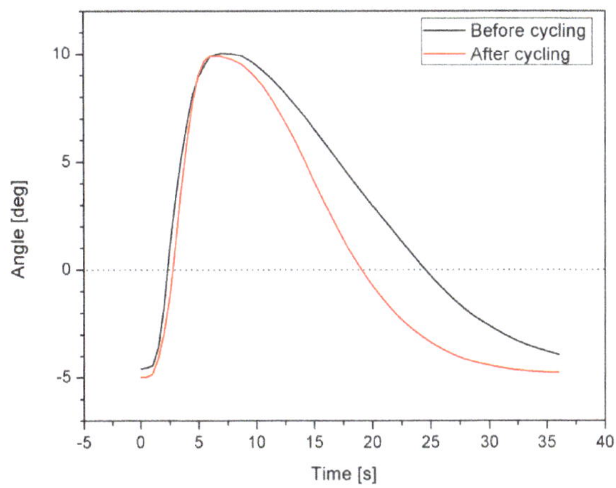

Figure 5. Angular stroke of the SMA-based device for ankle dorsiflexion: first activation and after 4000 cycles against a resisting torque of 400Ncm.

The closed loop control strategy was tested on three healthy volunteers (28.17±6.08 years old). Ag/AgCl electrodes were placed on the belly of the *tibialis anterior* muscle of the dominant side, the corresponding distal muscle-tendon junction, and the internal malleolus (driven electrode). Subjects were asked to perform a maximal isometric contraction at the ankle neutral position, then to sustain the minimum voluntary activation they could manage. Subsequently, values were set for the lower (110% of minimum individual contraction) and upper (60% of individual maximum isometric contraction) thresholds. Then, the control routine was launched and subjects were asked to follow on-screen instructions (graphic and written) trying to respond with just a supra-minimal contraction when cued to dorsiflex the ankle.

The measured joint angle and sEMG time courses for a representative subject are shown in Figure 6. When the lower threshold was crossed, the system triggered the powering of the orthosis, which completed the movement of dorsiflexion (assisted active session). It can be appreciated how passive mobilisation can be triggered by a very subtle muscular contraction, which may correspond only to a very slight movement (~1°). The movement produced by the orthosis as a consequence of a minimal contraction brings along some degree of reflex sEMG activity: this may also be thought of as an interesting result to the effect of rehabilitative exercise.

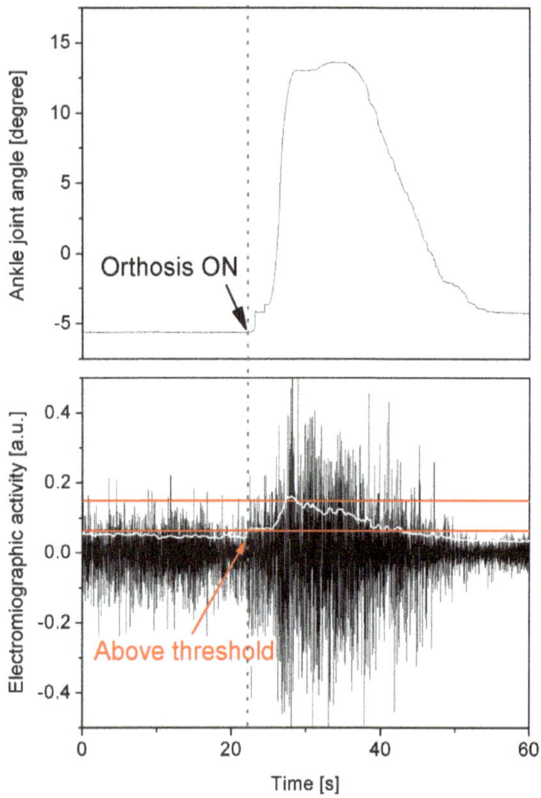

Figure 6. Top: angle timecourse measured during an EMG-triggered activation of the device for a healthy volunteer. Bottom: EMG recording during the trial. When the subject's muscular activity crosses the lower threshold, actuators are triggered to complete the movement of dorsiflexion.

3. SMA actuators for neuroscience

The development of SMA devices for use in Neuroscience is a very challenging task as diagnostic equipment utilised in this field of research is generally very sensitive to electromagnetic noise. Whereas some techniques (e.g. electroencephalography or near-infrared spectroscopy) are less affected by environmental conditions, magnetic resonance imaging (MRI, fMRI) and magnetoencephalography (MEG) acquisitions can be carried out only in highly shielded Faraday cages, to abate electromagnetic noise. Therefore, materials and devices must comply with a number of requirements, in order to be allowed into the acquisition room. In the next paragraph, these two diagnostic techniques will be discussed separately, as their peculiar characteristics demand different compatibility constraints.

3.1. Electromagnetic constraints

3.1.1. Functional magnetic resonance imaging (fMRI)

fMRI is a magnetic resonance procedure utilised in Neuroscience for measuring brain activity by detecting associated haemodynamic responses. MRI equipment uses a strong static magnetic field to polarise the nuclear spins of atoms in living tissues: the higher the magnetic field, the higher the polarisation of the spins, thus, the higher signal-to-noise ratio. Commercial scanners can generate polarisation fields up to 3T, but efforts are being made to increase the available magnetic field far beyond [13]. In addition to this static magnetic field, the MR scanner uses rapidly-varying magnetic field gradients for spatial encoding during the imaging sequence. The use of strong polarisation magnetic fields entails safety precautions, as high forces can be exerted especially on objects made of ferromagnetic materials, which may become dangerous projectiles for the patient, medical personnel, and the instruments. Besides safety issues, ferromagnetic materials generally affect image quality, as they alter the homogeneity of the static magnetic field. Another potential source of artefacts is the presence of objects made of conductive materials inside the acquisition room. In fact, the magnetic field gradients can induce electrical fields and currents flowing through the conductive materials (eddy-currents): these currents, in turn, induce magnetic fields which inevitably interact with the MR fields and affect the quality of the images. In much the same way as eddy-currents, it can be inferred that any current flowing inside conductive materials could generate image artefacts.

Given these strict constraints, a possible solution for moving the limbs of a patient during fMRI investigations by an actuator is to keep the actuator and control system outside the scanning room and transfer the mechanical energy to the subject via pneumatic, hydraulic, or mechanical (pulleys, ropes, etc.) means. These systems are usually complex and suffer from long transmission lines accompanied by dissipative and delay effects. It would be desirable to place actuators directly inside the acquisition room, but this limits the possibilities of using many categories of actuators. Of course, electromagnetic actuators are, in general, not compatible with the MRI environment, but also pneumatic or hydraulic motors could hardly get into the acquisition room, because of their bulky metallic components. Generally speaking, in order to be utilised in an MRI acquisition room, actuators should not include ferromagnetic and conductive materials. However, it is reported that small parts made of MR-incompatible materials do not compromise safety or generate image artefacts as long as they are sufficiently small and appropriately positioned relative to the imaged area [13]. This clears the way to the use of SMA actuators in this field, provided that some specific design rules are respected and that they are applied to body segments sufficiently distant from the head, which is the area of principal interest in Neuroscience.

3.1.2. Magnetoencephalography (MEG)

MEG is a technique for investigating neuronal activity in the living human brain by recording magnetic fields produced by the electrical currents flowing in cortical neurons. These weak magnetic fields (ranging 10^{-14}T-10^{-12}T), can be detected by employing arrays of

SQUID (Superconducting Quantum Interference Device) gradiometers, which convert the magnetic flux threading a pick-up coil into voltage. Since the magnetic fields to be measured are extremely small as compared to the Earth's magnetic field (10^{-5}-10^{-4}T), MEG measures are carried out in a shielded room that minimises interference. Of course, any electromagnetic noise should be avoided inside the acquisition room: there is no safety issue for the patient connected with this constraint, but if not respected the measure would be impossible because SQUID channels would saturate rapidly and would remain unusable as long as the pick-up coils are in the superconductive state. Unlike the case of MRI, theoretically any material could enter an MEG acquisition room. In practice, objects made of conductive materials should remain still inside the shielded room, as any movement would generate an artefact. These constraints are very demanding for most actuation technologies, but design solutions are possible for SMA actuators.

3.2. Compatible design guidelines

Some design indications could be useful for devices intended for either fMRI or MEG studies. In particular, the SMA element should have a shape or should be arranged in a way that limits to a great extent the magnetic fields induced by eddy-currents and by the current used for Joule's heating. For example, the coil of a spring is not particularly indicated, as the intensity of the magnetic field is proportional to the number of turns. Moreover, the SMA actuator should be supplied with a dc-current rather than other types of time-varying currents (pulse-width modulation, sinusoidal current…). In fact, this could help reduce the magnetic field induced by currents flowing into the SMA elements. Power generators and control systems should be positioned outside the acquisition room and wires passed through. As these cables could act as antennas that radiate electromagnetic noise, some countermeasures should be adopted: shielded or twisted cables can limit pick-up of stray frequencies, while using proper filters when connecting the wires inside and outside the acquisition room can help rejecting the time-varying components of the power signal. Coming to the control strategy, an open-loop may be preferred (for its simplicity), as both fMRI and MEG data analysis generally need a precise windowing of signals that matches different phases of movement, in order to extract features of interest with statistical significance. So, closed-loop architectures could be used e.g. to control precise abidance to set movement speed evolution, but might not be strictly necessary if the experimental protocol only requires repeatable ranges of motion in a given time. In fact, within the very protected environments of the shielded rooms, often no major disturbances are expected to intervene and perturb the movement.

Additional recommendations apply only to devices for use in the MRI acquisition room: in particular, ferromagnetic materials should be avoided for safety reasons, even if small parts could be tolerated in some cases. On the other hand, as we said before, during MEG acquisition conductive materials should not move inside the shielded room. Translating this information into design specifications, a first recommendation would be that the moving parts should not mount conductive materials, including the SMA actuators. However, the principle of SMA actuation is that metal moves! The geometry of the SMA element should be conceived so that macroscopic changes of space occupancy are avoided, thus limiting the amount of movement. For example, in SMA helical springs the linear length can vary considerably during actuation, spanning in many cases more than 300% the unloaded length. On the other hand, traction wires, having a limited elongation, change their space occupancy less dramatically. Interestingly, it should be noticed that the slow actuation rates typical of SMA could be an advantage for limiting artefacts influencing MEG measurements. In fact, it is reasonable that artefacts have mostly the same time frequency components as the movement that generated them, i.e. typically less than 1Hz for SMA actuation. Cortical oscillations range 1-100Hz, but when investigating sensorimotor representation (as expected in the case of Neuroscience for Neuromuscular Rehabilitation), the range of interest reduces to 8-100Hz, giving the possibility to devise suitable signal filtering stages.

3.3. Case study: amagnetic rotary actuator for ankle dorsiflexion during MEG and fMRI acquisitions

3.3.1. Concept

A possible device for ankle dorsiflexion could be devised as a leg part and a foot part connected by planar hinges parallel to the ankle joint, in a similar way to the device presented in section 2.4.2. Two SMA rotary actuators mounted externally with respect to the hinges promote the dorsiflexion of the foot part with respect to the leg part, while plantarflexion and position reset is left to viscoelastic resistance and foot weight (Figure 7). It should be noticed that the requirements for the ranges of movement are different in this application, as the purpose is to produce a clearly detectable movement without muscular stretching. This can be done maintaining joint angle negative or slightly positive (i.e. extended). The objective of the study was to test healthy subject, so it can be estimated that maximum resisting torque will be in the range 200-250Ncm (see Figure 2).

A rotary SMA actuator can be generally described as a structure in which the SMA wire connects parts that can rotate relative to one another about a central axis: when the elongated SMA wire is heated above A_f and recovers its deformation, the linear stroke ΔL is converted into a rotation $\Delta\theta$ of the moving parts. However, it is likely that a conspicuous length of wire is needed to produce a suitable amount of rotary stroke $\Delta\theta$ for ankle dorsiflexion. As we already said, in such cases coiling the wire could improve conveniently overall compactness of the actuator. Unfortunately, this design decision contrasts with the aim of fabricating an amagnetic actuator, as the electric current flowing in coils of wire produces, according to Ampère's Law, an overall magnetic field along the central axis of the solenoid, which is proportional to the number of turns. A very simple solution can be found exploiting the same physical laws of electromagnetic fields. In fact, the magnetic field vector is directed according to the right hand rule with respect to the direction in which the current flows that induced it. Thus, theoretically, given two identical and concentric solenoids traversed by the same electric current in opposite directions, the magnetic fields generated by the coils will mutually annihilate. Moreover, no ferromagnetic materials should be employed in the implementation of all other components of such actuator, as well as the assembled device.

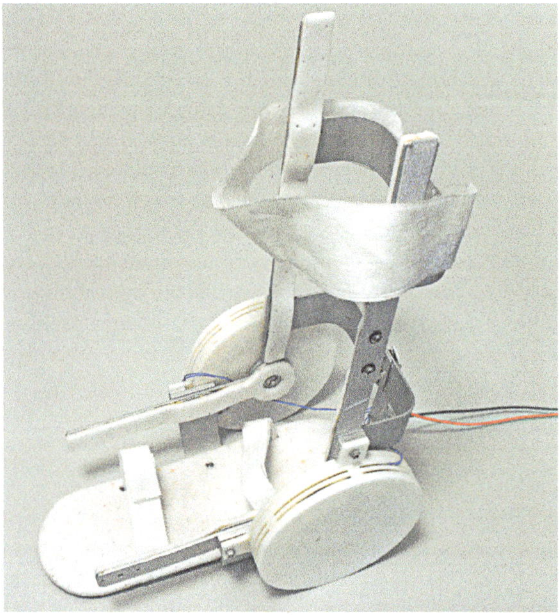

Figure 7. Implementation of an amagnetic device for ankle dorsiflexion with rotary actuators

3.3.2. Actuator implementation and mechanical characterisation

An implementation of the invented amagnetic actuator is shown in Figure 8. Two discs of approximately 10cm in diameter are connected through a central shaft allowing reciprocal rotation, leaving a clearance of 2cm between the two discs. One of the two discs mounts, along its circumference, 6 pins bearing 3 pulleys each, to create a spiralling sequence, along which the SMA wire can be coiled suitably. These pulleys have a double triangular groove, which makes the double coiling as compact as possible. The two ends of the SMA wire are fixed to one disc, while the mid-point of the wire connects to the other disc. The two halves of the wire are would along the pulleys thus creating two concentric coils very close to one another. By providing sufficient current flowing between the two ends of the SMA wire, shape recovery occurs generating a linear stroke ΔL that is converted into a reciprocal rotation $\Delta\theta$ of the two discs. Notice that the torque generated is given by twice the cross-section of the wire, while the induced magnetic field is self-compensated to a large extent, as the two coils are traversed by the same electric current in opposite directions [40].

As already described for the linear actuator in paragraph 2.4.3, wire length depends on the expected angular stroke and the working strain level, while torque output is connected to the wire diameter and mean stress on the cross-section. Moreover, the geometrical parameters of the assembled actuator affect both wire length and torque output. The rotary actuator was designed to provide an angular stroke up to 40° against resisting torques in the range 120Ncm-250Ncm. By limiting the linear strain to 3.8%, the required length of NiTi wire can be calculated in 219cm. The selected NiTi wire is 250μm in diameter, corresponding to stresses of 130-270MPa in the above-mentioned range of torques. Taking into account the localized strain on the wire resting on the pulleys, the total maximal strain reaches 4.42%, which in combination with the chosen stress level should guarantee suitable fatigue life (several thousands of cycles). Commercial stabilised NiTi wire was utilised, displaying large deformability (4.5%) at room temperature for stress levels as low as 150MPa [58]. The power dimensioning procedure was similar to the one described in paragraph 2.4.3 and led to utilizing 0.7A at 30V for each actuator. Control is achieved with an open-loop strategy, by which acquisition windows can be synchronised to the ankle movement. Power supply and control appliances are left outside the acquisition room, and shielded cables are let into the shielded room through suitable access vents.

Figure 8. Amagnetic SMA rotary actuator, consisting of two coils of wire traversed by the same current in opposite directions. This arrangement produces self-annihilation of induced magnetic fields.

In the present implementation, ferromagnetic materials were avoided. However, it was not possible to exclude completely metallic components, such as the bearing balls, the shafts and the pins. After careful consideration, it was decided that the rotating discs and the frame of the device could be made of an aluminium alloy, in order to minimise structural sections. This choice did not affect the acquisition of biosignals, as will be discussed in the next paragraphs.

Technical tests were carried out on the assembled device, to assess its characteristics. Increasing loads were attached to the foot part of the orthosis 13cm from the axis of rotation, while the device was held aloft by a static support. The weights used in this test produced resisting torques in the range 28-250Ncm. A direct current injection at 30V for 7s was applied to the actuators, connected in parallel; then 30s were allowed for position reset through natural cooling and the action of the weights. The resulting angular upward and downward strokes were measured by means of electrogoniometer SIM-HES-EG 042 (Signo Motus, Messina, Italy). Figures 9 shows the results. At lower values of resisting torques (i.e. in the range 28Ncm-120Ncm), angular stroke is not stabilised and steadily increases from 24° to 36°. For torques above 120Ncm, angular stroke is quite stable at 36°. Curves steadily shift towards negative angles (zero being the horizontal position, and negative in the direction of plantarflexion) with increasing torques. This increase in stroke is caused by the incomplete detwinning of martensite at lower torques: it is worth noticing that anyway, even at the lower values of measured angular stroke, suitable angular strokes are obtained.

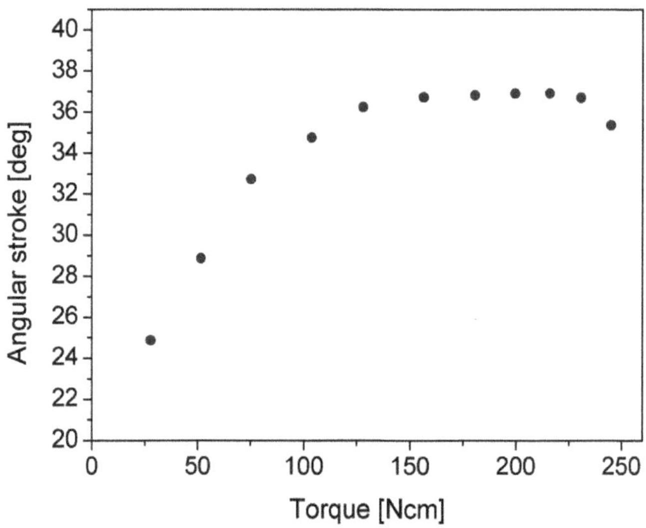

Figure 9. Angular stroke of the SMA-based exerciser for different resisting static torques. The dotted line is just a guide for the eye.

3.3.3. *Tests in a MEG shielded room*

In order to assess compatibility with the MEG technology, neural activity from a healthy subject's brain during repeated passive mobilisation of the ankle by the device was carried out with an MEG system composed of 153 SQUID channels covering the whole surface of the scalp. The electric power generator and the programming computer were kept out of the shielded room, and cables were passed through suitable vents in the shims. The NiTi wire was activated by a current pulse (ramp to 0.7A in 1s, then flat for 9s).

These tests revealed no significant noise on SQUID channels. Figure 10 (left) shows the signal recorded by one representative channel during MEG testing of the healthy volunteer.

At t=0s, the actuators on board the orthosis were switched on. There is no artefact at that moment, indicating that the level of noise in the acquisition room has not changed. Furthermore, the right graph in the same Figure demonstrates that the frequency components in the actuator-OFF and actuator-ON states are the same. The variation in intensity of the signal power spectral density (PSD) can be totally accounted for by a change in the cortical reactivity of the subject under testing: this variation is an important part of the measured quantities of interest.

Incidentally, the use of conductive materials in the construction of the device did not affect the acquisition. This is probably due to the slow movement provided by the actuators, which generates low-frequency artefacts that can be suitably eliminated during routine filtering and windowing of the MEG data.

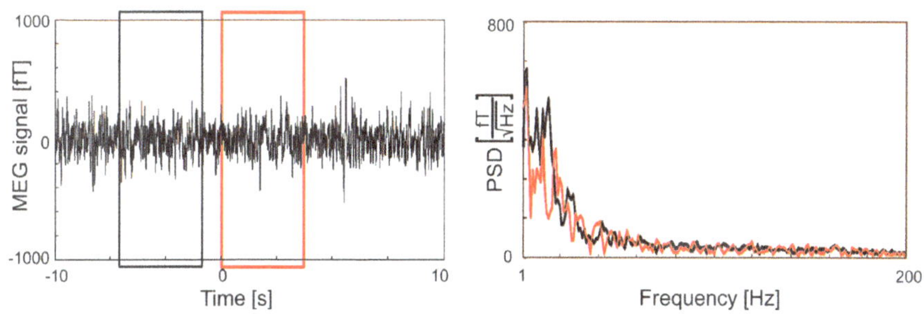

Figure 10. Left: MEG signal recorded by a representative SQUID channel from a healthy subject during passive mobilisation of the ankle by the amagnetic exerciser. Right: spectral components of the recorded MEG signal, before and after switching on the actuators at t=0s (black and red line, respectively).

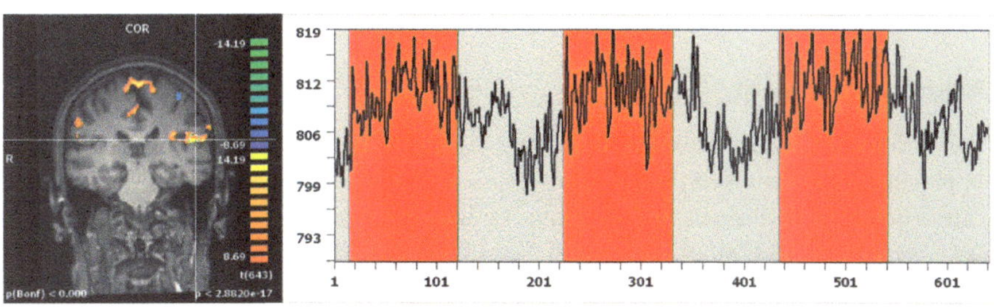

Figure 11. Left: fMRI image of metabolic activity in a healthy brain during passive mobilisation of the ankle. Right: time-course of the BOLD signal (arbitrary units) in the area of the brain highlighted in the left figure.

3.3.4. Tests in a MRI acquisition room

Magnetic resonance images were collected in a Philips 3T Achieva X-Series Magnetic Resonance scanner. fMRI signals were treated to extract the Blood Oxygen Level Dependent (BOLD) image, which depicts changes in neural metabolism related to neural activity. fMRI images (Figure 11) were not affected by any artefact. The BOLD signal was clearly collected from all parts of the brain and in some areas seemed to be temporally dependent on the movements of the ankle. The use of conductive materials in the implementation of the device did not affect the acquisition. The distance between the gantry (or head coil) and the ankle probably helped limit any influence on image encoding.

4. Conclusions

This Chapter showed some innovative applications of Shape Memory Alloys requiring deep understanding of the interaction of the material characteristics with the complex constraints imposed by the human body. In particular, it was explained how the design plan should be laid considering the many aspects connected to the state of the target patient, and the technical requirements should be chosen to meet very well identified needs. The field of Medical Rehabilitation is an interesting domain for exploiting the functional properties of NiTi-based alloys in making new lightweight and portable actuators. The Neuroscience applications introduced in this Chapter, on the other hand, albeit representing a niche sector *per se*, both make the most of the typical design techniques employed for rehabilitation devices, and provide a development ground for interesting industrial actuators with amagnetic characteristics. It is hoped that the SMA-based design strategies presented here will be of inspiration to engineers interested in utilising shape memory actuation for biomedical, robotic, aerospace or automotive applications with tight and mandatory external constraints where compactness, light weight or wearability are desired features of the device.

Author details

Simone Pittaccio* and Stefano Viscuso
National Research Council of Italy, Institute for Energetics and Interphases, Lecco, Italy

Acknowledgement

The authors would like to thank the staff of ITAB, University of Chieti, where the tests employing MEG and fMRI were conducted: in particular Gian Luca Romani, Filippo Zappasodi, Vittorio Pizzella and Gabriella Tamburro for the MEG measurements, Cosimo Del Gratta, Antonio Ferretti and Mauro Gianni Perrucci for the fMRI trials. The financial support from Regione Lombardia (Mind in Italy Project) is most gratefully acknowledged.

5. References

[1] Gracies J-M (2005) Pathophysiology of spastic paresis I: paresis and soft tissue changes, Muscle Nerve 31: 535–551.
[2] Gracies J-M (2005) Pathophysiology of spastic paresis II: emergence of muscle overactivity, Muscle Nerve 31: 552–571.

* Corresponding Author

[3] Simon O, Yelnik AP (2010) Managing spasticity with drugs. Eur J Phys Rehabil Med. 46: 401-410.

[4] European Stroke Organisation (ESO) Executive Committee, ESO Writing Committee (2008) Guidelines for management of ischaemic stroke and transient ischaemic attack, Cerebrovasc. Dis. 25: 457-507.

[5] Duncan PW, Zorowitz R, Bates B, Choi JY, Glasberg JJ, Graham GD, Katz RC, Lamberty K, Reker D (2005) Management of Adult Stroke Rehabilitation Care: a clinical practice guideline, Stroke. 36: e100-e143.

[6] Hesse S (2008) Treadmill training with partial body weight support after stroke: a review. NeuroRehabilitation. 23: 55-65.

[7] Krebs HI, Rossi S, Kim S, Artemiadis PK, Williams D, Castelli E, Cappa P. (2011) Pediatric anklebot. IEEE Int. Conf. Rehabil. Robot 1-5.

[8] Staubli P, Nef T, Klamroth-Marganska V, Riener R (2009) Effects of intensive arm training with the rehabilitation robot ARMin II in chronic stroke patients: four single-cases. J Neuroeng Rehabil. 17: 6:46.

[9] Fasoli SE, Krebs HI, Stein J, Frontera WR, Hogan N (2003) Effects of robotic therapy on motor impairment and recovery in chronic stroke. Arch. Phys. Med. Rehabil. 84: 477-482.

[10] Alary F, Simões C, Jousmäki V, Forss N, Hari R (2002) Cortical activation associated with passive movements of the human index finger: an MEG study. Neuroimage. 15: 691-696.

[11] Boonstra TW, Clairbois HE, Daffertshofer A, Verbunt J, van Dijk BW, Beek PJ (2005) MEG-compatible force sensor. J Neurosci Methods. 144: 193-196.

[12] Suminski AJ, Zimbelman JL, Scheidt RA (2007) Design and validation of a MR-compatible pneumatic manipulandum. J Neurosci Methods. 163: 255-266.

[13] Tsekos NV, Khanicheh A, Christoforou E, Mavroidis C (2007) Magnetic resonance-compatible robotic and mechatronics systems for image-guided interventions and rehabilitation: a review study. Annu Rev Biomed Eng. 9: 351-387.

[14] Onishi H, Oyama M, Soma T, Kubo M, Kirimoto H, Murakami H, Kameyama S (2010) Neuromagnetic activation of primary and secondary somatosensory cortex following tactile-on and tactile-off stimulation. Clin Neurophysiol. 121: 588-593.

[15] Hollnagel C, Brügger M, Vallery H, Wolf P, Dietz V, Kollias S, Riener R (2011) Brain activity during stepping: a novel MRI-compatible device. J Neurosci Methods. 201: 124-130.

[16] Funakubo H (1987) Shape memory alloys. New York: Gordon and Breach Science Publishers. 275p.

[17] Otsuka K, Wayman CM (1998) Shape Memory Materials, Cambridge: Cambridge University Press. 287 p.

[18] Otsuka K, Ren X (2005). Physical metallurgy of Ti–Ni-based shape memory alloys, Progr. in Mater. Science. 50: 511-678.

[19] Lagoudas DC (2008) Shape Memory Alloys: Modeling and Engineering Applications. Berlin: Springer. 435 p.

[20] Viscuso S, Pittaccio S, Caimmi M, Gasperini G, Pirovano S, Besseghini S, Molteni F (2009) Pseudoelastic alloy devices for spastic elbow relaxation. In: Lim CT, Goh JC, editors. ICBME 2008 – 13th International Conference on Biomedical Engineering, IFBME Proceedings 23. Berlin: Springer. pp. 1584–1587.
[21] Pittaccio S, Viscuso S (2009) The use of muscle "creep" as opposed to relaxation in stretching braces: a pseudoelastic device. In: Dössel O, Schlegel WC, editors. World Congress on Medical Physics and Biomedical Engineering 2009, IFMBE Proceedings 25/9. Berlin: Springer. pp. 178-181.
[22] Pittaccio S, Viscuso S (2009) Customizable neuro-mechanical model of a hemiplegic elbow interacting with a pseudoelastic dynamic orthosis. In: Dössel O, Schlegel WC, editors. World Congress on Medical Physics and Biomedical Engineering 2009, IFMBE Proceedings 25/9. Berlin: Springer. pp. 182-185.
[23] Viscuso S, Pittaccio S, Caimmi M, Gasperini G, Pirovano S, Villa E, Besseghini S, Molteni F (2009) Pseudoelastic Nitinol-Based Device for Relaxation of Spastic Elbow in Stroke Patients. J Mater Eng Perform 18: 805–813.
[24] Patent Number: WO/2011/137999. Joint For Articulations With Pseudo-Elastic Elements.
[25] Pittaccio S, Viscuso S, Beretta E, Turconi AC, Strazzer S (2010) Pilot studies suggesting new applications of NiTi in dynamic orthoses for the ankle joint. Prosthet Orthot Int 34: 305-318.
[26] Hanafusa A, Isomura T, Sekiguchi Y, Takahashi H, Dohi T (2002) Orthosis Design System for Malformed Ears Based on Spline Approximation. In: Dohi T, Kikinis R, editors. Medical Image Computing and Computer-Assisted Intervention — MICCAI 2002. Berlin: Springer. pp. 227-234.
[27] Shimizu Y, Kobayashi (2011) The Development of Foot Orthotics using Shape Memory Alloy for Preventing Falls. In: Werner B, editor. 2011 IEEE 11th International Conference on Bioinformatics and Bioengineering. Washington: Conference Publishing Services. pp. 117-120.
[28] Takami M, Fukui K, Saitou S, Sugiyama I, Terayama K (1992) Application of a shape memory alloy to hand splinting. Prosthet Orthot Int. 16: 57-63.
[29] Patent Number: US 2008/0294079 A1. Orthotic Apparatus And Method Of Operation.
[30] Patent Number: US 2011/0131838 A1. Dynamically Adjustable Orthotic Device.
[31] Machado LG, Savi MA (2003) Medical applications of shape memory alloys. Braz J Med Biol Res. 36: 683-691.
[32] Lai YJ, Yeh LJ, Chiu MC (2010) An Assessment of the Body Joint Bending Actuator using a Shape Memory Alloy. J Appl Sci, 10: 1973-1977.
[33] Lai YJ, Yeh LJ, Chiu MC (2012) An experimental investigation on shape memory alloy dynamic splint for a finger joint application. Sens Actuator A-Phys 173: 210–218.
[34] Torri M, Viscuso S, Pittaccio S, Nespoli A, Besseghini S (2006) Biomechanical design of a shape memory alloy spring for the activation of a flaccid hand rehabilitation device. In: Venugopalan R, Wu MH, editors. Medical Device Materials III: Proceedings of the Materials & Processes for Medical Devices Conference. Materials Park: ASM International. pp.237-242.

[35] Van Kuren MB, Gillette S, Mejia P, Stoever T, Walker A,(2005) Design considerations for a wearable pediatric rehabilitative boot. In: ICORR 2005, 9th International Conference on Rehabilitation Robotics. pp. 400- 403. Available: http://ieeexplore.ieee.org/stamp/stamp.jsp?tp=&arnumber=1501128. Accessed 2012 Mar 28.
[36] Pittaccio S, Viscuso S, Rossini M, Magoni L, Pirovano S, Villa E, Besseghini S, Molteni F (2009) SHADE: A shape-memory-activated device promoting ankle dorsiflexion. J Mater Eng Perform 18: 824–830.
[37] Viscuso S, Pittaccio S, Tecchio F, Zappasodi F, Rossini M, Magoni L, Pirovano S, Besseghini S, Molteni F (2009) Assessment of the peripheral performance and cortical effects of SHADE, an active device promoting ankle dorsiflexion. In: Lim CT, Goh JC, editors. ICBME 2008 – 13th International Conference on Biomedical Engineering, IFBME Proceedings 23. Berlin: Springer. pp. 961-965.
[38] Viscuso S, Pittaccio S, Zappasodi F, Tecchio F (2009) A magnetically quasi-transparent tool for ankle passive mobilization in investigations on cortical involvement using MEG. In: Dössel O, Schlegel WC, editors. World Congress on Medical Physics and Biomedical Engineering 2009, IFMBE Proceedings 25/7. Berlin: Springer. pp. 410-413.
[39] Pittaccio S, Zappasodi F, Viscuso S, Mastrolilli F, Ercolani M, Passarelli F, Molteni F, Besseghini S, Rossini PM, Tecchio F (2011) Primary sensory and motor cortex activities during voluntary and passive ankle mobilization by the SHADE orthosis. Hum Brain Mapp 32: 60-70.
[40] Patent Number: WO/2011/141183. Rotational Actuator.
[41] Pittaccio S, Viscuso S (2011) An EMG-Controlled SMA Device for the Rehabilitation of the Ankle Joint in Post-Acute Stroke. J Mater Eng Perform 20: 666-670.
[42] Merrett GV, Metcalf CD, Zheng D, Cunningham S, Barrow S, Demain SH (2011) Design and Qualitative Evaluation of Tactile Devices for Stroke Rehabilitation. Presented at IET Assisted Living 2011, London, UK. Available: http://eprints.soton.ac.uk/271802/. Accessed 2012 Mar 28.
[43] Patent Number: US 2001/0029224 A1. Exercise Kit And Method Of Using Same.
[44] Esfahani ET (2007) Developing an Active Ankle Foot Orthosis Based On Shape Memory Alloys. M.Sc. thesis. Available: http://etd.ohiolink.edu/send-pdf.cgi/Tarkesh%20 Esfahani%20Ehsan.pdf?toledo1197561946. Accessed 2012 Mar 28.
[45] Esfahani ET, Elahinia MH (2007) Stable Walking Pattern for an SMA-Actuated Biped IEEE/ASME Trans Mechatron 12: 534-541.
[46] Stirling L, Yu CH, Miller J, Hawkes E, Wood R, Goldfield E, Nagpal R (2011) Applicability of Shape Memory Alloy Wire for an Active, Soft Orthotic. J Mater Eng Perform 20: 658-662.
[47] Patent Number: WO 2011/008934 A2. Actively Controlled Orthotic Device.
[48] Patent Number: US 5103807. Shape Memory Alloy Orthotic Device.
[49] Abolfathi PP (2007) Development of an Instrumented and Powered Exoskeleton for the Rehabilitation of the Hand. Ph.D. thesis. Available: http://ses.library.usyd.edu.au/bitstream/2123/3690/1/01pp-abolfathi-2008-thesis.pdf. Accessed 2012 Mar 28.

[50] Patent Number: US 6379393 B1. Prosthetic, orthotic, and other rehabilitative robotic assistive devices actuated by smart materials.
[51] Jayatilake D, Gruebler A, Suzuki K (2010) Robot Assisted Smile Recovery. In: Kordic V, editor. Cutting Edge Robotics 2010. Rijeka: InTech. pp. 333-350.
[52] Esteki A, Mansour JM (1996) An experimentally based nonlinear viscoelastic model of joint passive moment. J Biomech. 29: 443-450.
[53] Given JD, Dewald JPA, Rymer WZ (1995) Joint dependent passive stiffness in paretic and contralateral limbs of spastic patients with hemiparetic stroke. J Neurol Neurosurg Psychiatry 59: 271-279.
[54] Lin CCK, Ju MS, Chen SM, Pan BW (2008) A Specialized Robot for Ankle Rehabilitation and Evaluation. J Med Biol Eng 28: 79-86.
[55] IEC 60601-1. Medical electrical equipment - Part 1: General requirements for basic safety and essential performance.
[56] ISO 14971. Medical devices - Application of risk management to medical devices.
[57] ISO 14155. Clinical investigation of medical devices for human subjects - Good clinical practice.
[58] Fumagalli L, Butera F, Coda A (2009) SmartFlex® NiTi Wires for Shape Memory Actuators. J Mater Eng Perform 18: 691-695.
[59] Liu SH, Huang TS, Yen JY (2010) Tracking Control of Shape-Memory-Alloy Actuators Based on Self-Sensing Feedback and Inverse Hysteresis Compensation. Sensors 10: 112-127.

Chapter 5

Modeling, Compensation and Control of Smart Devices with Hysteresis

Daniele Davino, Alessandro Giustiniani and Ciro Visone

Additional information is available at the end of the chapter

1. Introduction

The chapter aims to provide a review on the basic issues concerning the techniques to handle innovative devices employing materials with hysteresis and the related control strategies.

Nowadays, the great availability of materials designed to gather different physical characteristics (electric, magnetic, mechanic, etc.) working together to provide improved capabilities, has determined a huge rise of devices able to integrate different "coupled" functionalities into the same device. In this respect, the set of these "smart" or "multi-functional" materials and the devices employing them is really huge in number and quality. With no claim of exhaustivity, we can recall, for sake of example, thermo-electric, [11], or magneto-caloric [33] effects, where a coupling between Entropy/temperature and electric or magnetic fields is observed, and which promise new devices for heat recovery or refrigeration tasks, through the exploitation of new and more effective materials.

Other materials, conversely, are able to couple mechanical quantities (i.e. stress and strain) to electric or magnetic fields so amplifying the application range for actuation or sensing aims. The most known and widespread are the piezo-electric alloys ($Pb[Zr_xTi_{1-x}]O_3$, $PbTiO_3$), which show increased piezoelectric properties so allowing to make smart devices, otherwise unrealizable, [34], [32], [2]. A similar behavior is shown by the electro-active polymers, [3] which, exploiting the characteristics of polymers, allow to foresee really innovative applications. Materials with magneto-elastic coupling show a complementary behavior with respect to the former and enable to further increase the already huge set of potential smart applications. It is worth to recall the magnetostrictive materials, such as Terfenol-D or Galfenol, [21], exploited for actuation or sensing purposes, or the Ni-Mn-Ga alloys, which presents different magneto-elastic behavior and, even if less assessed than classical magnetostrictives, promise very interesting application, [37].

Omitting electro- or magneto-rheological materials and many others, we would finally mention Ni-Ti alloys showing a thermo-mechanical coupling by the exploitation of the Austenite-Martensite phase transition that enables to realize further and interesting smart devices, [31], [4]. In summary, the interest and attention of researchers and inventors is so focused on these materials and their applications, that everyone can experience, navigating on the web, an impressive "blooming" of ideas and proposals. The key element is the consciousness that smart materials achieve new functionalities going well beyond the "sum" of the individual properties and allowing to develop new and really innovative devices.

Among them, those materials coupling mechanical to other physical quantities (thermal, electric or magnetic) received a special attention since their suitability for sensing or actuation goals. The latter involved many researchers in a multi-disciplinary frame in designing and producing really smart actuators, able to provide high forces, or high precision *micropositioning*, or high speed responses, in dependence of the selected application.

In particular, as we can find in the scientific literature, the employ of Piezo-electric materials, [23], or magnetostrictives, [21], which are able to cover complementary sets of applications, play a primary role and many devices suited for micropositioning, active vibration control, smart actuation, ultrasonic generators, are available.

Similar conclusions can be drawn for Shape Memory Alloys (SMA) which show huge deformations driven by temperature variations and therefore are generally quite slow [45]. The Ni-Mn-Ga materials, also referred to as Ferromagnetic Shape Memory Alloys (FSMA), [40] overcome this limitation still preserving high deformations as SMAs.

However, all of them, share *rate independent* memory properties which strongly affect the global behavior of the device and its performances. Rate independence means that the observed memory behavior doesn't arise from "dynamics" and therefore is still kept also for "quasi-static" input variations, as happens in the well-known behavior of ferromagnetic materials. The common way to refer to it is *hysteresis*.

In the sequel, a strategy to "handle" devices employing smart materials with hysteresis in general working conditions will be outlined and several applications employing magnetostrictives will be discussed to check the validity of the proposed techniques.

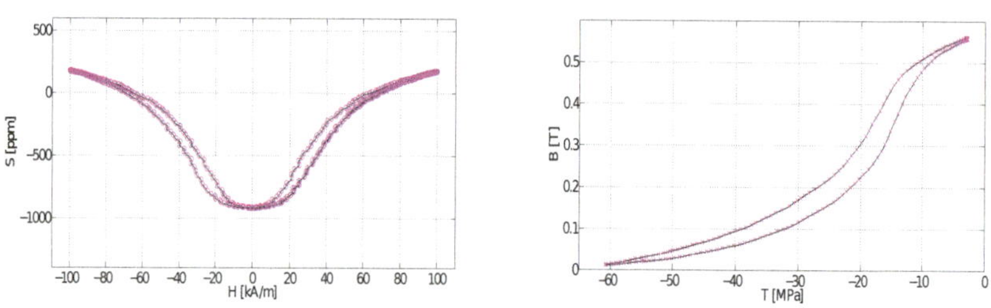

(a) Magnetostriction at constant stress

(b) Inverse Magnetostriction at constant field

Figure 1. Elastic and magnetic response of a Terfenol-D sample to a magnetic (a) or mechanical (b) input

2. Hysteresis operators and their basic properties

The general approach to materials showing nonlinear coupling and hysteresis between the involved variables, can be written as

$$y_1 = \mathcal{F}_1(x_1, x_2), \tag{1}$$

$$y_2 = \mathcal{F}_2(x_1, x_2), \tag{2}$$

being (x_1, x_2) the couple of input variables, and (y_1, y_2) the couple of output variables. Of course, the choice of input and output variables is, in principle, arbitrary and is performed according to the specific application needs. Moreover, $\mathcal{F}_1$ and $\mathcal{F}_2$ represents arbitrary and general operators with memory, which "link" the input to the output functions. This is a quite generic picture, by which we would just acquaint the reader on the complex phenomena lying below the behavior of such materials. In particular, we would evidence that, generally speaking, equations above define a *multi-variate operator* with memory which could be handled by means of *vector operators* with hysteresis, [28], [25]. Unfortunately, they require huge efforts in identification procedures and high computational weight, so resulting unsuitable for *sensing* or *actuation* tasks. As a consequence, other approaches have been considered with the aim to gather good generality and modeling capabilities to a relative simple handling and computational effort, [28], [1], [42]. There, the basic concern was to fully exploit the "machinery" of the classical hysteresis operators, well suited to link together *one input* to *one output*. They are usually referred to as SISO (Single Input-Single Output) systems. As a starting point, any functional material can be modeled as a SISO system whenever only one of the output variables is of concern and, moreover, one of the inputs is kept constant. For the sake of example, Fig.1-(a) and -(b) sketches the typical elastic response driven by magnetic field or the flux density induced by applied stress, respectively. In each of these pictures, a specific characteristic with memory is described and can easily be modeled through a classical SISO hysteresis operator. In a more general framework, as shown in Fig.2-(a) and -(b), a SISO modeling is still applicable whenever the interest is only on one of the output variables (i.e. the *strain* in picture -(a)), and when *stress* can be assumed as stationary (same picture). Of course, the same reasoning can be applied when the *flux density* as output variable is of concern (picture -(b)). A quite unifying approach to define a wide class of hysteresis operator

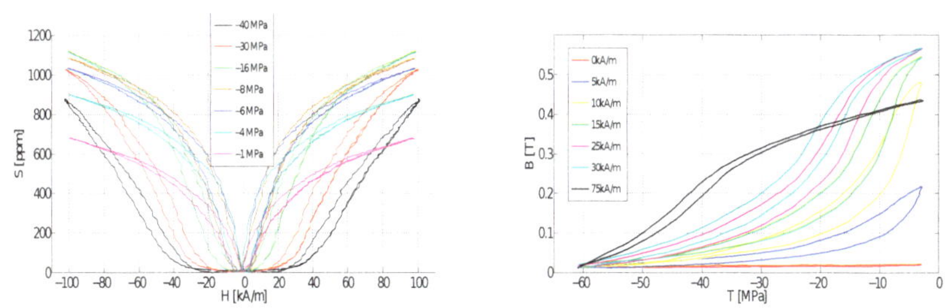

(a) Magnetostriction as function of stress

(b) Inverse Magnetostriction as a function of magnetic field

Figure 2. Elastic and magnetic response of a Terfenol-D sample to a magnetic (a) or mechanical (b) input at various magnetic fields and stresses

is based on the *Preisach memory* updating rules, [28], [25] and we refer to them in the sequel. For the sake of simplicity and assuming some basic knowledge on the Preisach operator, we can introduce the Preisach hysteresis operator as the superposition of a continuous set of ideal relay operators, $\hat{\gamma}_{rs}$ defined by the parameters $r \in \mathbb{R}^+$ and $s \in \mathbb{R}$ as specified in the Fig. 3. Here it should be noticed that the *relay* can be represented by means of the r and s parameters, or by its switching fields a and b ($a \geq b$), equivalently. In the latter case the formalism is the one exploited by I.D. Mayergoyz in [28], while the former yields to the formalism used in the books [7, 26] and one can be derived from the other by a simple 45° axis rotation, [44]. In the sequel we decided to use the second approach which allows the reader to refer to those books so exploiting several results there addressed. On the contrary, this choice has the drawback to discuss some tool of the Preisach formalism (i.e. identification, or inversion algorithms) in a less natural way. For this reason we would further evidence the aim of this chapter, focused in providing a sufficiently complete discussion concerning the modeling and compensation of systems with hysteresis and referring for details to the cited references.

Each *relay*, referred to $R_{r,s}[x]$, has $r \geq 0$ and $s \in \mathbb{R}$. We assume now that they are distributed into the $\mathbb{R}^+ \times \mathbb{R}$ half plane with a prescribed distribution function $\mu(r,s)$ and the output of this set is the linear weighted superposition of the response of each *relay*. In symbols, [28], [25], [26]:

$$y(t) = \iint_{\mathbb{R}^+} \mu(r,s) R_{rs}[x](t) dr ds \tag{3}$$

Of course the past input history of each relay of the distribution results in a non trivial way to compute the integral. However, this problem can be overcome by giving a geometric interpretation to it. First of all, let us associate to each point of the half plane (r,s) the relay having the same parameters. Moreover, recall that each relay of the distribution switches from the state -1 to $+1$ if the input field exceeds the upper switching point, that is $x(t) \geq s+r$, while the opposite transition from positive to negative output is performed if the field is below the lower switching point or, in symbols: $x(t) \leq s-r$. This yields to the following conclusion: the relay operators switched to the $+1$ state are those below the straight line of equation $s+r = x$; conversely, those switched from the state $+1$ to -1 lie above the line $s-r = x$. If all the relays start from the state -1 (usually referred as the state of *negative saturation*) and applying the input field $x(t)$, the line separating the $+1$ from the -1 elements is sketched in Fig. 3-(b) and represents the state of the system. When, from such initial state,

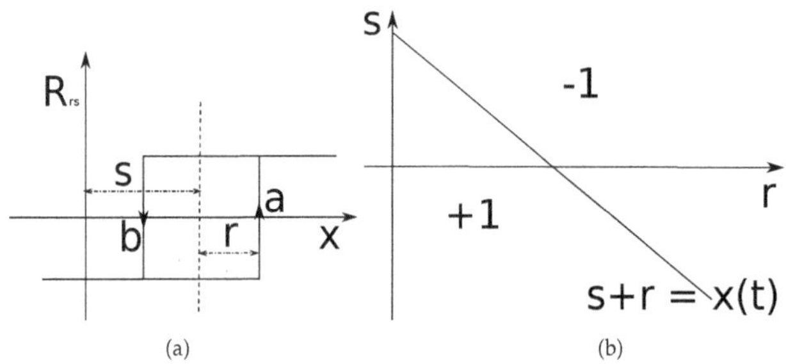

Figure 3. (a) The ideal relay with parameters r and s; (b) geometrical interpretation of the relay switching

the input is decreased, the relays above the line $s - r = x(t)$ switch down to -1. The result is sketched Fig. 4-(a), while in Fig. 4-(b) it is represented the state corresponding to an arbitrary *non-monotone* input variation. The *demagnetized* or *virgin state* is represented in the $\mathbb{R}^+ \times \mathbb{R}$ half plane as the horizontal axis ($s = 0$). We will refer to it as the h_0 state.

After this brief resume, it should be quite easy to realize that the above mentioned rules define a specific relation between the set of continuous input functions, $x(t)$ and the piece-wise linear function $h(r,t)$, as follows:

$$\Theta : x \in C^0[0,T] \rightarrow h(r,t) \in \Psi_0, \tag{4}$$

with, h the staircase of the classical Preisach model (cfr.[28]), and Ψ_0 the set of admissible Preisach states. The properties of this operator, referred to as Preisach Hysteresis Generator (PHG), can be easily sketched. To this aim, let us consider the functions $x(t)$ and $z(t)$, shown in Fig. 5-(a), sharing the same sequence of input extrema and linked by the equation $x(t) = z \circ \rho(t)$. The function ρ is a time rescaling and $\circ$ the composition operator. In other words the two functions differ only by the rate of variation. It is easy to verify that the staircases corresponding to x and z coincide (Fig. 5-(b)). The memory operator, therefore, is not affected by the input rate, but only by the sequence of the input extrema.

Let us now consider the memory operator with initial state equal to the h_{-1} *virgin state*, i.e. $h_0 = h_{-1}$ undergoing the input $x(t)$, shown in Fig. 5-(a). The vertex of the staircase (x_3, x_4) are wiped out as soon as the input exceeds x_3 as they were never existed, as shown in Fig. 5-(b). The final state generated by the sequence of input extrema $(x_1, x_2, x_3, x_4, x_5)$ coincides with the one generated by the *reduced* input sequence (x_1, x_2, x_5), [41]. The latter is the basic rule for memory updating which is universally referred to as *Preisach deletion rule*. In conclusion, the Preisach-memory operator is *rate independent* and fulfills the *wiping-out* property. What we have just introduced is the machinery enabling the classical Preisach operator, defined in eqn. (3), to take into account *rate-independent* memory phenomena. To

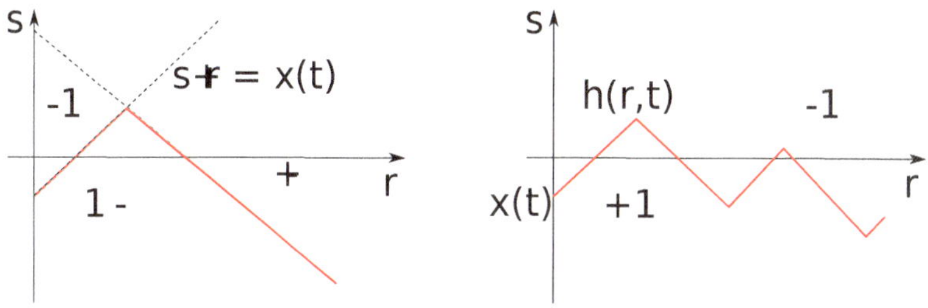

(a) (b)

Figure 4. (a) staircase in the case of non-monotone input variation (b) Staircase corresponding to a non-monotone input $x(t)$

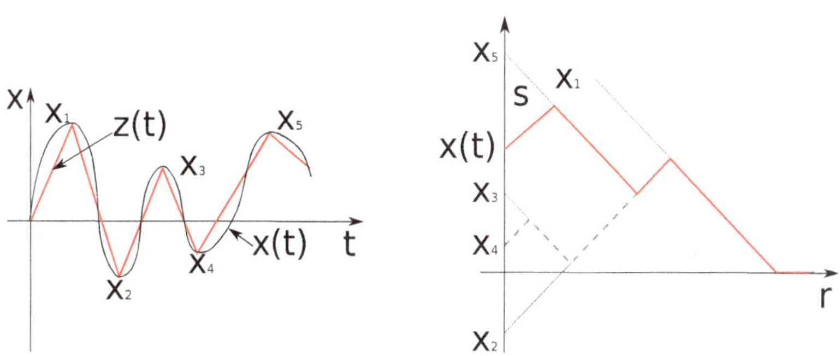

(a) (b)

Figure 5. (a) staircase in the case of non-monotone input variation (b) Staircase corresponding to a non-monotone input $x(t)$

fully exploit such operator, it is possible to define a larger class of operators with hysteresis, based on the Preisach memory updating rules, as proposed in [7], where any operator with Preisach memory can be represented as follows:

$$\mathcal{W} = \mathcal{F} \circ \Theta, \tag{5}$$

where Θ is the PHG, while

$$\mathcal{F} : h(r, x) \in \Psi_0 \rightarrow y(t) \in \mathbb{R}. \tag{6}$$

represents a memoryless functional by which a wide class of hysteresis operators, based on the Preisach memory, can be defined. It should be noted that this formalism, adopted in [7, 26], is slightly different with respect to [28], representing the Preisach plane by means of a rotated coordinate system, [44]. For the sake of example, the following functional

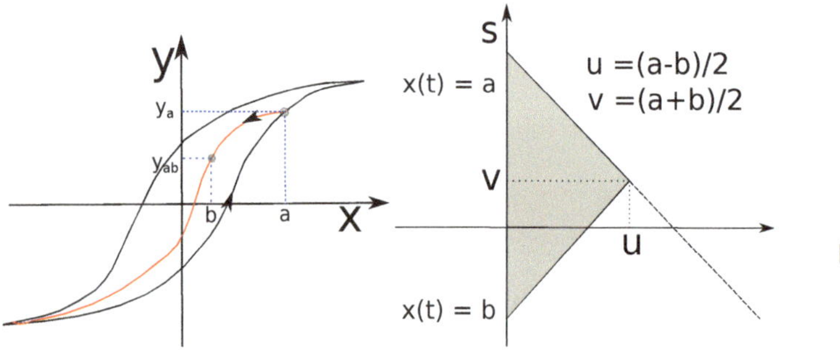

(a) Definition of first order reversal curve (b) Triangular domain over which the Everett integral is defined

Figure 6. Identification procedure of a Preisach model, based on the exploitation of first order reversal curves

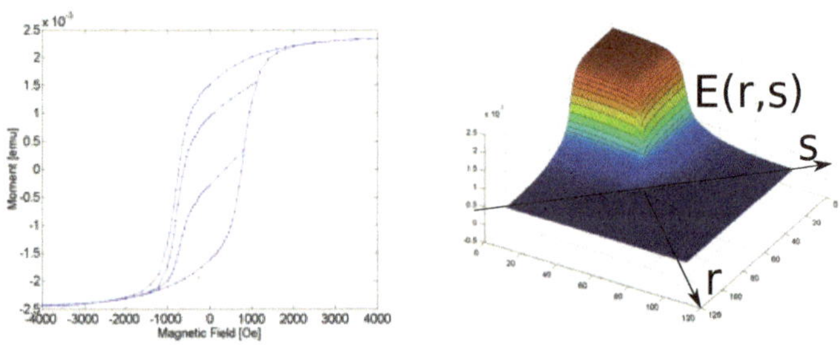

(a) Characteristic of a floppy disk (b) Everett function of a floppy disk

Figure 7. Example of the identification of the Everett function from experimental data coming from a floppy disk

$$\mathcal{F} : h \in \Psi_0 \rightarrow \int_0^{+\infty} dr \left(\int_{-\infty}^{h} \mu(r, s) ds - \int_h^{+\infty} \mu(r, s) ds \right), \tag{7}$$

defines the Classical Preisach operator, and $\hat{\mu}(r,s)$ is a density function to be determined. The functional

$$\mathcal{F} : h \in \Psi_0 \to \int_0^{+\infty} \mu(r,s)h(r,t)dr, \tag{8}$$

conversely, defines the well known Prandtl-Ishilinskii operator [7]. Such models have been widely exploited in last years by several authors to compensate the memory of real devices, so allowing a model-based control approach. A non exhaustive review can be found in [44]. It should be quite evident that these models are phenomenological in nature and so require well-defined identification procedures to link their parameters (i.e. the mapping functional, $\mathcal{F}$) to the real material, through a suitable set of measured data. It should be said that such a procedure is not addressed for any arbitrary functional, while, in the case of the Preisach operator, a simple and well-behaved identification procedure is available and, for convenience it is reported with some detail hereafter.

In particular, it could be observed that in order to compute the output of a classical Preisach model, it is not necessary to know the Preisach Distribution Function, $\mu(r,s)$ [28], but it is sufficient the knowledge of the integrals of $\mu(r,s)$ over any triangle of the $\mathbb{R}^+ \times \mathbb{R}$ half plane and vertex (u,v), according to the procedure described below. Preliminary, let us define the following function known as *Everett function*, as follows:

$$\mathcal{E}(u,v) = \int_0^u dr \left(\int_{v-u+r}^{v} \mu(r,s)ds + \int_v^{v+u-r} \mu(r,s)ds \right). \tag{9}$$

The link between the actual measured data and the Everett function can be outlined as follows. Let us start from negative saturation and increase the input until the value $x(t) = a$ and the corresponding output is y_a are attained. Then decrease the input until the value $x(t) = b$ is reached; the corresponding output is referred to as y_{ab}. The hysteresis branch (Fig. 6-(a)) traced from the reversal point y_a until, again, the negative saturation is known as *first order reversal curve*, or *for*-curve. It is easy to realize (cfr. also [28]) that the following relation linking the output measured data and the Everett integral holds:

$$\mathcal{E}(u,v) = \frac{y_a - y_{ab}}{2} \tag{10}$$

with

$$u = \frac{a-b}{2} \tag{11}$$

$$v = \frac{a+b}{2}. \tag{12}$$

Let us subdivide the input interval into N equal intervals of width $\Delta x = x_{max} - x_{min}$ and assume $x_1 = x_{max} = N\Delta x$, $x_k = (N-k+1)\Delta x$. The corresponding reversal curve (Fig. 6-(b)) so has $N-k+1$ measured points, $y_{a_k b_n}$. Hence the total number of measured points to be used for the identification procedure, are $N \times (N-1)/2$, which correspond to the number of nodes in the grid in the $\mathbb{R}^+ \times \mathbb{R}$ half plane. Now, if we associate the value of the Everett integral to each point of the grid, a discrete version of the Everett function is easily available. The details on its construction and use can be found in [28]. An example of the reversal curves for a real material, and the corresponding Everett function is shown in Fig. 7.

3. Compensation algorithms

The definition of suitable control algorithms to be embedded in actuators employing smart materials is sensibly affected by the memory phenomena involved. This because, roughly speaking, the controller has no memory of past history which, conversely, affects the plant behavior, and will be not able to provide suitable correction to track a desired trajectory. The system therefore will show poor performances. A simple way to overcome this problems, proposing quite a simple control strategy, is based on the following well known issues. As a preliminary step, let us assume that the plant is adequately modeled by the series connection of a hysteresis operator and a linear dynamic component. In Fig. 8-(a) it is shown a control scheme employing a generic nonlinear controller which gather both dynamics and *rate-independent* properties, even if specifying its details is non-trivial. Let us assume that the *rate independent* behavior of the plant can be modeled by the aid of a hysteresis operator $W[x]$, as defined in section 2. If the inverse of the hysteresis operator partly modeling the plant were available, their series connection in the control scheme bears to a linear system, allowing the employment of all standard linear control strategies, as shown in Fig. 8-(b). In summary, in order to simplify the controller design, it is suitable to compensate nonlinearity and *rate-independent* memory of the actuator. This approach is generally referred to as *model-based* control strategy. The issue is now to address the definition of the compensator of a hysteresis

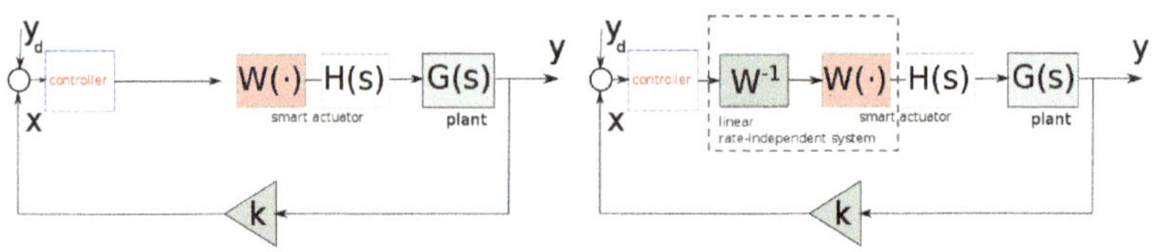

(a) The general controller should take into account memory effects

(b) Model-based control scheme: a suitable modeling of the plant is performed which enable to provide a compensator

Figure 8. General control scheme (a) - Model-based control scheme (b)

operator and to define the conditions under which such an operator admits an inverse or *compensator*,[7, 27]. In particular:

Definition 1. *A hysteresis operator* $\mathcal{W}^{-1}$ *with initial state* $\phi_{-1} \in \Psi_0$, *is called a* compensator *(or inverse) of the operator* $\mathcal{W}$, *with initial state* ψ_{-1} *if, for any state* $\phi \in \Psi_0$, *there is a state* $\psi \in \Psi_0$, *such that* $\mathcal{W}_\psi \circ \mathcal{W}_\phi^{-1}\, x(t) = \mathcal{W}_\phi^{-1} \circ \mathcal{W}_\psi\, x(t) = x(t)$, *for every input function x(t).*

Of course, the above definition lead us asking the conditions making this issue sound. This requires to specify the hysteresis operator with more detail. To this aim, we refer to the class of operators based on the Preisach deletion rule, stated in eqn. (5). We refer to this operators, to as $\mathcal{P}_0$-operators. In particular, we can define a piece-wise strictly monotone operator of this class, [7], according to the following

Definition 2. *The operator* $\mathcal{W} = \mathcal{F} \circ \Theta$ *with initial state* ψ_{-1} *mapping the set of continous functions,* $C[0, T]$, *into itself, is piece–wise strictly increasing if*

$$[\mathcal{F}(\Theta(x, \psi) - \mathcal{F}(\psi)]\,[x - \psi(0)] > 0, \quad x \neq \psi(0), \tag{13}$$

where $\psi(0)$ is the past input, while the symbol $\Theta(x, \psi)$ is the staircase attained by starting from ψ and applying the value of input x at time t.

Such a property guarantees the invertibility of $\mathcal{P}_0$-operators and is connected to the specific memoryless mapping functional F. It turns to the classical definition of monotonicity for the (single valued) hysteresis branch when the state (i.e. the staircase in the Preisach plane) is specified. In the particular case of the classical Preisach operator, with mapping functional specified in eqn. (7), the following theorem holds [6]:

Theorem 1. *The Preisach operator, $\mathcal{H} \in \mathcal{P}_0$, in eqn. 3 with distribution function μ, admits an inverse if and only if $\mu \geq 0$, and $E(u,v) \geq 0$, for every $(u,v) \in \mathbb{R}^+ \times \mathbb{R}$.*

Before discussing an efficient algorithm to implement the inverse of Preisach operator, it is important asking why the inverse operator cannot be employed by just exchanging the role of input and output.[1] In principle, this is not accurate, due to a specific property of the Preisach mapping functional, known as *congruency*. Let us refer to Fig. 9-(a), where it is sketched the state updating corresponding to the same *up and forth* input variation (from x_- to x_+) for two different states h_1 and h_2. It is quite evident that two minor loops are generated, each corresponding to an initial state, and they are *congruent* by vertical translation, Fig. 9-(b). The inverse operator describe a characteristic which is placed symmetrically to the direct characteristic, with respect to the $y = x$ line. As a consequence, any minor loop described by the model is transformed into one which is symmetrically placed with respect the same line.This implies that two congruent minor loops generated by the model, are transformed by the inverse operator into two minor loops which are congruent by a horizontal translation. Fig. 10-(a) illustrates such a situation. Conversely the operator which just exchanges the

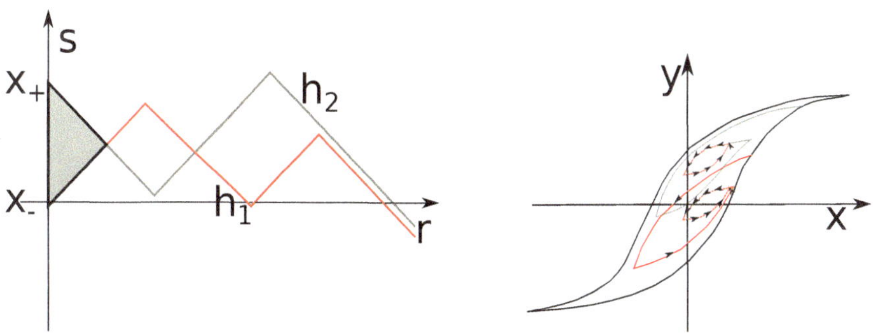

(a) States corresponding to the same up and forth input variation

(b) Congruent minor loops

Figure 9. Illustration of the Congruency property of the Classical Preisach Problem

role of the variables that we call *pseudo-compensator* is still a Preisach operator and therefore generates vertically congruent minor loops, as shown in Fig. 10(b). It is so manifest that this operator cannot be strictly considered as the inverse of a Preisach operator, due to the limitations imposed by the congruency property. However this observation doesn't limit the possibility to adopt pseudo-compensator as a hysteresis model by itself which is able to

[1] In reality this is a procedure to define and approximation of the inverse, referred to as *pseudo-compensator* as detailed in [30]

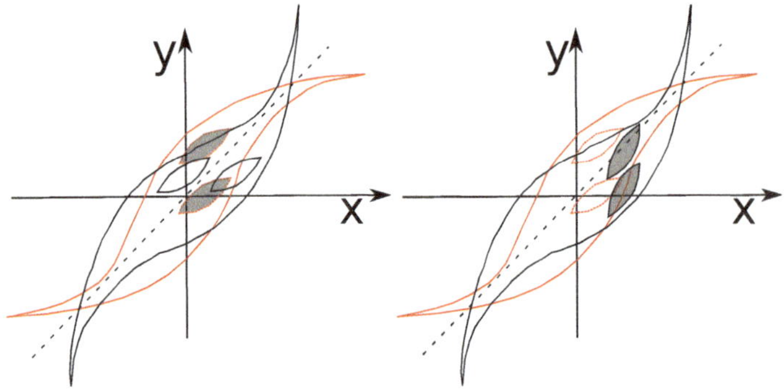

(a) States corresponding to the same up and forth input variation

(b) Congruent minor loops

Figure 10. Illustration of the Congruency property of the Classical Preisach Problem

guarantee good performances, as widely described in [20, 29]. In summary, the *inverse* of a classical Preisach operator *is not* a Preisach operator; the pseudo-compensator, conversely, *is* a Preisach operator and cannot rigorously represent the Preisach compensator.

It should be stressed that even if the inverse of the Preisach operator exists, it is not possible, generally speaking, to represent the compensator in a closed form, except than in very specific cases, [44], [26], [27], [43]. So it is mandatory the definition of well-defined and computationally efficient inversion algorithms in general working conditions. To this aim we will briefly sketch the issue, by referring to [12, 44] for details. When the conditions of invertibility of the Preisach operator are fulfilled the problem of finding its inverse has sense and the inversion procedure is based on the algorithm of state updating which assumed the output $y(t)$ as known. In other words, the algorithm described hereafter, allows to provide the input field of the operator at time instant t, corresponding to a given and known output $y(t)$, once the initial state is assigned. This procedure, since no numerical iterations are needed, allows to compute the inverse with the same computational efficiency and accuracy of the classical algorithm used to compute the Preisach operator. For this reason we refer to it as a "fast" inversion algorithm, [13] [14]. If the Preisach plane is divided into a $\Delta x \times \Delta x$ mesh, with $\Delta x = \frac{x_{max} - x_{min}}{N}$ then the Everett function is uniformly sampled into a $(N+1) \times N/2$ mesh, as sketched in the Fig. 11. Therefore the solution of the above equation can be performed by a simple inspection of a look-up table, according to the rules described below. Moreover, let us assume that the Everett integral is limited to the triangle ABC. This is a quite reasonable hypothesis, since the function must vanish at infinity. In the same plane, it can be evidenced the initial state of the operator, $h_0(r,t)$. Let us further assume the output $y(t) = y_0 + \Delta y$, is known. In the ordinary fashion of handling the Preisach operator, the output variation Δy coincides with the integral of the distribution function over the area S, that is:

$$\iint_S \mu(r,s) dr\, ds = \mathcal{E}(r,s) - \iint_{S^*} \mu(r,s) dr\, ds. \tag{14}$$

By exploiting the wiping out property, all the vertex of the staircase such that the constraint

$$\Delta y \geq \mathcal{E}_k(r,s) - \iint_{S^*} \mu(r,s) dr\, ds. \tag{15}$$

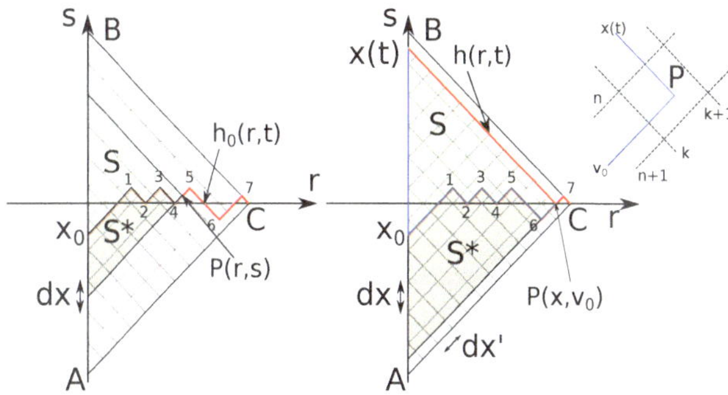

(a) State updating algorithm

(b) Input computation by interpolation over the grid points

Figure 11. The algorithm of inversion of the Preisach operator

is fulfilled, are cancelled. The suffix k of he Everett function evidence the vertex of the triangle. In the case reported in Fig.11-(a) the initial state h_0 represented by the sequence of dominant extrema $\{0,1,2,3,4,5,6,7\}$ is updated and the vertex $\{0,1,2,3,4\}$ cancelled. So the state updating algorithm consists in applying the above condition and cancel all the vertex until it is verified. Once the last vertex has been cancelled, the following condition hold:

$$\Delta y = \mathcal{E}(x, v_0) - \iint_{S^*} \mu(r,s)dr\,ds. \tag{16}$$

being x the unknown value of the input field to be determined. In Fig. 11-(b) it is shown how $x(t)$ wipes out the vertex of the state function, and the operator attains the new state $h(r,t) = \{x, P, 7, C\}$ evidenced in red in the same figure where it is also shown a zoom of the new state around the point $P(x, v_0)$ lying within a square determined by the points $(n+1, k)$, $(n+1, k+1)$, (n, k) and $(n, k+1)$. As a first step, the Everett integrals at the points $(k+1, v_0)$ and (k, v_0) are found by linear interpolation:

$$\mathcal{E}(m, v_0) = \mathcal{E}(m, n) - (\mathcal{E}(m,n) - \mathcal{E}(m, n+1))\,\frac{v_0 - n\Delta x'}{\Delta x'} \tag{17}$$

being $m = k+1$ and $m = k$. Finally, once these values have been determined, and exploiting eqn.(16) the final output value can be written as:

$$x' = \frac{\Delta y + \iint\limits_{S} \mu(r,s)dr\,ds - \mathcal{E}(k, v_0)}{\mathcal{E}(k+1, v_0) - \mathcal{E}(k, v_0)}\Delta x' + k\Delta x', \tag{18}$$

where $x' = x/\sqrt{(2)}$ and $\Delta x' = \Delta x/\sqrt{(2)}$. For this reason the equation does not change when the variables without index are concerned. Moreover, it should be stressed that the integral in the previous equation is a known quantity. Such an approach requires only three linear interpolations at most and does not need any lengthy iteration. The procedure only requires the knowledge of the matrix storing the samples of the Everett integrals and its precision is linked to the mesh adopted for the Preisach plane or, in other words, to the number of measured samples used for the identification. Moreover, the algorithm, as already mentioned, does not increase the computational weight of the algorithm which is equivalent to that of the direct operator.

4. Multi-variate systems with hysteresis

Smart materials, in general application conditions, cannot be considered as SISO systems, since each output quantity is a function of at least two input functions. For sake of example, in a magnetostrictive material the strain depends both on the mechanical load (stress) and magnetic field. A similar behavior can be experienced for the magnetic *flux density*. The behavior is illustrated in Fig. 12-(a) where the strain is affected by both the stress and field variation. Moreover, not only the H-field affects the internal state of the system, which is also altered by the applied stress, as shown by the hysteresis branches sketched in Fig. 12-(b). The experimental evidence witnessed by Figs. 2 and 12, allows to conclude

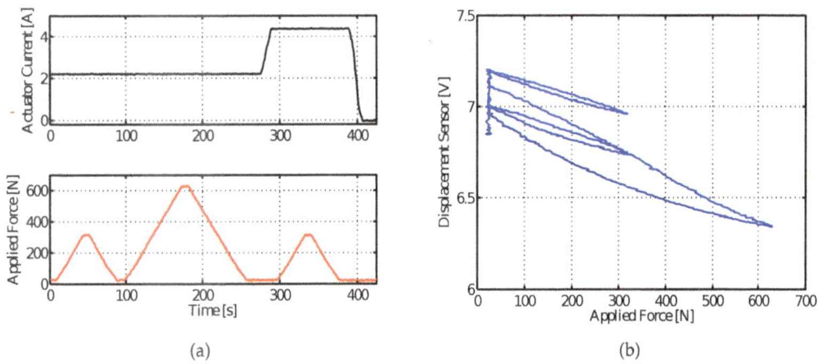

Figure 12. Applied current and force vs. time to a magnetostrictive sample (a); displacement/force hysteresis branches (b).

that the modeling of these systems requires a generalization of all the classical approaches to hysteresis phenomena. This need started two decades ago [1], [46] and yielded to models able to couple together two different input variables (i.e. ε and σ). This effort become of fundamental importance whenever the coexistence of non stationary stress and fields cannot be overlooked.

The simplest way to define more general models is to keep the usual SISO structure of systems with hysteresis, as the ones discussed in the previous section. To this aim, let us consider the mapping:

$$\zeta : (u, v) \in C^0[0, T] \times C^0[0, T] \to x \in C^0[0, T], \tag{19}$$

where ζ is the operator relating the couple of continous functions (u, v), affecting the system behavior, to the input, x, of the memory operator, Θ. For the sake of example, in the case of a magnetostrictive material, ζ "links" the stress σ and magnetic field H to the input of the memory operator. Therefore

$$\mathcal{W} = \mathcal{F} \circ \Theta \circ \zeta, \tag{20}$$

represents a hysteresis operator with two inputs which generalizes the definition in [5] to a wider class of hysteresis operators. Before proceeding in the description of multi-variate systems with hysteresis in control applications, it is mandatory to provide some information concerning its identification. To this aim, let us now consider the following operator, specified for a magneto-elastic material

$$y = \Gamma \circ \Theta \circ \zeta(x_1, x_2) + q(x_2), \tag{21}$$

where $y = \varepsilon$, $x_1 = H$, and $x_2 = \sigma$, being ε, H, and σ the strain, magnetic field and stress experienced by the material, respectively. Moreover, $q(\sigma)$ is a pure elastic response to be identified with almost trivial mechanical measurements at zero magnetic field. Even it is not strictly necessary, the latter assumption allows to take into account separately the effects of vertical translation of the magnetostrictive response due to pure mechanical reasons. Let us preliminary assume that the set of $j = 1, .., M$ anhysteretic[2] curves, e_j shown in Fig. 13-(b) is available. Moreover, assume that the interval of input variation ($x_2 = \sigma$) is $[\sigma_{min}, \sigma_{max}]$ and that the magnetostrictive response experiences a monotonic decrease for increasing compressive stresses, that is $\sigma_{min} = \sigma^*$. With reference to the figure, this corresponds to a 1.05 MPa applied mechanical stress. Finally, each curve is discretized into N samples such that $H_{i+1} - H_i = \Delta H$, with $i = 1, ..N$. Assuming further that $\zeta(H, \sigma^*) = H$ and $\lambda(\sigma^*) = 0$, if

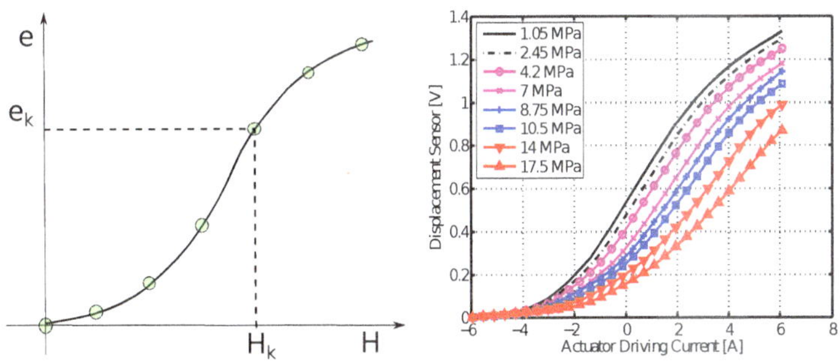

(a) (b)

Figure 13. Example of *Anhysteretic* curve (a) Experimental Anisteretic curves for a magnetostrictive material (b).

$\sigma = \sigma^*$, then the operator takes the following classical form

$$e(t) \equiv \varepsilon(t) - q(\sigma) = \Gamma \circ \Theta[H], \tag{22}$$

which could be easily identified, according to the procedure sketched in sect. 2 and detailed in [28]. Once the Preisach distribution function is known, the procedure to identify the $\zeta(\cdot,\cdot)$ function can start. Recalling now that in suitable conditions $\Gamma \circ \Theta$ admits an inverse and that an efficient inversion algorithm is available, as shown in section 3, the eq. (22) can easily be re-arranged as:

$$\Gamma^{-1}[\varepsilon - q(\sigma)] = f(H, \sigma). \tag{23}$$

By this procedure, a set of $M \times N$ measured samples is now available, which enable us finding the samples $\zeta(H_i, \sigma_k)$ of the unknown function.

[2] It is defined as *anhysteretic*, the curve in the input-output plane, as in Fig. 13-(a), each point of which corresponds to the output of the $\mathcal{P}_0$ operator in the state represented by the horizontal curve $s = H_k$, that can be constructed by applying to the system the input $x(t)$ slowly converging to H_k

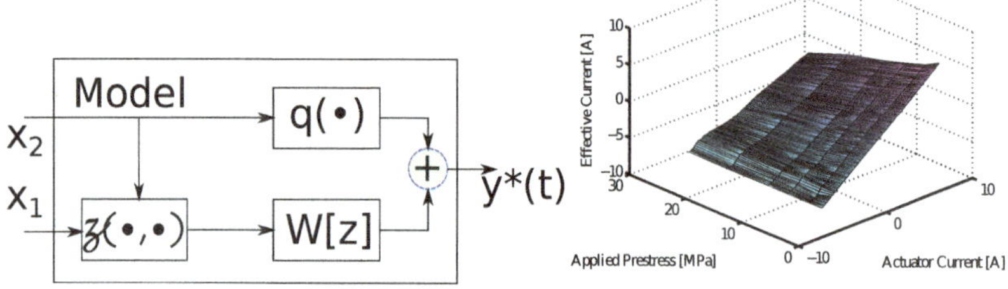

(a) (b)

Figure 14. Block scheme of the 2-variables operator (a); Example of reconstruction of the ζ function, named as "effective current" in the vertical axis of the plot (b).

In order to show the effectiveness of the described procedure, it is important now to show the performances of a multi-variable model as the one described in eqn. (23). In particular, in Fig. 14-(b) it is shown the reconstructed ζ function, while in Fig. 14-(a), the scheme of the model is shown. It undergoes the input variable x_1 and x_2, according to eqn. (23), which in the specific case are the magnetic field and the stress, respectively. The output $y^*(t)$ is compared to the actual measured strain function and the results are sketched in Fig. 15. In particular, in Fig. 15-(a) it is shown the time variation of the input variables, i.e. $x_1(t) = H(t)$ and $x_2(t) = \sigma(t)$, in the first two frames, while in the third, it is described the output of the ζ-function, to which the hysteresis operator $\mathcal{W}$ is subject. In Fig. 15-(b) it is shown the system response, compared to the time behavior of the measured strain. Here the performances of the model described so far are compared to those of a model proposed in [10], where the dependence of memory on the stress is not taken into account.

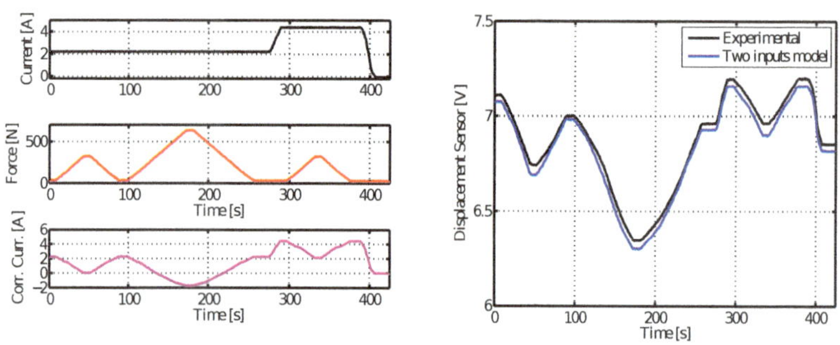

(a) (b)

Figure 15. Comparison of the model's behavior to experiments for a prescribed history. Stress (load), magnetic field (curr) and ζ (corr-curr) vs. time (a); output displacement compared to the experimental data (b).

In conclusion of this section, it is worth to be further stressed that the procedure described so far is of general breath and can be exploited in modeling and compensation of any material/system with multi-variable hysteresis as most of multifunctional materials behave.

5. Model-based control strategies for smart actuators

After a complete review on the modeling and compensation strategies of actual systems with hysteresis, in general working conditions, has been settled, it is possible to discuss different approaches to their control through the definition of specific application frameworks. As a first step, let us start from the classical *model based* control approach, already sketched in Fig. 8, sect. 3

There, the employment of the compensation algorithm allows a easy design of the controller, which could be, as a first attempt, a classical PI system, with all its limits, but with the great advantage of its implementation neatness. This viewpoint has been widely used in smart actuation tasks where Piezo, SMAs or magnetostrictives were employed, [23, 24, 27, 35, 39]. In Fig. 16-(a) the whole control system is shown. In this figure, the block $D(s) = H(s)G(s)$ merge together the linear dynamic part of the actuator and the linear dynamic plant. So the series connection of $\mathcal{W}$ and $D(s)$ represents the *smart actuator* and the plant (SmA). This is a quite general scheme allowing to discuss the control system of the actuator both when loaded by the plant or *unloaded*. The part (b) of the same Fig. 16 shows the same system actuator-plant when it is affected by two inputs (x_1, x_2). The detailed discussion of this scheme will be provided later, at the end of this section.

In the example shown in Fig. 17 a triangular waveform with period $T = 0.02$ s is tracked by the control system sketched in Fig. 16-(a), with and without the compensation algorithm described in the previous sections, [12]. The effects of compensation evidence the increase of the tracking performances of the actuator. In [9] it is also discussed the improvements of the control performances also in term of stability.

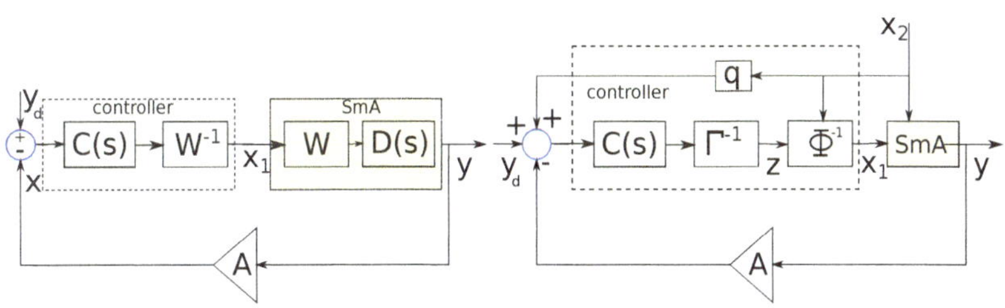

(a) Model-based control scheme (b) Control scheme with two variables feeding the Sma

Figure 16. Model-based control scheme (a) and its generalization to a multi-variable, model-based controller (b). In both cases, the smart actuator and the plant (SmA) are represented by the connection of $\mathcal{W}$ and $D(s)$

The examples considered until now refers to *slow* actuation applications, such as micro positioning tasks (cfr. also [8]). When the assumption of low input variation cannot be assumed as the general working condition, a more accurate discussion should be provided. To this aim, let us again consider the Fig 2, showing how the mechanical response of the material is strongly affected by the applied constant stress (i.e. the *pre-stress*). Normally the magnetostrictive material is subject to a prescribed (and optimal) stress, with the aim to maximize such response. When the actuator is working to actuate a prescribed mechanical load, the total stress experienced by the material has very low fluctuations with respect to the applied stress, whenever slow variations of the applied field are concerned. For high frequencies application tasks this assumption no longer holds and the material experiences a simultaneous variation of stress and field. In that case the employment of a compensation algorithm with two variable is necessary. To this aim the *multi-variable* hysteresis modeling described in the section above could be an interesting starting point to define suitable multi-variable compensation and control strategies. A first attempt to tackle such a quite demanding problem was settled in [10], where a simple generalization of the Preisach model, taking into account the presence of a second variable (i.e. the stress) in the distribution function was stated and discussed. Such approach had the drawback that only a correction of the operator's output was concerned while no effects on the state were considered. It should be said, however, that even so the system showed promising characteristics, which yielded to further improvements and to the definition of the model described above (eqn. (21)). In the following we will sketch how the latter model can be employed to define suitable and well behaved multi-variable control strategies, [15–17]. However, a preliminary discussion to settle the existence conditions of the compensator for such a kind of multi-variable hysteresis operator is required. Let us now refer to the model specified in eqn. (21) which is re-arranged

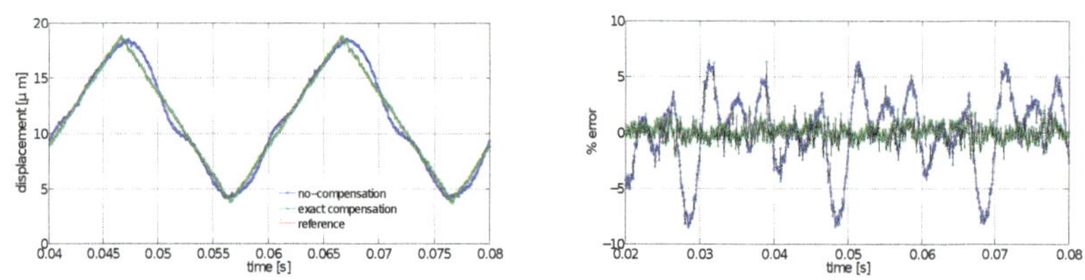

(a) Model-based control scheme

(b) Green: compensated control; Blue: uncompensated control

Figure 17. Model-based control scheme (a) and its performance in tracking a reference signal (b)

as follows:

$$z \equiv \mathcal{W}^{-1}[y - q] = \zeta(x_1, x_2), \tag{24}$$

where

$$\zeta : (x_1, x_2) \in \Omega \longrightarrow \mathbb{R}$$

is a continuous function and Ω the set where it is defined. Moreover, consider now the set where the above eqn. (24) admits at least one solution:

$$S = \left\{ (\bar{x}_2, \bar{z}) \in \mathbb{R}^2 : \exists x_1 \text{ with } \zeta(x_1, \bar{x}_2) \equiv \bar{z} \right\} \tag{25}$$

and the function:

$$\Phi : x_1 \in X \subseteq \mathbb{R} \longrightarrow (x_2, z) \in S. \tag{26}$$

It should be noticed that the latter defines a *single valued* function, being $\zeta(x_1, x_2)$ a *single valued* function. Moreover, recall that $e = y - q$ is the model's output without the pure mechanical response represented by $q(x_2)$.

This function could be either injective or not, depending on the characteristics of the real material, as evidenced hereafter. To this aim, let us consider the material, showing the relation between the output $y - q$ and the input variables x_1 and x_2 as shown in Fig. 18-(a). There with z it has been indicated the input variable corresponding to the output $y - q$ when $x_2 = x_2^*$. If now $x_2 = x_2'$ two distinct values of x_1 are obtained. This implies that the function Φ is *not injectve* (Fig 18-(b)). Conversely, if we refer to Fig. 19-(a), the value of x_1 corresponding

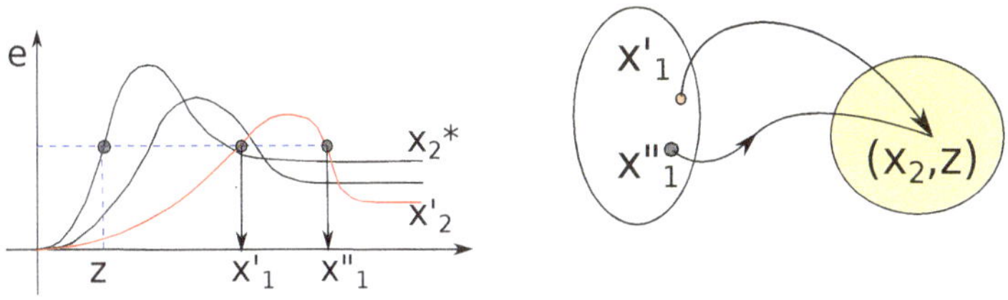

(a) Non monotone response characteristic of a general smart material.

(b) Example of non-injective Φ function

Figure 18. Example of the Φ function reconstructed from input-output available data. It can be stressed that the non-monotonicity of the response functions, as sketched in (a), yields to $\frac{\partial \zeta}{\partial x_1}$ that can assume any sign for assigned x_2

to a given z (which, as before, refers to the assigned value of output $y - q$ on the curve corresponding to $x_2 = x_2^*$), is now unique, when it exists. In fact in the same figure, the possibility that no x_1 values corresponding to a given couple (x_2, z) is also considered. This example evidences that not all values (x_2, z) belongs to the domain of definition of the function Φ. In this case the function is *injective* in S. Moreover, it would appear quite evident that the properties of the Φ function are strictly related to the monotonicity of the functions $(y - q) \leftrightarrow z$ or, in other words to the derivatives $\frac{\partial \zeta(x_1, x_2)}{\partial x_1}$, as detailed hereafter.

The function

$$z = \zeta(x_1, x_2)$$

can be rearranged by the aid of the function: $g : \mathbb{R}^2 \to \mathbb{R}^2$, having components:

$$\begin{cases} z = \zeta(x_1, x_2); \\ x = x_2, \end{cases} \tag{27}$$

and *Jacobian*:

$$J_g(x_1, x_2) = \begin{vmatrix} \frac{\partial \zeta}{\partial x_1} & \frac{\partial \zeta}{\partial x_2} \\ 0 & 1 \end{vmatrix} = \frac{\partial \zeta}{\partial x_1}. \tag{28}$$

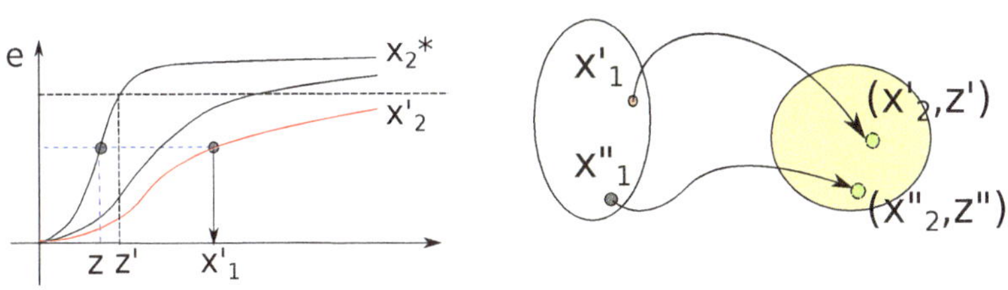

(a) Monotone response characteristic of a general smart material.

(b) Example of injective Φ function

Figure 19. Example of the Φ function reconstructed from input-output available data. It can be stressed that the strict monotonicity of the response functions, as sketched in (a), yields to $\frac{\partial \zeta}{\partial x_1} > 0$ for assigned x_2.

If now we define the set:

$$A = \{(P \equiv (x_1, x_2) \in \Omega : J(P) \neq 0\}, \tag{29}$$

it is possible to conclude that whenever A is a non-empty set, for every $P \in A$ the g function g is locally invertible and so the equation $\zeta(x_1, x_2) = z$ has only one x_1 solution. The sets

$$S^* = \left\{(\bar{x}_2, \bar{z}) \in \mathbb{R}^2 : (x_1, x_2) \in A\right\} \tag{30}$$

is a subset of S, i.e. $S^* \subseteq S$. As a consequence, Φ is injective in S^* and finally it admits the inverse function

$$\Phi^{-1} : S^* \longrightarrow X^*,$$

being X^* a subset of X. This line of reasoning shows the conditions which guarantee the existence of the Φ^{-1} function and hence the possibility to define the compensator of the 2-variables operator with hysteresis. This is a crucial point since only with this condition a generalization of the classical *model-based* control strategy can be settled for multi-variable systems with rate independent memory.

The availability of a procedure enabling the inversion of a hysteresis operator depending on two input variables opens the opportunity to extend *model-based* control strategies to systems having in realty two inputs (for example, desired strain and mechanical stress). This need arises in specific dynamic working conditions of magnetostrictives, (i.e. active vibration suppression of structures), where the device could experience forces of comparable magnitude of the applied stress, but could be extended to similar conditions involving different smart materials.

The simplest control strategy exploiting the results concerning the inversion of multi-variable hysteresis operators, is sketched in Fig. 16-(b). There we can observe the compensation algorithm described so far. In particular the operator Γ^{-1} which, in open loop, takes the variable $e = y - q$ and provides the corresponding input function for the reference case (i.e. $x_2 = x_2^*$), that is $x_1 = z$. The inverse of the Φ function, takes z and x_2 and provides the control variable x_1. If a given desired signal is provided, in presence of a time varying $x_2 = \sigma(t)$, the global behavior of the actuators drastically changes and, without a suitable *stress-dependent* compensation algorithm, the controller performances suddenly worsen. For the sake of example, in Fig. 20 it is shown the relative error in tracking a $\sin(t)/t$ function provided a time variation of the applied force over the full simulation time interval of about 400N. It is evident the improved behavior of the 2-variables control (2VC) system with respect to the case when stress corrections are not provided, [18]. The present approach allows to compensate the effects of nonlinearity and hysteresis in more general working conditions, guaranteeing a close-to-linear behavior of the actuator and making so easier the controller design. Unfortunately, the effects of the compensation algorithm together with the PI controller are beneficial for the whole control system whenever the desired signal is slowly varying, but its performances decrease for faster actuation purposes. Further, multi-variable compensation and control is mostly required exactly when faster signal to track are involved. Thus, different control schemes able to guarantee sufficient performances in a wider frequency range are necessary, [19, 22]. As a first attempt to address the issue, let us recall that closed loop control represents a necessary solution, in presence of disturbances model uncertainties, guaranteeing performances otherwise unachievable by open loop solutions. On the contrary, feedforward controls are easier to implement, do not affect the stability of the overall system

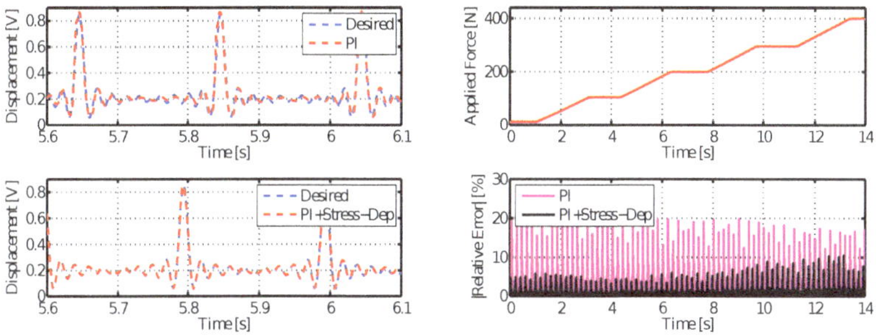

(a) Desired and measured output vs. time. Up: PI controller; bottom; PI+2VC (b) Applied stress and relative tracking error vs. time

Figure 20. Example of real time control task, by considering the effects of compensation with one or two variables

and show faster behavior with respect to feedback solutions since its action depends only on the feedforward controller and not on process dynamics. Modern control systems often employ both actions with the aim to obtain the advantages of both configuration, avoiding, as far as possible, their drawbacks so yielding to a controller with higher performances. Such a configuration is called *two-degrees-of-freedom*, 2-DoF, control system since in this way it is possible to design a controller that weights independently the reference signal and the measurements coming from the process [36, 38]. Examples of 2-DoF control systems for

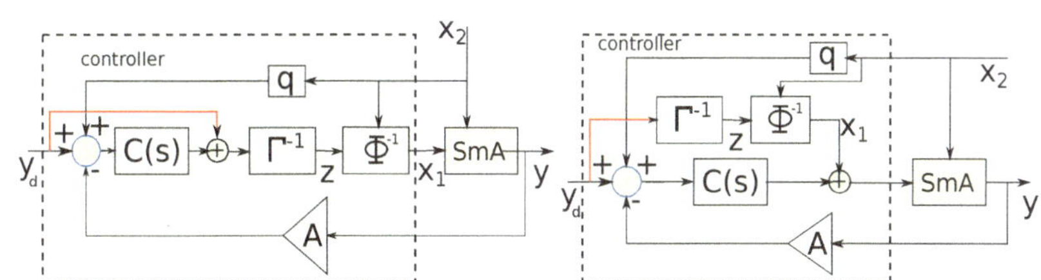

(a) (b)

Figure 21. Block schemes of 2-DoF control systems. Hysteresis compensation on the feedback loop (a); Hysteresis compensation on the feedforward action (b)

actuators with memory, proposed in [19], are shown in Fig. 21. In both of the diagrams it is possible to locate a feedback and a feedforward action. In part -(a) of the figure, the feedforward action trivially consists in adding the desired output to the output of the linear controller $C(s)$. The compensation, is located on the feedback action. In that case both the compensator and actuator with hysteresis experiences exactly the same input history and so they share the same internal state. Conversely, the scheme reported on the part -(b) of the same figure shows the compensation located on the feedforward action, while the feedback loop is without compensation. In such a scheme it is quite evident that the real actuator and its compensator doesn't experiences the same input history. For this reason we could expect a different behavior with respect to the previous system.

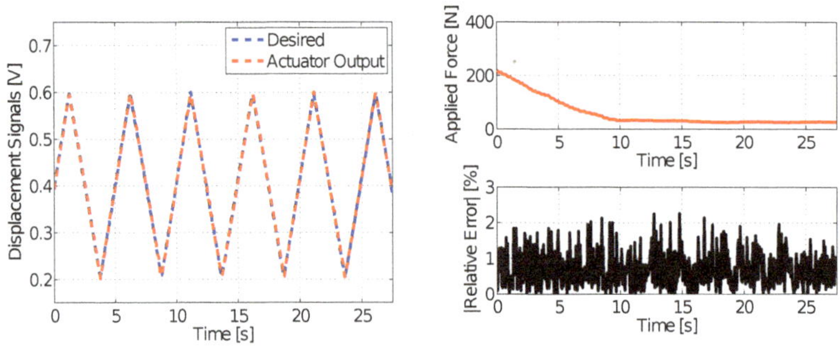

(a) (b)

Figure 22. Tracking performances of a PI control system with 2VC for low frequency application. In this case the period of the desired output is $T = 5\mathrm{s}$

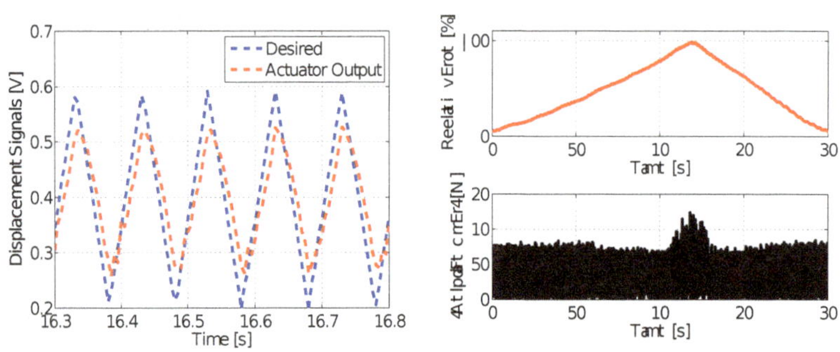

(a) (b)

Figure 23. Tracking performances of a PI control system with 2VC for higher frequency application. In this case the period of the desired output is $T = 0.1\mathrm{s}$

In order to evidence the drawbacks of a classical feedback loop employing a simple PI controller, and the improvements provided by the 2-DoF control system, let us discuss the experiments performed by a *real-time* environment where the compensation algorithm and the PI controller are implemented in a PC with Matlab, and a xPC Target toolbox with a 16 bit acquisition/generation card are also employed. Further details are discussed in [19]. In particular, in Fig. 22 it is shown the performances of the feedback system, sketched in Fig. 16, in tracking a triangular waveform, with period of $5\,s$. Here the relative error lies below 3%. When the frequency is increased to 10Hz, the problems in tracking the same waveform dramatically arise and tracking error goes well beyond 15%, as evidenced in Fig 23. The performances results of the scheme reported in Fig. 21-(b) are shown in Fig. 24, with an increase of tracking performances. In Fig. 25, conversely, the performances of the 2-DoF control system presented in Fig. 21-(a) are shown. It can be observed a further increase of the tracking performances of the system with quite low tracking error, since the system provides the same state evolution of both the actuator and compensator.

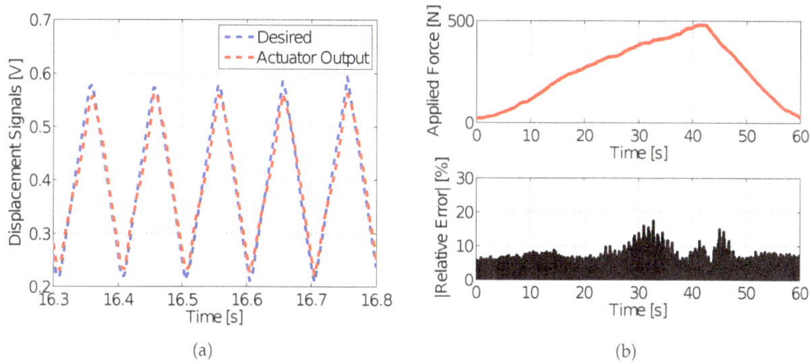

Figure 24. Tracking performances of a 2-DoF control system with 2VC for higher frequency application. In this case the period of the desired output is $T = 0.1$s and the controller is specified in Fig. 21-(b)

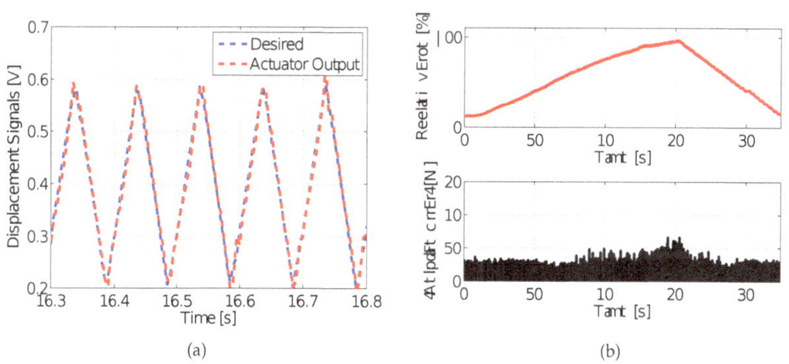

Figure 25. Tracking performances of a 2-DoF control system with 2VC for higher frequency application. In this case the period of the desired output is $T = 0.1$s and the controller is specified in Fig. 21-(a)

In conclusion the availability of suitable and well assessed tools for the modeling and compensation of actuators with hysteresis enables to define control algorithms that fully exploits this high modeling capabilities. The result is the availability of a model-based control system which gather good performances and quite low computational effort, guaranteeing its suitability for real time and fast actuation tasks.

6. Conclusion

The chapter aimed to provide a view on the basic problems involved when modeling and control of smart materials, normally affected by memory phenomena, are concerned. So, after a brief review of some of the contributions provided in this field, the attention was focused specifically on materials showing a magnetostrictive behavior, such as Terfenol-D and Galfenol. However, all the ideas and tools introduced and discussed could be widely exploited for other smart materials with non-negligible memory effects.

In Sect. 2 a brief description of the phenomena involved in magnetostrictive materials is drawn and the ideas of their modeling are also provided. In this respect, the basic operators describing systems with hysteresis is presented, with a specific attention to the classical Preisach model. Although the most known and applied approach is the one in the 'rotated' system of coordinates, according to Mayergoyz's book, [28], the adopted formalism allover the chapter is that preferred in the books [7] and [26], showing some interesting characteristic from the modeling viewpoint. In any case, the relationship between them is explicitly pointed out.

In the next section 3 the attention is focused on the main modeling and compensation issues which are normally adopted for control purposes. Here, a specific attention is devoted to the congruency property of Preisach operator, which forces a specific attention in the inversion of the operator. Finally, a basic point is discussed with some detail, that is, the definition of compensation algorithms which doesn't require further computational effort with respect to the 'direct' operator, i.e. the Preisach operator. This specific point becomes crucial when a *model-based* real time control approach is of concern.

In order to put the issues discussed in the chapter in the current framework of smart materials and devices, in section 5 the need to manage two independent variables in controlling the device is emphasized, by proposing a well-behaved procedure to handle the stress and magnetic field simultaneously. The effectiveness of this approach for a more precise description and control of smart devices is also discussed by comparison to measured data. Such discussion paves the way to the last section of the chapter, where some application in *real time* controlling smart materials are presented. In particular, 'standard' and 2-DoF control strategies are presented, all fulfilling the constraint to keep a low computational complexity of the whole control system, still ensuring good tracking and stability performances.

Author details

Daniele Davino, Alessandro Giustiniani, Ciro Visone
University of Sannio - Engineering Department, 82100 Benevento (BN), Italy

7. References

[1] Adly, A.A., Mayergoyz, I.D. & Bergqvist, A. (1991). Preisach modeling of magnetostrictive hysteresis *AIP Journal of Applied Physics* 69, 5777.

[2] Anton, S.R. & Sodano, H.A. (2007). A review of power harvesting using piezoelectric materials (2003-2006), *IOP Smart Materials and Structures*, vol. 16, 3, pp. R1-R21.

[3] Bar-Cohen, Y. (2002). Electro-active polymers: Current capabilities and challenges, *Proceedings of SPIE*, San Diego, CA.

[4] Bellouard, Y. (2008). Shape memory alloys for microsystems: A review from a material research perspective *ELSEVIER Materials Science and Engineering A* 481-482, pp. 582-589.

[5] Bergqvist, A. & Engdahl, G. (1991). A stress-dependent magnetic Preisach hysteresis model, *IEEE Transactions on Magnetics*, Vol. 27, No. 6, pt 2, pp. 4796-4798.

[6] Brokate, M. (1989). Some mathematical properties of the Preisach model of hysteresis, *IEEE Transactions on Magnetics*, 25, pp. 2922-24.

[7] Brokate, M. & Sprekels, J. (1996) *Hysteresis and Phase Transitions*,Springer.

[8] Cavallo, A., Natale, C. Pirozzi, S. & Visone, C. (2003). Effects of Hysteresis Compensation in Feedback Control Systems, *IEEE Transactions on Magnetics* Vol. 39, No. 3, pp.1389-1392.

[9] Cavallo, A., Natale, C., Pirozzi, S., Visone, C. & Formisano, A. (2004). Feedback Control Systems for Micropositioning Tasks With Hysteresis Compensation, *IEEE Transactions on Magnetics*, Vol. 40, No. 2, pp. 876-879.
[10] Cavallo, A., Davino, D., De Maria, G., Natale, C., Pirozzi, S. & Visone, C. (2008). Hysteresis compensation of smart actuators under variable stress conditions, *ELSEVIER Physica B*, 403, 2-3, pp. 261-265.
[11] Chung, D.Y., Hogan, T., Brazis, P., Rocci-Lane, M., Kannewurf, C., Bastea, M., Uher, C. & Kanatzidis, M. G. (2000). $CsBi_4Te_6$: A High-Performance Thermoelectric Material for Low-Temperature Applications, *Science* 287, pp. 1024-27
[12] Davino, D., Natale, C., Pirozzi, S. & Visone, C. (2005). A fast compensation algorithm for real-time control of magnetostrictive actuators, *ELSEVIER Journal of Magnetism and Magnetic Materials*, 290-291, pp. 1351-54.
[13] Davino, D., Giustiniani, A. & Visone, C. (2008). Fast Inverse Preisach Models in Algorithms for Static and Quasistatic Magnetic-Field Computations, *IEEE Transactions on Magnetics*, 44 (6) , pp. 862-865.
[14] Davino, D., Giustiniani, A. & Visone, C. (2008). Properties of Hysteresis Models Relevant in Electromagnetic Fields Numerical Solvers, *ELSEVIER Physica B: Condensed Matter*, 403 , pp. 414-417.
[15] Davino, D., Giustiniani, A. & Visone, C. (2008). A Magnetostrictive Model with Stress Dependence for Real-Time Applications, *IEEE Transactions on Magnetics*, 44 (11-2) , pp. 3193-3196.
[16] Davino, D., Giustiniani, A. & Visone, C. (2009). Experimental Properties of an Efficient Stress-Dependent Magnetostriction Model, *AIP Journal of Applied Physics*, 105 (07D512) , pp. 1-3.
[17] Davino, D., Giustiniani, A. & Visone, C. (2010). Design and Test of a Stress Dependent Compensator for Magnetostrictive Actuators, *IEEE Transactions on Magnetics*, 46 (2) , pp. 646-649.
[18] Davino, D., Giustiniani, A. & Visone, C. (2011). Compensation and control of two-inputs systems with hysteresis, *IOP Journal of Physics: Conference Series*, 268, doi:10.1088/1742-6596/268/1/012005.
[19] Davino, D., Giustiniani, A., Iannelli, L. & Visone, C. (2012). Comparison of real-time control strategies with hysteresis compensation for magnetostrictive actuators, *IOS International Journal of Applied Electromagnetics and Mechanics*, in press.
[20] Dlala, E. & Arkkio, A. (2006). Inverted and forward Preisach models for numerical analysis of electromagnetic field problems, *IEEE Transactions on Magnetics*, Vol. 42, No. 8, pp. 1963-1973.
[21] Engdhal, G. (editor) (2000) *Handbook of Giant Magnetostrictive Materials*, Academic Press.
[22] Esbrook, A., Guibord, M., Tan, X. & Khalil, H.K. (2010). Control of systems with hysteresis via servocompensation and its application to nanopositioning, *Proceedings of the 2010 American Control Conference, ACC 2010*, Article number 5531422, Pages 6531-6536.
[23] Ge, P. & Jouaneh, M. (1996). Tracking Control of a Piezoceramic Actuator, *IEEE Transactions On Control Systems Technology*, 4, pp. 209.
[24] Hasegawa, T. & Majima, S. (1998). A control system to compensate the hysteresis by Preisach model on SMA actuator, *Proceedings of IEEE International Symposium on Micromechatronics and Human Science*, 171.
[25] Krasnoselskii, M. A. & Pokrovskii, A. V. (1989). Systems with Hysteresis, Springer-Verlag.

[26] Krejci, P. (1996). Hysteresis, convexity and dissipation in hyperbolic equations, *Gakuto Int. Series Math. Sci. & Appl.*, 8, Gakkotosho, Tokyo.
[27] Krejci, P. & Kuhnen, K. (2001). Inverse control of systems with hysteresis and creeps, *Proceedings of Control Theory and Applications*, Vol. 148, No. 3, pp. 185-192.
[28] Mayergoyz, I. D. (1991) *Mathematical Models of Hysteresis*, Springer.
[29] Miano, G., Serpico, C. & Visone, C. (1997). A new model of magnetic hysteresis, based on Stop hysterons: an application to magnetic field discussion, *IEEE Transactions on Magnetics*, Vol. 32, No. 3, pp. 1132-1135.
[30] Natale, C., Velardi, F. & Visone, C. (2001). Identification and compensation of Preisach hysteresis models for magnetostrictive actuators, *ELSEVIER Physica B*, 306, pp. 161-165.
[31] Otsuka, K. & Wayman, C.M. (1998) *Shape Memory Materials*, Cambridge University Press.
[32] Panda, P. K. (2009). Review: environmental friendly lead-free piezoelectric materials, *SPRINGER Journals of Matererials Science*, 44, pp. 5049-5062, DOI 10.1007/s10853-009-3643-0.
[33] Pecharsky, V.K. & Gschneidner, K.A. (2006). Advanced magnetocaloric materials: What does the future hold? *International Journal of Refrigeration* 29 pp. 1239-1249.
[34] Polla, D.L. & Francis, L.F. (1998). Processing And Characterization Of Piezoelectric Materials And Integration Into Microelectromechanical Systems, *Annual Review Matererials Science*, 28, pp. 563-597.
[35] Schafer, J. & Janocha, H. (1995). Compensation Of Hysteresis In Solid-State Actuators, *ELSEVIER Sensors And Actuators A* 49, pp. 97-102.
[36] Skogestad, S. & Postlethwaite, I. (2005). Multivariable Feedback Control: Analysis and Design, *Wiley-Interscience.*
[37] Söderberg, O., Ge, Y., Sozinov, A., Hannula, S.P. & Lindroos, V.K: (2005). Recent breakthrough development of the magnetic shape memory effect in Ni-Mn-Ga alloys *IOP Smart Materials and Structures* 14, 5.
[38] Sugie, T. & Yoshikawa, T. (1986). General solution of robust tracking problem in two-degree-of-freedom control systems *IEEE Transactions on Automatic Control*, Vol. 31 (6), pp. 552-554.
[39] Tan, X., Venkataraman, R. & Krishnaprasad, P.S. (2001). Control of hysteresis: Theory and experimental results, SPIE Modeling, Signal Processing, and Control in Smart Structures (Rao V. S., Ed.), 4326, pp. 101-112.
[40] Tellinen, J., et al., (2002). Basic properties of magnetic shape memory actuators, *Proceedings of 8th ACTUATOR Conference*, Bremen, Germany.
[41] Visintin, A. (1991) *Differential Models of Hysteresis*, Springer.
[42] Visone, C. & Serpico, C. (2001). Hysteresis operators for the modeling of magnetostrictive materials, *ELSEVIER Physica B: Condensed Matter*, 306 (1-4) , pp. 78-83.
[43] Visone, C. & Sjostrom, M. (2004). Exact invertible hysteresis models based on play operators, 2004, *ELSEVIER Physica B: Condensed Matter*, 343, pp. 148Ð52.
[44] Visone, C. (2008). Hysteresis Modelling and Compensation for Smart Sensors and Actuators, *IOP Journal of Physics Conference Series.*
[45] Webb, G.V., Lagoudas, D.C. & Kurdila, J. (1998). Hysteresis modeling of SMA actuators for control applications, *SAGE Journal of Intelligent Material Systems and Structures*, Vol. 9, No. 6, pp. 432-48.
[46] Webb, G., Lagoudas, D. & Kurdila, A. (1999). Adaptive hysteresis compensation for SMA actuators with stress-induced variations in hysteresis, *SAGE Journal of Intelligent Material Systems and Structures*, Vol. 10, No. 11, pp. 845-854.

Chapter 6

1D Phenomenological Modeling of Shape Memory and Pseudoelasticity in NiTi Alloys

Carmine Maletta and Franco Furgiuele

Additional information is available at the end of the chapter

1. Introduction

A one-dimensional phenomenological approach to simulate both the mechanical and functional properties in shape memory alloys (SMAs) is described in the following sections. In fact, shape-memory alloys exhibit unique mechanical and functional features, due to reversible transformations in crystal structure. In particular, on the macroscopic scale, SMAs are able to remember a geometrical shape and can return to that shape by activating the phase transition mechanisms. Many kinds of SMAs have been exploited in the last decades, such as the copper-zinc-aluminum (ZnCuAl), copper-aluminum-nickel (CuAlNi), nickel-manganese-gallium (NiMnGa), nickel-titanium (NiTi), and other ones made by alloying zinc, copper, gold, iron, etc. Among these alloys the near equiatomic NiTi binary system shows the most exploitable characteristics due to the high stress and strain recovery capabilities associated with their functional properties, namely pseudoelastic effect (PE) and shape memory effect (SME). These properties are due to a reversible solid state phase transformation between a parent phase (austenite) and a product phase (martensite), the so called thermoelastic martensitic transformation (TMT), that can be activated either by temperature (Thermally Induced Martensite, TIM), or by applied stress (Stress Induced Martensite, SIM) [1]. Due to these features NiTi alloys are currently used in an increasing number of applications in many fields of engineering [2], for the realization of smart sensors and actuators, joining devices, hydraulic and pneumatic valves, release/separation systems, consumer applications and commercial gadgets. However, thanks to their good mechanical properties and biocompatibility the most important applications of NiTi alloys are in the field of medicine, where pseudoelasticity is mainly exploited for the realization of several components, such as cardiovascular stent, embolic protection filters, orthopedic components, orthodontic wires, micro surgical and endoscopic devices. As a direct consequence of their interesting features NiTi alloys have attracted the interest of scientific and engineering community in the last years. However, despite the increasing interest and the efforts of many researchers to better understand these unusual mechanisms, the use of NiTi alloys is currently limited to high-value applications (i.e. medical devices, MEMS, etc.), due to the high cost of the raw material as well as to the complex component manufacturing; in fact, an accurate control of the processing parameters must be carried out as the functional and mechanical properties of NiTi alloys are significantly affected by the thermo-mechanical loading history experienced during manufacturing [3–8]. On the other hand, the design of complex shaped NiTi-based components needs an accurate knowledge of the mechanical and functional response of the material, as well as how this evolves during subsequent thermo-mechanical processes. Within this context the use of numerical modeling techniques, to simulate both mechanical and functional behavior of SMAs, is of major concern and, consequently, many studies have been focused on this topic in the last few years [9, 10], with the aim to model the non-linear hysteretic behavior that describes the phase transformation, and the related functional properties. Some of these models are based on microscopic and mesoscopic approaches [10], where the thermo-mechanical behavior is modeled starting from molecular level and lattice level, respectively; other models are based on macroscopic approaches, where only phenomenological features of the SMAs are used [11–24]. In this field, some authors proposed one-dimensional models based on an assumed polynomial-free energy potential [11, 12] while other models are based on an assumed phase transformation kinetic and consider simple mathematical functions to describe the phase transformation behavior of the material [13–15]. These models are probably the most popular in the literature due to their phenomenological approaches, which allow easy developments without considering the underlying physics of the transformation kinetic. Furthermore, other models are based on the elastoplasticity theory [16–22] which are capable of describing the functional behavior of the material using plasticity concepts. Finally, some researchers used the Galerkin method to describe thermo-mechanical behaviors of shape memory alloys [23, 24]. More recently, a 1-D phenomenological approach to simulate both the shape memory effect [27–29] and pseudoelastic effect [30] in NiTi-based shape memory alloys has been developed and it is described in the following sections. In particular, the temperature-strain and stress-strain hysteretic behavior of SMAs, associated with the thermally induced and stress-induced phase transition mechanisms, are modeled from a phenomenological point of view, *i.e.* without considering the underlying physics of the problem, by using Prandtl-Ishlinksii hysteresis operators [25, 26]. The main features of this approach is a simple implementation together with a good accuracy and efficiency in predicting the stress-strain hysteretic behavior of 1D components. Unfortunately, the one dimensional nature of the proposed model, represents one of the major drawback with respect to some of the pre-existing phenomenological models, which are based on more thermodynamically consistent frameworks and, consequently, are able to capture several behaviors of NiTi alloys, such as detailed stress-strain distribution in 2D and 3D components. However, the high computational efficiency of the proposed model allows its use for real time simulation and control o 1D SMA components. The parameters of the phenomenological model are identified by simple and efficient numerical procedures, starting from a set of experimentally measured hysteresis loops. The identification procedures have been developed in the commercial software package $Matlab^{TM}$, while the computed parameters are used in $Simulink^{TM}$ models, which are able to simulate the whole path dependent hysteretic behavior of the SMAs, *i.e.* for generic complete and incomplete stress-induced and/or thermally induced phase transition mechanisms. The models are also able to capture the hysteresis modifications due to complex loading conditions, *i.e.* they are able to predict the change of the transformation stresses and temperatures according to the

Clausius-Clapeyron relation [1]. The unique thermo-mechanical features of SMAs are firstly illustrated in the following section 2, while the numerical approach is described in section 3 together with some case studies, involving both shape memory and pseudoelasticity, and the results show good accuracy and small computational time.

2. Thermo-mechanical properties of NiTi alloys

2.1. Thermoelastic martensitic transformation

Nickel-Titanium (NiTi) based shape memory alloys exhibit unique thermomechanical properties due to a reversible solid state phase transformation between a high temperature parent phase (B2 - austenite) and a low temperature product phase (B19′ - martensite), the so called Thermoelastic Martensitic Transformation (TMT). In particular, TMT can be activated by a temperature variation (TIM, Thermally Induced Martensite) or by the application of an external mechanical load (SIM, Stress Induced Martensite), and it allows the crystal lattice structure to accommodate to the minimum energy state for a given temperature and/or stress. Figure 1 schematically shows the crystal structures of the two phases.

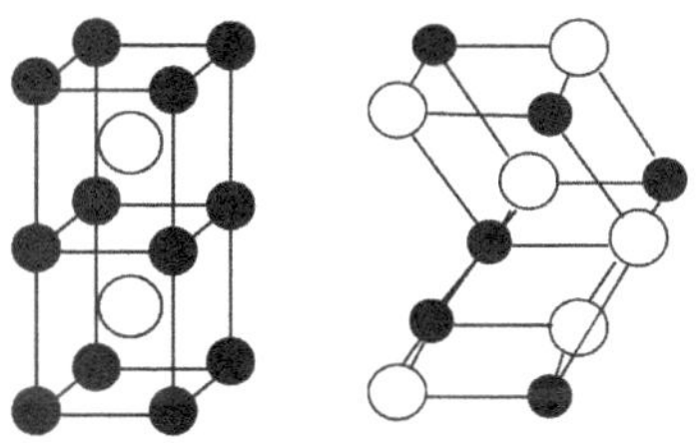

Figure 1. Schematic depiction of the lattice structures of austenite and martensite.

The austenitic phase is characterized by a Body Centered Cubic structure (BCC), with a nickel atom at the center of the crystallographic cube and titanium atoms at the cube's corners, while the lattice structure of the martensitic phase consists of a rhombus alignment with an atom at each of the rhombus corners. On the macroscopic scale the two crystal structures exhibit different engineering properties, such as Young's modulus, electrical resistance, damping behavior, etc. As a consequence, the transition between the two phases gives the possibility to obtain variable and/or tunable properties, *i.e.* NiTi alloys are able to change and or/adapt their response as a function of external stimuli. In addition, phase transition mechanisms are also associated with high strain recovery capabilities resulting from both thermally-induced and/or stress-induced transformations as described in the following section.

2.1.1. *Thermally-induced martensitic transformation*

When cooling the austenitic structure a thermally-induced martensitic transformation (B2→B19′) occurs in the temperature range between martensite start temperature (M_S), and martensite finish temperature (M_f). When the martensitic structure is heated the reverse transformation between martensite and austenite (B19′→B2) occurs in the range between austenite start temperature (A_s) and austenite finish temperature (A_f). These characteristic temperatures, the so called phase transition temperatures (PPTs), can be regarded as material parameters, which depend on the alloy composition and on the thermomechanical processing conditions [1], and can be easily detected by Differential Scanning Calorimetry (DSC) investigation, as schematically shown in Figure 2.

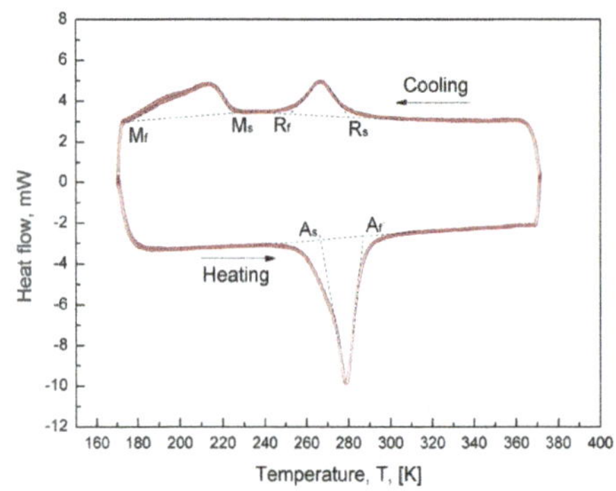

Figure 2. Differential scanning calorimetry thermogram of a NiTi alloy

In addition, transformation from B2 cubic austenite into monoclinic B19′ martensite could occur either directly or via an intermediate rhombohedral phase (R-phase), as shown in Figure 2. In particular, the R-phase transformation (B2→R) can be observed during cooling from A_f to M_s prior to martensitic transformation, resulting in a sequential transformation B2→R→B19′. However, it is worth noting, that the B2→R transformation is observed only under specific processing conditions of the alloy [33]. In addition, marked differences are normally observed between direct and reverse transformation temperatures as a direct consequence of the thermal hysteretic behavior of the alloy, as illustrated in Figure 3. In particular, this figure shows the thermal hysteresis describing the evolution of the volume fraction of martensite (ξ_M) together with the characteristic transformation temperatures.

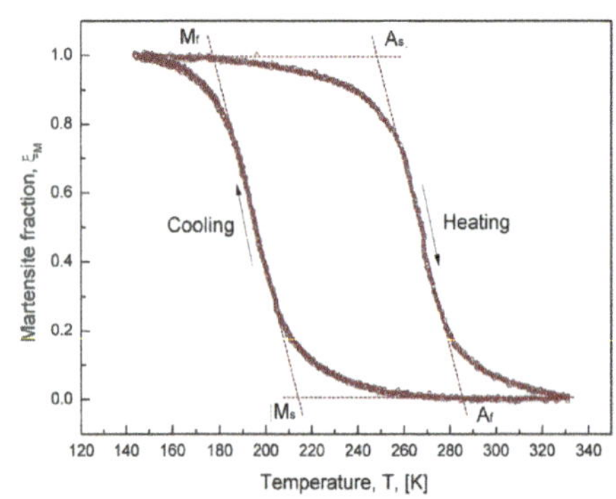

Figure 3. Thermal hysteresis of a NiTi alloy describing the evolution of volume fraction of martensite

2.1.2. Stress-induced martensitic transformation

When a mechanical load is applied to the austenitic structure the stress-induced B2→B19′ transformation occurs, corresponding to a plateau in the stress-strain curve of the alloy. If the mechanical load is removed the reverse B19′→B2 transformation occurs which is related to another stress plateau and allows an almost complete strain recovery. Figure 4 illustrates an example of stress-strain curve of a NiTi alloy exhibiting stress-induced phase transformation mechanisms, together with the characteristic transformation stresses of the alloy, *i.e.* the stresses for direct B2→B19′ transformation (σ_s^{AM}, σ_f^{AM}) and the stresses for reverse B19′→B2 transformation (σ_s^{MA}, σ_f^{MA}).

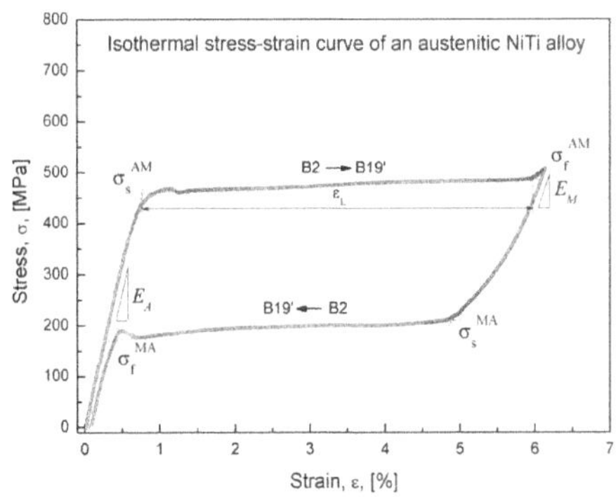

Figure 4. Stress-strain curve of an austenitic NiTi alloy with characteristic transformation stresses.

Figure 4 also illustrates the recovery strain ε_L due to the stress-induced transformation mechanisms, together with the Young's moduli of the two phases (E_A and E_M). Another stress-induced microstructural change occurs when a mechanical load is applied to the martensitic structure, *i.e.* for $T < M_f$, the so called detwinning. This mechanism can be regarded as a variant reorientation process and, on the macroscopic scale, it causes large plastic-like deformations which corresponds to a plateau in the stress-strain curve of the alloy. This mechanism is responsible for the shape memory effect as described in the following section. In addition, it is worth noting that NiTi SMAs exhibit a marked temperature dependent stress-strain response, as schematically depicted in Figure 5. In particular, the temperature dependence of transformation stresses is given by the Clausius-Clapeyron relation of equations 1:

$$\frac{d\sigma^{AM}}{dT} = C_M; \frac{d\sigma^{MA}}{dT} = C_A \tag{1}$$

where C_M (direct martensitic transformation) and C_A (reverse austenitic transformation) are in the range between 5 and 10 $MPaK^{-1}$.

2.2. Shape memory effect

Shape Memory Effect (SME) is the ability of a SMA to remember a predetermined shape and to recover this shape even after being subjected to large mechanical deformations (up to 10%). In NiTi alloys this property is observed under martensitic conditions, *i.e.* when $T < M_f$,

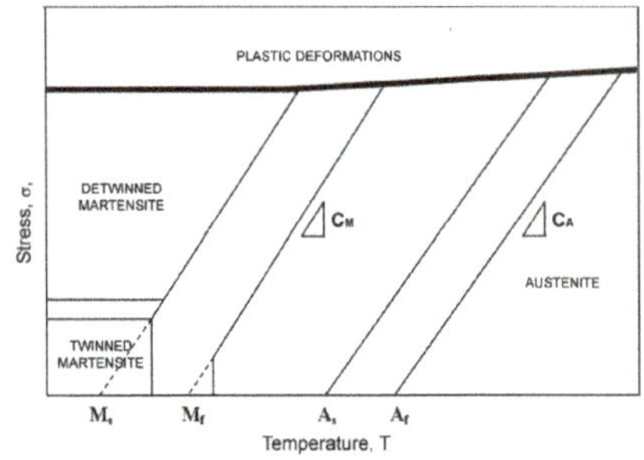

Figure 5. Relation between transformation stresses and temperature according to the Clausius-Clapeyron relation.

and it can be attributed to the combination of two microstructural changes: *i*) detwinning of martensitic variants and *ii*) thermally induced phase transformation. Figure 6 shows a schematic depiction of the SME together with the associated phase transition mechanisms (Figure 6.a) and the corresponding stress-strain-temperature response (Figure 6.b).

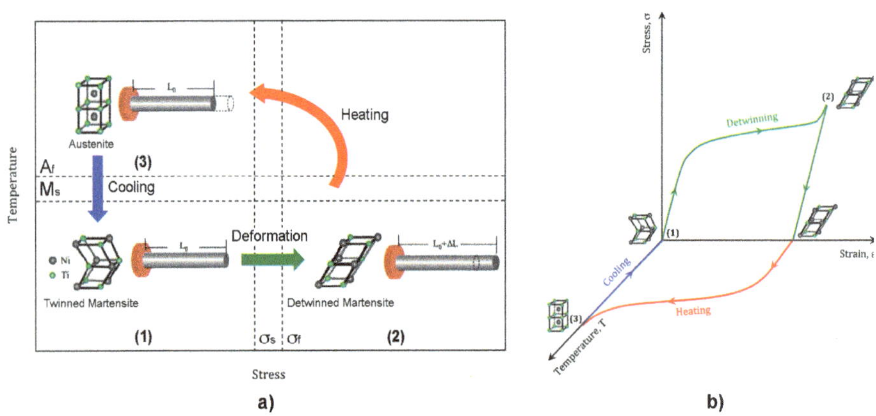

Figure 6. Schematic depiction of the shape memory effect: a) phase transition mechanisms and b) stress-strain-temperature response

In particular, Figs. 6 show that if a mechanical load is applied to the twinned martensitic structure (1), *i.e.* for $T < M_f$, detwinning occurs at a given critical stress value which corresponds to large plastic-like deformations (up to 10%) through a plateau in the stress-strain curve (2). In fact, these deformations persist after complete unloading as only elastic recovery of the detwinned structure is observed. However, if the material is heated up to the austenite finish temperature ($T > A_f$) a complete thermally induced phase transformation occurs from the detwinned martensitic structure to the austenitic one (3) and, on the macroscopic scale, this transformation allows a complete shape recovery. Finally, if the material is cooled down to the martensite finish temperature ($T < M_f$) it is able to remember its original twinned martensitic structure (1). This unusual functional property is also known as one-way shape memory effect (OWSME) as it defines the ability of material to remember just one shape, the cold one ($T < M_f$), and to recover this shape after being mechanically deformed. However, under specific thermo-mechanical treatments NiTi alloys could exhibit another shape memory mechanism, the so called two-way shape memory effect (TWSME), *i.e.* they are able to remember a cold shape, linked to the martensitic structure, and a hot shape, linked to the austenite. As a consequence, during repeated heating and cooling, the material changes its shape in a reversible way, through a hysteresis loop, as schematically illustrated in Figure 7.

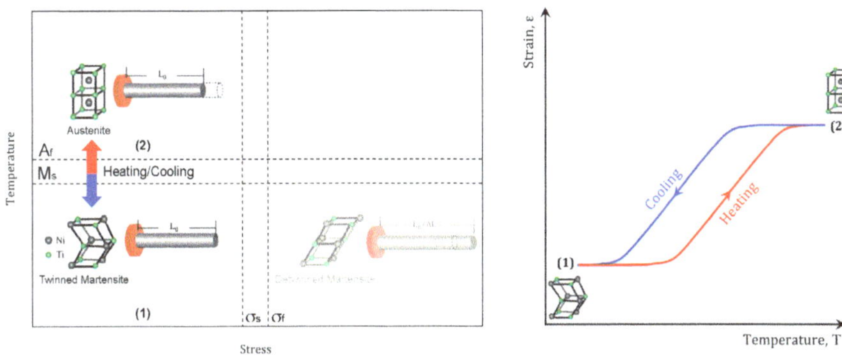

Figure 7. Schematic depiction of the two-way shape memory effect: a) phase transition mechanisms and b) strain-temperature hysteretic response

In particular, TWSME can be induced by proper thermomechanical procedures, the so-called training, which usually involve repeated deformations and transformations between austenite and martensite. This thermomechanical process produces a dislocation structure and, consequently, creates an anisotropic stress field that benefits the formation of preferentially oriented martensite variants [32], resulting in a macroscopic shape change between the phase transition temperatures.

Figure 8 shows an example of the thermomechanical cycle, which is composed of four subsequent steps: 1) strain controlled uniaxial loading up to a training deformation ε_{tr}, 2) complete unloading at the same rate and recording of the residual strain ε_r, 3) heating up to the austenite finish temperature A_f, in stress free conditions, to activate SME and measuring the recovery deformation ε_{re} and permanent strain ε_p, and 4) cooling down to the martensite finish temperature M_f, in stress free conditions, and recording the induced two-way shape memory strain ε_{tw}. Experimental measurements have been carried out in [28] where several training cycles have been executed with a training deformation $\varepsilon_{tr} = 5.5\%$. Each training cycle starts from the end of the cooling stage of the previous one, so that the total deformation at the $i-th$ cycle, $\varepsilon_{tot(i)}$, can be defined as follows:

$$\varepsilon_{tot(i)} = \varepsilon_{tr(i)} + \varepsilon_{p(i-1)} + \varepsilon_{tw(i-1)} \tag{2}$$

Figure 9 reports the measured ε_{tw}, ε_p, ε_{pe}, and ε_{tot} *vs* the number of training cycles. The figure clearly shows that the two-way shape memory strain ε_{tw} increases with increasing the number

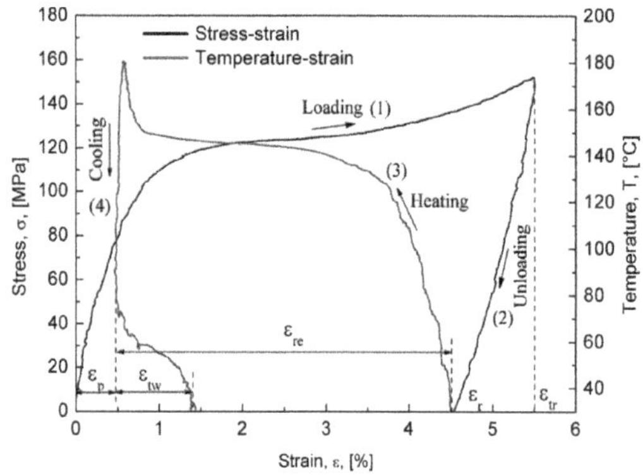

Figure 8. Example of training cycle: 1) loading, 2) unloading, 3) heating up to A_f, and 4) cooling down to M_f [28].

of training cycles, and a similar behavior is observed for the permanent deformation ε_p, the strain recovery ε_{re}, and the total deformation ε_{tot}. In particular, ε_{tw} increases from 1.0% at the first training cycle to 2.8% after six cycles; ε_p and ε_{re}increase from 1.5% to 4.2% and from 4% to 6%, respectively, and, consequently, ε_{tot} increases from 5.5% to 11.8%. In Figure 10, the

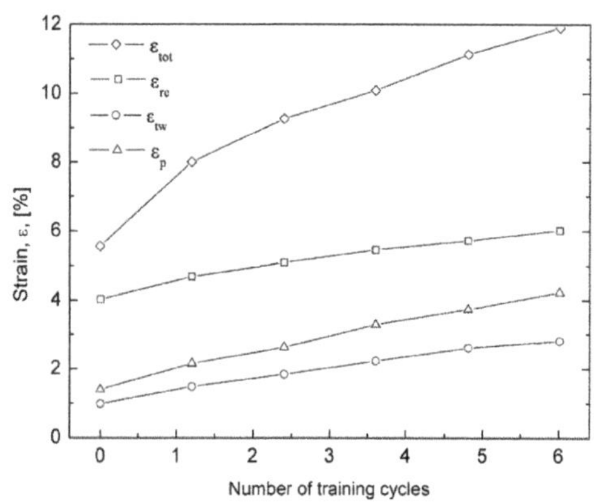

Figure 9. Measured deformations versus number of training cycles: ε_{tw}, ε_p, ε_{pe}, and ε_{tot} [28].

measured thermal hysteresis behavior strain *vs* temperature, describing the TWSME of the trained material, is shown. In particular, Figure 10.a illustrates the stress-free hysteresis loop, together with the PTTs, while in Figure 10.b the stress-free thermal hysteresis loop is compared with those obtained under a tensile stress of 50 MPa and 100 MPa. The comparison clearly shows and a systematic increase in ε_{tw}, as well as in all PTTs, when increasing the applied stress σ. In particular, the increase of ε_{tw} is attributed to *i*) the variation of Young's modulus in the thermal hysteresis behavior between martensite and austenite and *ii*) the increased volume fraction of favorably oriented martensite variants with increasing external stress.

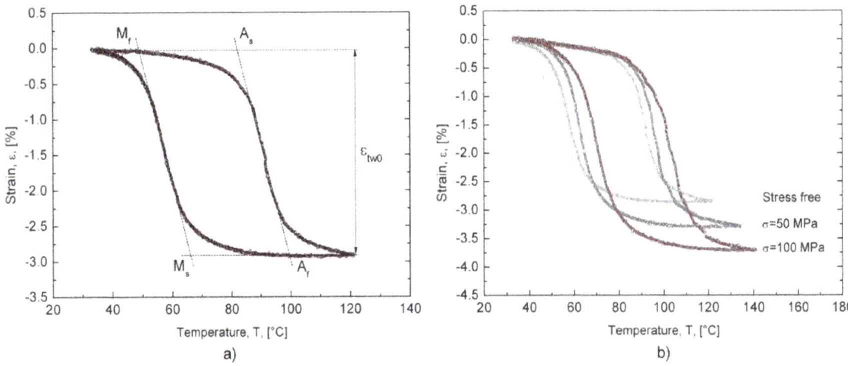

Figure 10. Thermal hysteresis behavior of the trained material strain *vs* temperature: a) stress-free hysteresis loop with an highlight of the PTTs and b) effects of the applied tensile stress [28].

2.3. Pseudoelastic effect

The pseudoelastic (PE) effect in NiTi alloys consists in the high strain recovery capability (up to 10%) observed during isothermal loading/unloading cycles carried out at temperature $T > A_f$, *i.e.* when the alloy is in austenitic conditions. This functional property can be directly attributed to the reversible stress-induced martensitic transformations as discussed in section 2.1.2. In particular, Figure 11 illustrates that if a mechanical load is applied to austenitic structure (1) the B2→B19′ transformation occurs and, on the macroscopic scale, large mechanical deformation are achieved through a stress-strain transformation plateau (2). However, if the mechanical load is removed the reverse B19′→B2 transformation occurs and, consequently, the material is able to recover its original shape through an unloading plateau in the stress strain curve. However, the reverse transformation occurs at lower stress values resulting in a marked stress-strain hysteretic behavior.

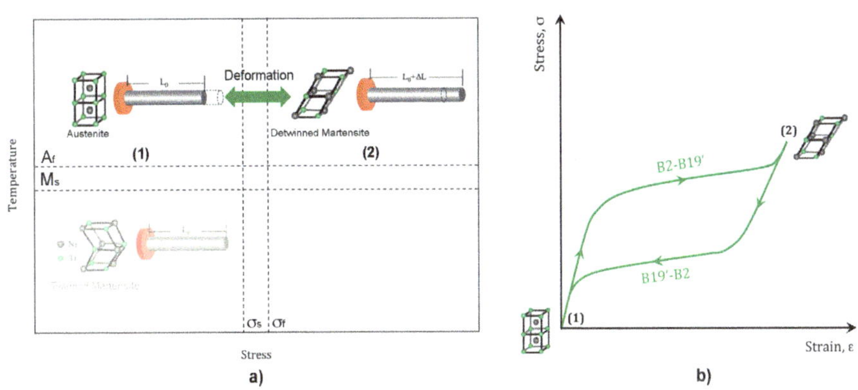

Figure 11. Schematic depiction of the pseudo elastic effect: a) stress-induced phase transition mechanisms and b) stress-strain hysteretic response

It is worth noting that Figure 11.b illustrates the nominal stress-strain behavior of a pseudoelastic SMA, while NiTi alloys exhibit a marked evolution of the stress-strain hysteretic behavior in the first mechanical cycles, up to a stable response, due to the formation of stabilized martensite [32], which causes a reduction of the pseudoelastic recovery of the SMA. In particular, Figure 12 illustrates the evolution of the stress-strain hysteretic behavior of the material in the first 25 mechanical cycles for a fixed value of total strain $\varepsilon_{tot} = 3.5\%$. These data were obtained from experimental testing of a commercial pseudoelastic NiTi alloy [30]; the figure clearly shows a marked reduction of the pseudoelastic recovery, from 3.5% to about 3%, but the stress-strain loops becomes stable after 20 cycles. Furthermore, as reported in section

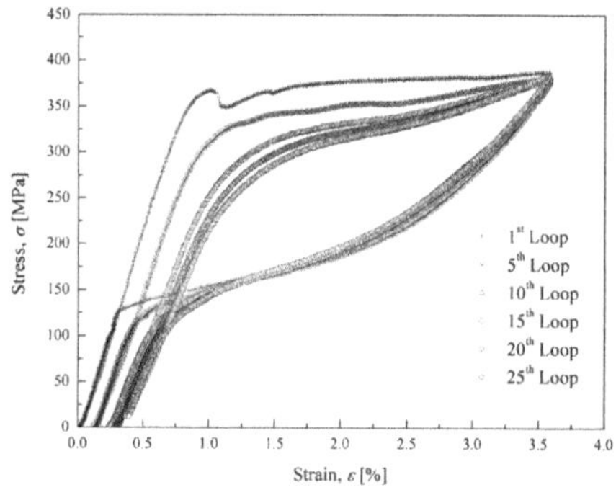

Figure 12. Evolution of the stress-strain response of a commercial pseudoelastic NiTi alloy during the first 25 mechanical cycles for a maximum applied deformation $\varepsilon_{tot} = 3.5\%$ [30]

2.1.2, the critical stresses for direct and inverse phase transformation are significantly affected by the temperature, according to the Clausius-Clapeyron constant (Equation 1), as illustrated in the experimentally measured curves in Figure 13 [30]. These curves have been obtained from isothermal loading unloading cycles, carried out at increasing values of the testing temperature for $T > A_f$ ($303K - 328K$), by using a SMA with a stable pseudoelastic response. The analysis of the data in Figure 13 allows to obtain the value of the Clausius-Clapeyron constant ($C_M = C_A = 8.7MPaK^{-1}$).

3. Hysteresis modeling: basics

A one-dimensional numerical approach to simulate the stress-strain and strain-temperature hysteresis behavior of SMAs is described in this section. In particular, in a purely phenomenological way, the hysteresis loop is modeled by a Prandtl-Ishlinskii hysteresis operator H [26]; the basic idea of this approach consists in modeling the non-linear hysteretic behavior by a weighted superposition of many elementary hysteresis operators, such as the backlash operators Hr, as schematically illustrated in Figure 14.

$$H = \{w\}^T \{H_r\} \tag{3}$$

where $\{H_r\}$ is the vector of backlash operators and $\{w\}$ is the corresponding vector of weights. As shown in Figure 14.a, each backlash operator H_{ri} is characterized by its dead band width dw_i, while the corresponding weight w_i represents the slope of the oblique lines of the operator. As illustrated in Figure 14.b, which represent a generic $x(t) - y(t)$ hysteretic

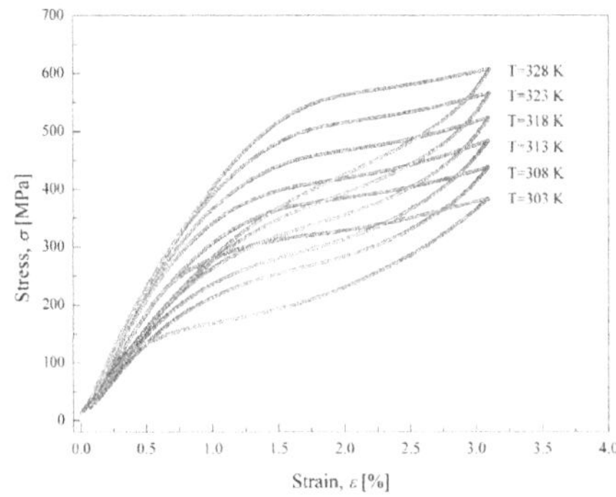

Figure 13. Isothermal stress-strain hysteresis loops of a commercial pseudoelastic NiTi alloy as a function of the testing temperature [30].

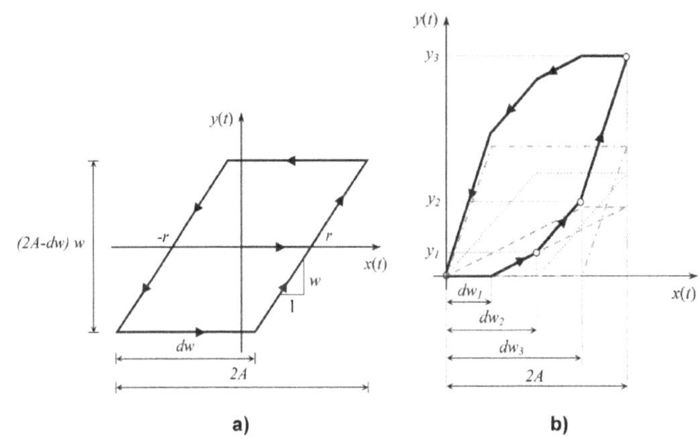

Figure 14. Schematic depiction of the hysteresis operators: a) elementary backlash operator H_r and b) Prandtl-Ishlinskii hysteresis operator H given by weighted superposition of elementary hysteresis operators.

behavior, the proposed approach consists of modeling the hysteretic loop by a linear piecewise discretization. The accuracy of the model can be improved by increasing the total number of linear pieces, which represent the total number of the backlash operators. The problem of modeling the hysteretic behavior, starting from the experimental measurements, is now reduced to the determination of the deadband width vector $\{dw\}$ of the backlash operators and the associated gain vector $\{w\}$. In particular, the parameters of the model can be easily identified by the outer loop of the hysteretic region by using the following simple relation:

$$y_k = \sum_{i=1}^{k}(dw_{k+1} - dw_i)w_i \tag{4}$$

where y_k is the output value of the lower branch of the loop in the generic point of discontinuity k, as shown in Figure 14.b. The vector $\{dw\}$ is a user defined discretization of the total amplitude of the input signal. Equation 4 can be rewritten in matrix form as follows:

$$\{y\} = [A]\{w\} \tag{5}$$

where the matrix $[A]$ is constructed, for a given $\{dw\}$ vector, by using equation 4; the unknown vector $\{w\}$ can be found by solving a system of N linear equations, where N is the total number of backlash operators, as follows:

$$\{w\} = [A]^{-1}\{y\} \tag{6}$$

The main drawback of the Prandtl-Ishlinskii approach consists in the fact that only loops with an odd symmetry to the relative center can be modeled; in fact the symmetry of the elementary hysteresis operator, with respect to the center of the loop, persists under linear superposition. However, this drawback can be overcome by using a modified Prandtl-Ishlinksii operator, as described in details in [26], in which a weighted superposition of saturation operators is combined with the hysteresis operator. The parameters of this sub model, such as saturation limits $\{S\}$ and associated gains $\{w_s\}$, can be identified by using a procedure similar to that described above.

3.1. Modeling of pseudoelastic effect

The numerical model described in this section is able to simulate the pseudoelastic effect of a shape memory alloy [30], *i.e.* the the stress-strain ($\sigma - \varepsilon$) hysteretic behavior, based on the Prandtl-Ishlinksii operator and on the assumptions reported in the following.

3.1.1. Basic assumptions

Figure 15 shows the stabilized stress-strain hysteretic behavior of a commercial NiTi alloy, *i.e.* the response of the material after the first training cycles (see Figure 12), for different values of the applied deformation. The figure also illustrates the Young's moduli of austenite and detwinned martensitic structures, E_A and E_M, together with the generic young's modulus of the alloy, $E(\xi_M)$, corresponding to an incomplete stress induced martensitic transformation, *i.e.* as a function of the martensite fraction ξ_M $(0 < \xi_M < 1)$. In particular, E_A represents the slope of the early stage of the stress-strain loading curve, E_M is measured from the unloading curve of a complete martensitic transformation ($\xi_M = 1$), while $E(\xi_M)$ is obtained from the unloading path of an incomplete phase transformation.

The total strain ε can be decomposed in elastic and a transformation strain components, ε_{el} and ε_{tr}, respectively:

$$\varepsilon = \varepsilon_{el} + \varepsilon_{tr} \tag{7}$$

where the elastic strain can be expressed as a function of the applied stress, σ, and of the Young's modulus, $E(\xi_M)$, of the material:

$$\varepsilon_{el} = \frac{\sigma}{E(\xi_M)} \tag{8}$$

As schematically shown in Figure 15 the Young's modulus changes during stress-induced phase transformation between austenite and martensite, *i.e.* it decreases from E_A to E_M, and

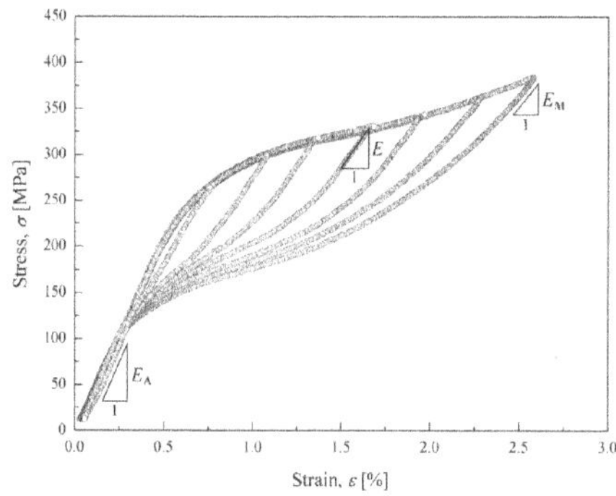

Figure 15. Stress-strain hysteretic behavior of a commercial NiTi alloy together with the Young's moduli of austenite and detwinned martensite [30].

it is assumed to be dependent on the volume fraction of martensite ξ_M according to the Reuss formula [20]:

$$\frac{1}{E(\xi_M)} = \frac{\xi_M}{E_M} + \frac{1-\xi_M}{E_A} \tag{9}$$

The evolution of martensite is assumed to be a linear function of the stress in the stress-strain transformation curves, *i.e.* in the range $(\sigma_s^{AM}, \sigma_f^{AM})$ in the loading stage and $(\sigma_s^{MA}, \sigma_f^{MA})$ in the unloading stage. In particular, the evolution rule for a complete transformation, can be expressed as follows:

$$\xi_M = \begin{cases} \frac{\sigma-\sigma_s^{AM}}{\sigma_f^{AM}-\sigma_s^{AM}} & \text{Loading path} \\ 1 - \frac{\sigma-\sigma_s^{MA}}{\sigma_f^{MA}-\sigma_s^{MA}} & \text{Unloading path} \end{cases} \tag{10}$$

The assumptions given by equations (9) and (10) have been validated by experimental measurements of the Young's modulus, as reported in Figure16. In particular, the figure shows the measured values of the Young's modulus as a function of the applied stress, on the direct stress-strain transformation plateau, together with the predictions of the Reuss formula, and a satisfactory agreement is observed.

However, it's worth noting that the evolution of martensite is characterized by a hysteretic behavior, *i.e.* it is stress path dependent, and equations (9) and (10) do not correctly predict the effects of incomplete transformations. These effects are simulated mathematically, by using a backlash operator, as schematically illustrated in Figure 17; in particular, Figure 17.a shows the stress *vs* time path, while Figure 17.b illustrates the evolution of ξ_M and $1/E(\xi_M)$ *vs* the applied stress. Furthermore, the continuous lines in Figure 17.b are relative to a complete martensitic transformation, *i.e.* ξ_M increases from 0 to 1, while the dashed lines show the effects of an incomplete transformation. Due to the modification in Young's modulus during the phase transformation between austenite and martensite, as shown in Figure 17.b, both elastic and transformation strain components are represented by a hysteretic behavior and it can be calculated using equations (7-10); as an example in Figure 18 a typical $\sigma - \varepsilon$ hysteresis loop for a complete phase transformation is compared with the corresponding computed elastic strain $(\sigma - \varepsilon_{el})$ and transformation strain $(\sigma - \varepsilon_{tr})$ hysteresis loops.

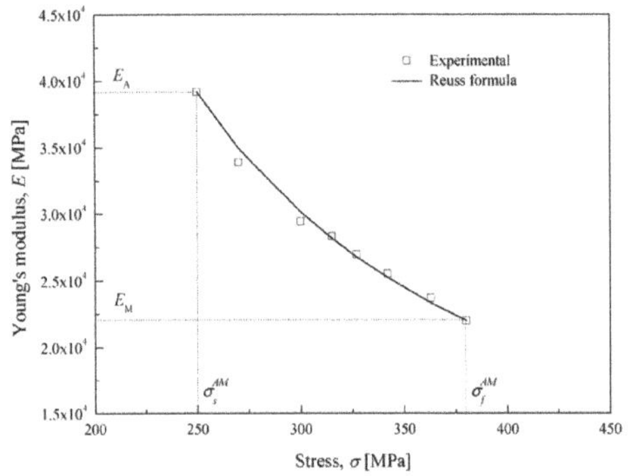

Figure 16. Evolution of the Young's modulus as a function of the applied stress: experimental measurements *vs* simulations [30].

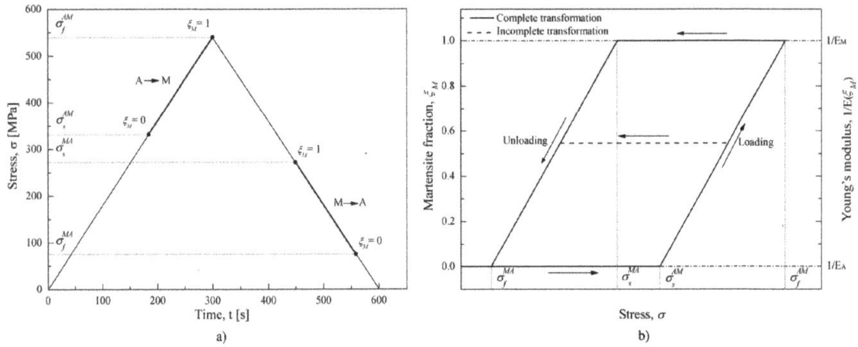

Figure 17. Evolution rule of the martensite fraction ξ_M in the tension cycle: a) example of stress-time path and b) simple hysteresis model to predict the Young's modulus $E(\xi_M)$ [30].

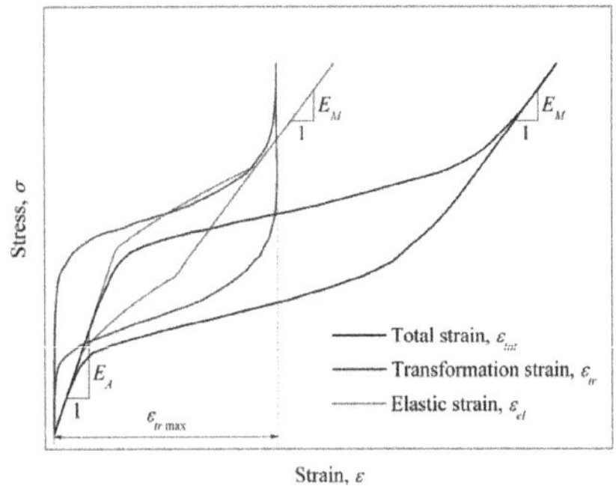

Figure 18. Stress strain ($\sigma - \varepsilon$) hysteresis loops: elastic strain ε_{el} and transformation strain ε_{tr} [30].

3.1.2. Numerical flowchart

Based on the Prandtl-Ishlinskii hysteresis modeling approach together with the model assumptions described in the previous section, a one-dimensional numerical model can be easily developed by using commercial software packages. In particular, Figure 19 illustrates an implementation of the model in the *MatlabSimulink*^{TM} platform, in which the two strain components, ε_{el} and ε_{tr}, are treated separately. As clearly illustrated in Figure 19, both stress

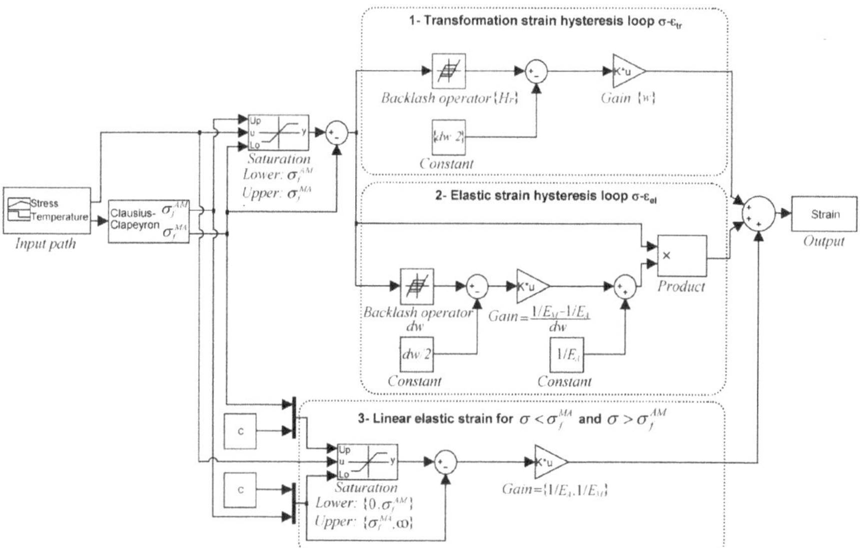

Figure 19. Flowchart of a *MatlabSimulink*^{TM} model to simulate the pseudoelastic effect in SMAs [30].

and temperature are taken as input variables, and the critical stresses are calculated, based on the current value of the temperature, by using the Clausius-Clapeyron relation (equation 1). In the model the saturation operators are used to identify the stress range where the material exhibits the hysteretic behavior ($\sigma_f^{MA} < \sigma < \sigma_f^{AM}$), and the dead zones of transformation where only elastic deformation of austenite ($\sigma < \sigma_f^{MA}$) or martensite ($\sigma > \sigma_f^{AM}$) occurs. Three different sub-models are highlighted in Figure 19 which simulate the $\sigma - \varepsilon_{tr}$ and $\sigma - \varepsilon_{el}$ hysteresis loops, and the linear elastic response of the material in austenitic and martensitic conditions. In particular, a Prandtl-Ishlinskii hysteresis operator, was used to model the $\sigma - \varepsilon_{tr}$ hysteretic behavior, a single backlash operator was adopted to model the $\sigma - \varepsilon_{el}$ loop (see Figure 17.b), and a gain operator was used to describe the linear elastic response for $\sigma < \sigma_f^{MA}$ and for $\sigma > \sigma_f^{AM}$. The parameters of the Prandtl-Ishlinskii operator describing the $\sigma - \varepsilon_{tr}$ hysteretic behavior, *i.e.* the deadband width vector $\{dw\}$ and the associated gain vector $\{w\}$ are determined from an experimentally measured stress-strain hysteresis loop by using the procedure described in section 3. In particular, as shown in Figure 20, the generic input and output variables (x and y) can be regarded as the stress and strain values (σ and ε), respectively. As a consequence, the vector $\{dw\}$ represents a user defined discretization of the stress amplitude while the vector $\{w\}$ can be determined by using equation (6), where the vector $\{y\}$ is the vector of strain values $\{\varepsilon\}$. The computational procedure to obtain the aforementioned model parameters from an experimentally measured hysteretic loop can be easily implemented, which generates the vector $\{dw\}$, by a partition of the input stress amplitude, and calculates the unknown vector of weight $\{w\}$ by solving a system of N linear equations, where N is the number of backlash operators.

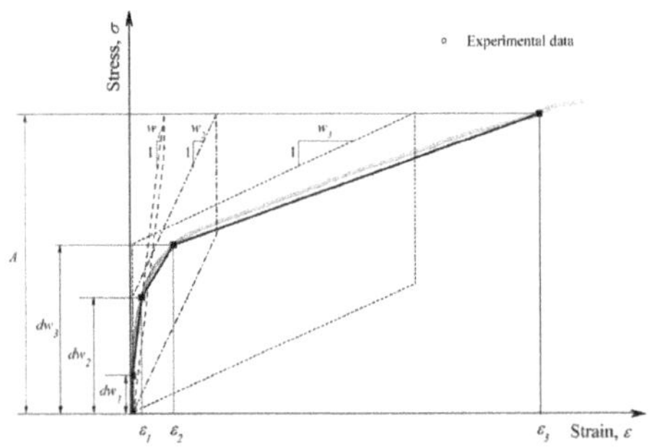

Figure 20. Loading branch of the stress-strain hysteresis loop and linear piecewise discretization obtained by a weighted superposition of three backlash operators [30].

3.1.3. Numerical results vs experiments

The accuracy of the numerical method is illustrated by comparisons with experimentally measured hysteresis loops, by using he thermo-mechanical parameters reported in [30] (E_A=39 GPa, E_M=20 GPa, σ_s^{AM}=260 MPa, σ_f^{AM}=385 MPa, σ_s^{MA}=250 MPa, σ_f^{MA}=125 MPa, $C_A = CM$=10.3 MPa/K). Figure 21 shows the stress-strain hysteretic behavior of the SMA for a stress path which involves several incomplete stress-induced martensitic transformations ($A \rightarrow M$), by repeated isothermal tension cycles ($T = 303K$) carried out between $\sigma_{min} = 0$ and decreasing values of $\sigma_{max} < \sigma_f^{AM}$. In particular, the comparison between numerical simulations and experimental measurements, illustrated in Figure 21.a, clearly shows a satisfactory accuracy of the model in predicting the non-linear stress-strain hysteretic behavior of the material; the figure also shows that the model is able to capture the modification of Young's modulus in the stress-strain transformation curve, as it correctly predicts the change in the slopes of the unloading curves. Furthermore, Figure 21.b shows the evolutions of the transformation strain, ε_{tr}, and elastic strain, ε_{el}, for the same input stress path of Figure 21.a. As illustrated in section 3.1.1, both ε_{tr} and ε_{el} are characterized by hysteretic behaviors, which are due to the mismatch between the critical stresses in the stress induced transformations and the modification of Young's modulus in the stress-strain transformation curve, respectively.

Figures 22 show comparisons between numerical predictions and experimental measurements for two input isothermal stress paths ($T = 303K$) which involve incomplete $M \rightarrow A$ transformations (Figure 22.a) and both incomplete $A \rightarrow M$ and $M \rightarrow A$ transformations (Figure 22.b). In particular, Figure 22.a shows the hysteretic response of the material for a stress path which is composed of some subsequent tension cycles between increasing values of $\sigma_{min} < \sigma_f^{MA}$ and $\sigma_{max} = const$, while Figure 22.b is relative to a stress

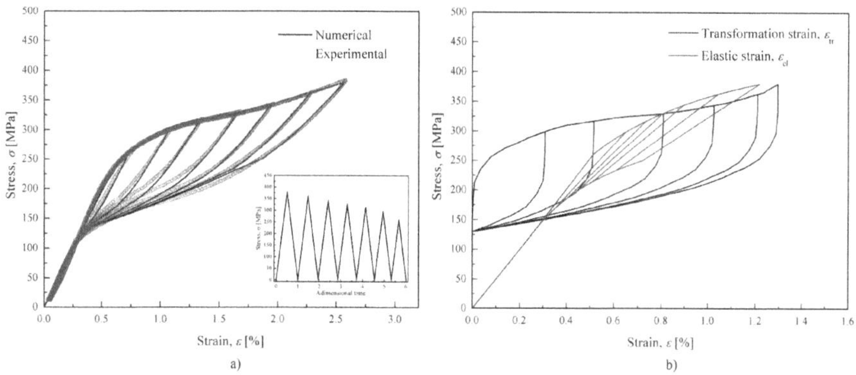

Figure 21. Numerical simulation for isothermal stress cycles with incomplete $A \rightarrow M$ transformations: a) comparison with experimentally measured loops and b) evolution of transformation strain, ε_{tr}, and elastic strain, ε_{el} [30].

path which involves different subsequent tension cycles carried out between increasing values of $\sigma_{min} > \sigma_f^{MA}$ and decreasing values of $\sigma_{max} < \sigma_f^{AM}$; both figures shows good agreements between experiments and numerical simulations.

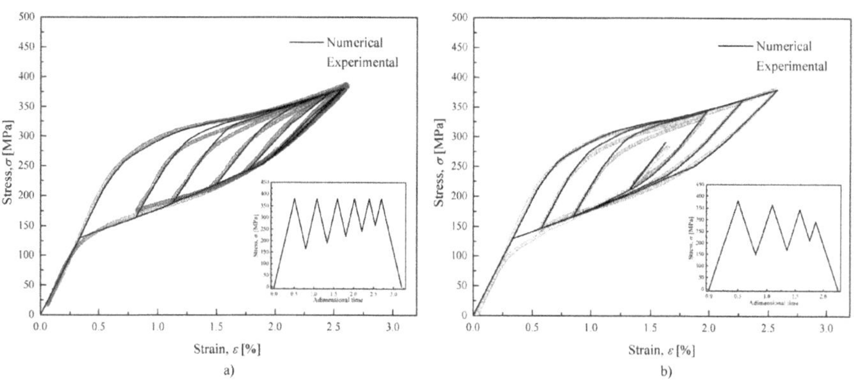

Figure 22. Comparison between numerical simulations and experimentally measured hysteresis loops for isothermal stress cycles with: a) incomplete $M \rightarrow A$ transformations and b) incomplete $A \rightarrow M$ and $M \rightarrow A$ transformations [30].

3.2. Modeling of two-way shape memory effect

The numerical model described in this section is able to simulate the two-way shape memory effect of a trained NiTi based shape memory alloy, *i.e.* the the strain-temperature ($\varepsilon - T$) hysteretic behavior [27–29]. Furthermore, the model is able to capture the effects of applied stress on the thermal recovery of the alloy.

3.2.1. *Basic assumptions*

In a pure phenomenological way, the variation of the two-way shape memory strain, ε_{tw}, with increasing the applied stress (see Figure 10.b) can be attributed to two different mechanisms: *i*) the variation of Young's modulus in the thermal hysteresis behavior between martensite and austenite, *ii*) the increased volume fraction of favorably oriented martensite variants with increasing external stress.

$$\Delta\varepsilon_{tw} = \varepsilon_{tw} - \varepsilon_{tw0} = \left(\frac{1}{E_M} - \frac{1}{E_A}\right)\sigma + c\sigma \qquad (11)$$

where subscript 0 indicate the stress-free condition, E_M and E_A represent the Young's moduli of martensite and austenite, respectively, while c can be assumed as a material constant. In particular, the first term at the right end side of equation 11, indicated as $\Delta\varepsilon_{mech}$, describe the first effect while the last terms, namely $\Delta\varepsilon_{mem}$, take into account the second effect. Young's moduli E_M and E_A can be measured by isothermal tensile tests carried out at temperatures $T < M_f$ and $T > A_f$, respectively, while the parameter c can be obtained from experimental measurements of the two-way shape memory strain carried out at different value of the applied stress. In the following the values of the parameters reported in [28] have been used ($E_M = 36 * 10^3 MPa$, $E_A = 67 * 10^3 MPa$, $c = 8.5 * 10^{-5} MPa^{-1}$). Two simple mathematical functions, obtained by a numerical fitting of the experimental data, are used to describe the phase transformation kinetics. In particular, as reported in the model by Tanaka and Nagaki [13], the heating and cooling branches of the hysteresis loop can be represented by two exponential curves:

$$\varepsilon = \begin{cases} \varepsilon_{tw} e^{a_A(A_f - T + \delta T_M)} & \mathrm{M} \rightarrow \mathrm{A} \\ \varepsilon_{tw}[1 - e^{a_M(M_f - T + \delta T_A)}) & \mathrm{A} \rightarrow \mathrm{M} \end{cases} \qquad (12)$$

where ε_{tw}, M_s, and A_s are functions of the applied stress, as reported in Equations 11 and 1; a_M, a_A, δT_M, and δT_A, which define the shape of the heating and cooling branches of the loop, can be identified by a numerical fitting of the experimental data. Figure 23.a shows a comparison between experimental measurement and exponential curves; in the figure, points P_1 and P_2 represent the range where the numerical fitting is executed to identify the parameters of the heating branch of the loop, while points P_3 and P_4 are relative to the cooling branch. In Figure 23.b, a linear fitting between points P_1 and P_2 of the experimental data in the $T - log\varepsilon$ plane is shown, where the slope of the line defines the parameter a_A and the intersection with the $log\varepsilon$ axis allows us to obtain the parameter δT_M. If the loop is characterized by an odd symmetry with respect to its center, as is quite well observed in the investigated material, the same values can be assumed for the constants a_A and a_M and δT_A, and δT_M. The two exponential curves describe the two branches of the hysteresis loop in a parametric way for a generic value of the applied stress by using Equations 1 and 11. Starting from the curves $\varepsilon - T$, the numerical method based on the Prandtl-Ishlinksii operator can be developed, which is able to predict the output response for a generic temperature path, as decribed in the following section.

3.2.2. *Numerical flowchart*

The numerical method described above can be easily implemented in a $MatlabSimulink^{TM}$ model, as shown in Figure 24, by a modified Prandtl-Hishlinkii hysteresis operator. The

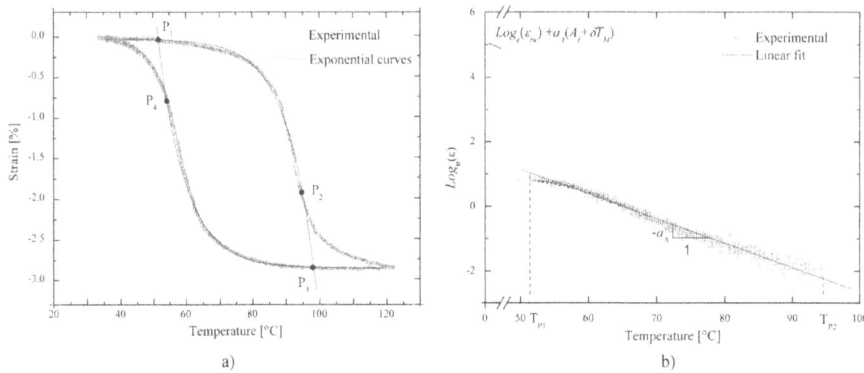

Figure 23. Numerical fitting of the experimental data: a) comparison between exponential curves and experimental measurements; b) numerical fitting in the $T - log\varepsilon$ plane to identify the parameters of the exponential curve in the heating branch of the hysteresis loop [28]

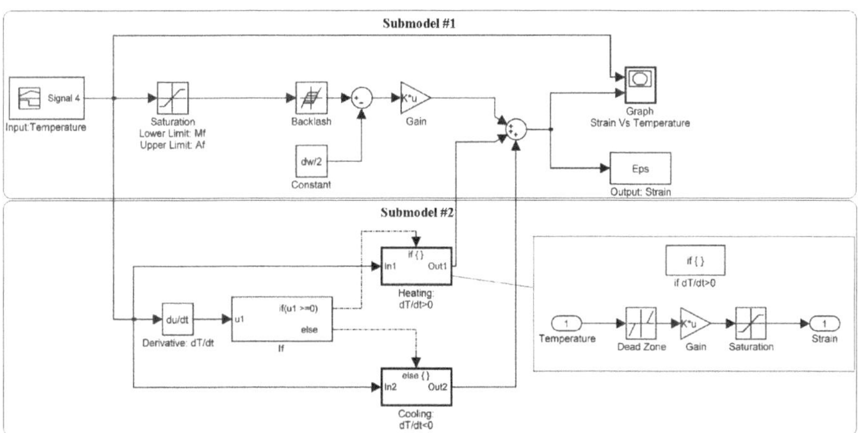

Figure 24. Flowchart of a $MatlabSimulink^{TM}$ model to simulate the two way shape memory effect in SMAs.

Prandtl-Hishlinkii operator is implemented in the submodel #1 of Figure 24, by a weighted superposition of several backlash operators, and the corresponding parameters, *i.e.* the deadband width vector $\{dw\}$ and the associated gain vector $\{w\}$, are determined from the exponential curves of equation 12, which in turn are obtained from a fitting of experimental data ($T - \varepsilon$) as illustrated in Figure 23. In particular, the generic input and output variables (x and y) can be regarded as the temperature and strain values (T and ε), respectively. As a consequence, the vector $\{dw\}$ represents a user defined discretization of the temperature amplitude while the vector $\{w\}$ can be determined by using equation (6), where the vector $\{y\}$ is the vector of strain values $\{\varepsilon\}$ obtained from equation 12 ($\varepsilon_i = \varepsilon(T_i)$). The saturation operator in the submodel #1 is used to simulate the dead zones of transformation, *i.e.* the material behavior when the temperature is above A_f during heating and below M_f in the cooling branch of the hysteresis loop. In particular, this operator imposes upper and lower bounds on the temperature, which are A_f and M_f, respectively, so that when the temperature is outside these bounds the signal is clipped to the upper or lower bounds. In Figure 25.a, the response of the submodel #1, in terms of strain-temperature loop, is illustrated for a thermal cycle between the temperatures $T_0 < M_f$ and $T_1 > A_f$; the figure clearly shows that the hysteretic behavior of the material is properly described in the range of temperatures between M_f and A_f, and the dead zone of transformation, when the temperature exceeds A_f or falls below M_f, are also simulated.

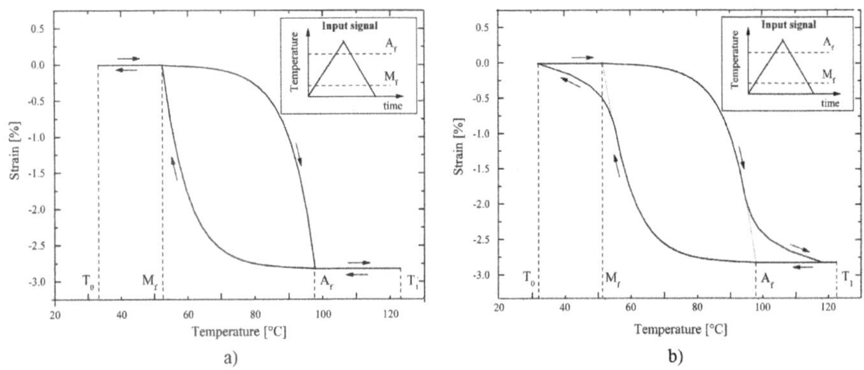

Figure 25. Numerically simulated loop for a thermal cycle between the temperatures $T_0 < M_f$ and $T_1 > A_f$ obtained by: a) Prandtl-Ishlinskii model and b) modified Prandtl-Ishlinskii model [28].

Unfortunately, when comparing the experimental results with the numerically simulated loops, high errors are observed in the extremity of the hysteretic region, *i.e.* when the temperature is below M_f during cooling and above A_f during heating. To overcome this limitation, a modified *Simulink* model can be implemented by including the submodel #2 of Figure 24. This latter uses two subsystems, for the heating and cooling branches of the loop, which modifies the output response of the system when the temperature is near M_f and A_f. In particular, each subsystem implements a weighted superposition of several dead band operators, which is executed by a series of a dead band block and a gain block, while the saturation block assures that the correction is carried out only in a limited range of temperatures near M_f and A_f. Figure 25.b shows a simulated hysteresis loop, obtained by the modified model, between the temperatures $T_0 < M_f$ and $T_1 > A_f$; the figure clearly shows that the model allows a better simulation of the extremity of the hysteretic region with respect to Figure 25.a.

3.2.3. *Numerical results vs experiments*

In this section, the accuracy and efficiency of the 1D numerical model are illustrated by comparing some experimentally measured hysteresis loops with the corresponding numerical predictions. The simulations have been carried out by using a model with 20 backlash operators and 5 dead zone operators to modify the loops in the extremity of the hysteretic region. Figure 26.a shows a comparison between the experimentally measured hysteresis loop for a stress-free martensitic transformation, between the temperatures $T_0 < M_f$ and $T_1 > A_f$, and the numerically simulated one; the figure clearly shows a good accuracy of the numerical model with very small errors. In Figure 26.b, a comparison between numerical predictions and experimental results when the material is subjected to a tensile stress $\sigma = 100MPa$ is shown. Also, in this case a satisfactory agreement is observed, but the errors increase with respect to the stress free condition. However, it is important to point out that the model parameters were identified by using the measured hysteresis loop under stress-free conditions, and by applying Equations 1 and 11 to modify both the PTTs and ε_{tw}.

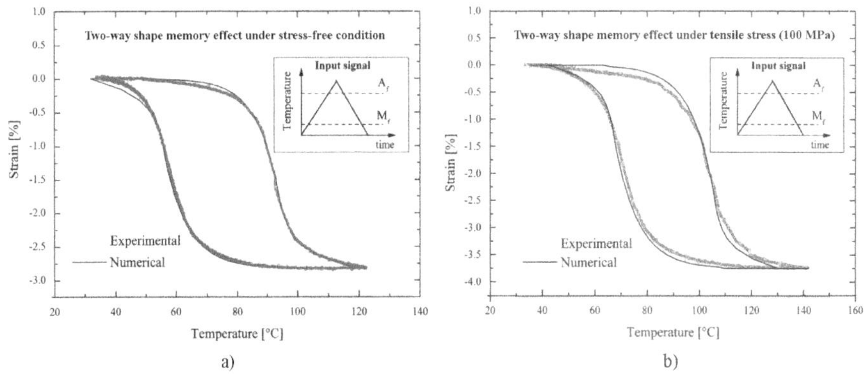

Figure 26. Comparison between experimental measurements and numerical predictions for a thermal cycle between the temperatures $T_0 < M_f$ and $T_1 > A_f$ under a) stress-free condition and b) tensile strees of 100 MPa [28].

The accuracy of the model was also analyzed when the material is subjected to partial thermal cycles, *i.e.* to incomplete martensitic transformations. Figures 27.a and 27.b show the hysteretic behavior of the material for two different temperature-time paths in stress-free conditions; in particular, Figure 27.a shows incomplete $A \rightarrow M$ transformations, while Figure

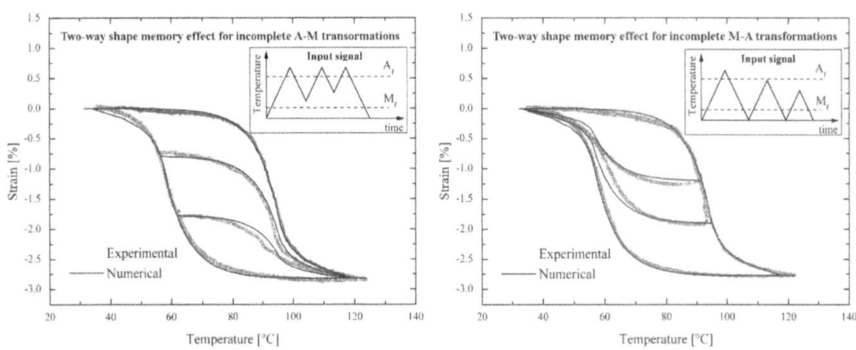

Figure 27. Comparison between experimental measurements and numerical predictions for two different temperature-time paths: a) incomplete $A \rightarrow M$ transformations and b) incomplete $M \rightarrow A$ transformations [28].

27.b illustrates incomplete $M \rightarrow A$ transformations. As shown in the figures, the comparison between experimental measurements and numerical predictions show a good accordance in both cases. It is worth noting that the same model parameters of the first example (Figure 26.a) were used in these numerical simulations.

Author details

Maletta Carmine

Assistant professor of Machine Design at University of Calabria, Department of Mechanical Engineering, Rende (CS), Italy

Furgiuele Franco

Full professor of Machine Design at University of Calabria, Department of Mechanical Engineering, Rende (CS), Italy

4. References

[1] Otsuka K, Ren X, (2005) Physical Metallurgy of TiÐNi-Based Shape Memory Alloys. Prog. Mater. Sci. 50: 511-678.
[2] Otsuka K, Wayman CM (1998) Shape Memory Materials. Cambridge University 559 Press, Cambridge.
[3] Wu Ming H (2002) Fabrication of Nitilon Materials and Components. Mater. Sci. Forum. 394-395:285-292.
[4] Schlossmacher P, Haas T, Shussler A (1997) Laser-Welding of a Ni-Rich Tini Shape Memory Alloy: Mechanical Behavior. J. Phys. IV. 7(5):251-256.
[5] Tuissi A, Besseghini S, Ranucci T, Squatrito F, Pozzi M (1999) Effect of Nd-YAG Laser Welding on the Functional Properties of the Ni-49.6 at. %Ti. Mater. Sci. Eng. A. 273-275:813-817.
[6] Theisen W, Schuermann A (2004) Electro Discharge Machining of Nickel-Titanium Shape Memory Alloys. Mater. Sci. Eng. A. 378:200-204.
[7] Falvo A, Maletta C, Furgiuele F M (2005) Laser Welding of a Niti Alloy: Mechanical and Shape Memory Behavior, Mater. Sci. Eng. A. 412:235-240.
[8] Falvo A, Maletta C, Furgiuele F M (2006) Functional Behavior of a Niti Welded Joint: Two-Way Shape Memory Effect. Mater. Sci. Eng. A. 481-482:647-650.
[9] Gall K A, Seitoglu H, Chumlyakov Y (2000) Niti Experiments Versus Modeling: Where Do We Stand?, Proc. SPIE. 3992:536Ð547.
[10] Paiva A, Savi M A (2006) An Overview of Constitutive Models for Shape Memory Alloys. Math. Probl. Eng. art. no. 56876.
[11] Falk F (1980) Model Free-Energy, Mechanics and Thermodynamics of Shape Memory Alloys. Acta Metall. 28(12):1773-1780.
[12] Falk F (1983) One-Dimensional Model of Shape Memory Alloys. Arch. Mech. 35(1):63-84.
[13] Tanaka K, Nagaki S (1982) Thermomechanical Description of Materials With Internal Variables in the Process of Phase Transitions. Ing. Arch. 51:287-299.
[14] Liang C, Rogers C A (1990) One-Dimensional Thermomechanical Constitutive Relations for Shape Memory Materials. J. Intell. Mater. Syst. Struct. 1:207-234.

[15] Brinson L C (1993) One Dimensional Constitutive Behavior of Shape Memory Alloys: Themomechanical Derivation With Non-Constant Material Functions and Redefined Martensite Internal Variable. J. Intell. Mater. Syst. Struct. 4:229-242.
[16] Bertran A (1982) Thermomechanical Constitutive Equations for the Description of Shape Memory Effects in Alloys. Nucl. Eng. Des. 74(2):173-182.
[17] Souza A C, Mamiya E, Zouain N (1998) Three-Dimensional Model for Solids Undergoing Stressinduced Phase Transformations. Eur. J. Mech. A/Solids. 17(5):789-806.
[18] Boydand J G, Lagoudas D C (1996) A Thermodynamical Constitutive Model for Shape Memory Materials. Part I. The Monolithic Shape Memory Alloy International Journal of Plasticity. 12(6):805-842.
[19] Auricchio F, Lubliner J (1997) A Uniaxial Model for Shape Memory Alloys. Int. J. Solids Struct. 34(27):3601-3618.
[20] Auricchio F, Sacco E (1997) A One-Dimensional Model for Superelastic Shape Memory Alloys With Different Elastic Properties Between Austenite and Martensite. Int. J. Non-Linear Mech. 32(6):1101-1114.
[21] Auricchio F, Taylor R L, Lubliner J (1997) Shape-Memory Alloys: Macro and Numerical Simulations of the Superelastic Behavior. Comput. Methods Appl. Mech. Eng. 146(3-4): 281-312.
[22] Marfia S, Sacco E, Reddy J N (2003) Superelastic and Shape Memory Effects in Laminated Shape-Memory-Alloy Beams. AIAA J. 41:100-109.
[23] Liew K M, Ren J Reddy J N (2005) Numerical simulation of thermomechanical behaviors of shape memory alloys via a non-linearmesh-free Galerkin formulation. Int. J. Numer. Meth. Engng 63:1014-1040.
[24] Amalraj J, Bhattacharyya A, Faulkner M G (2000) Finite element modeling of phase transformation in shape memory alloy wires with variable material properties. Smart Mater. Struct. 9:622-31.
[25] Krejci P, Kuhnen K, (2001) Inverse Control of Systems With Hysteresis and Creep, IEE Proc.: Control Theory Appl. 148(3):185-192.
[26] Kuhnen K, Janocha H (2001) Inverse Feedforward Controller for Complex Hysteretic Nonlinearities in Smart Material Systems. Control Intell. Syst. 29(3):74-83.
[27] Falvo A, Maletta C, Furgiuele F M (2007) Two-Way Shape Memory Effect of a Ti Rich Niti Alloy: Experimental Measurements and Numerical Simulations. Smart Mater. Struct. 16:771-778.
[28] Falvo A, Maletta C, Furgiuele F M (2008) A Phenomenological Approach for Real-Time Simulation of the Two-Way Shape Memory Effect in NiTi Alloys. ASME Journal of Engineering Materials and Technology. 130:771-778.
[29] Falvo A, Maletta C, Furgiuele F (2008) Hysteresis modeling of two-way shape memory effect in NiTi alloys. Meccanica. 43:165-172.
[30] Falvo A, Maletta C, Furgiuele F (2009) A phenomenological model for superelasticity in NiTi alloys. Smart Mater. Struct. 18:025005.
[31] Raniecki B, Lexcellent C H and Tanaka K 1992 Thermodynamic Models of Pseudoelastic Behavior of Shape Memory Alloys .Arch. Mech. 44 261-284.
[32] Hamilton R F, Sehitoglu H, Chumlyakov Y, Maier H J (2004) Stress Dependence of the Hysteresis in Single Crystal Niti Alloys. Acta Mater. 52:3383-3402.

[33] Sittner P, Landa M, Lukas P, Novak V (2006) R-phase transformation phenomena in thermomechanically loaded NiTi polycrystals. Mechanics of Materials 38:475-492

Section 2

Smart Polymers

Chapter 7

A Technology for Soft and Wearable Generators

Thomas G. McKay, Benjamin M. O'Brien and Iain A. Anderson

Additional information is available at the end of the chapter

1. Introduction

Never again needing to recharge the batteries of portable electronic devices is an exciting prospect. The uptake of portable electronics is increasing and recharging or replacing batteries is not only an inconvenience, but also contributes to an environmental hazard. Furthermore, with technologies such as GPS, pressure sensors, and cameras becoming smaller and cheaper, smart clothing such as the Adidas miCoach range are now a reality [1]. For seamless integration of smart devices into clothing, the inconvenience of battery replacement or recharging needs to be eliminated.

Conveniently there is an alternative source of energy in the exact location where portable and wearable devices operate: biomechanical energy from human movements. Interestingly, the device often credited as being the first ever power generator was a biomechanical energy harvester. 17th century engineer Otto von Guericke produced a sulfur globe which was charged through the triboelectric effect[1] when it was rubbed with a dry hand [2]. However, four centuries later the ability to harvest appreciable amounts of energy from biomechanical motions remains a research challenge. In this chapter we will discuss recent progress towards harvesting biomechanical energy using a soft and wearable electro-active polymer technology.

To provide sufficient power for portable devices a biomechanical energy harvester needs to supply in the order of a few Watts. For instance, an insulin pump consumes approximately 5 Watts [3] and the Nexus One smartphone (HTC Corporation, Taiwan) consumes 25 milli-Watts in standby, 330 milli-Watts in idle, and 750 milli-Watts during a phone call [4]. So to eliminate the need to recharge batteries, a similar or greater quantity of power needs to be sourced.

---

[1] The triboelectric effect refers to the transfer of electrons between two materials when they come into contact. Some materials are more susceptible to donating electrons when the contact is separated and some are more likely to hold on to electrons so the materials remain charged when separated.

Starner and Paradiso provide a good review of the energy available for scavenging from human movement and identify walking as a rich source of energy [5]. They highlight that 13Watts of power is available from the heel-strike of a 68Kg person if the sole of their shoe is compressed by 1cm when walking at 2 steps per second. Furthermore, a typical running shoe midsole dissipates a relatively large amount of energy as heat. Shorten's analysis of the energetics of a shoe midsole worn by a 76Kg runner suggests that a running shoe will dissipate between 2 and 10 Joules per step [6]. A well designed energy harvesting shoe could instead turn this energy into electricity, without altering the comfort or energy expenditure of the person wearing the shoe.

The most prevalent energy harvesting technology, electromagnetic generators, have been used to harvest energy from human gait, but additional mechanisms are required to condition the mechanical energy [5, 7]. Donelan et al. developed a knee-brace generator for harvesting energy from human gait. Their generator contained auxiliary components including a gear train, bearings, and a separate input shaft to convert the mechanical energy to suit their electromagnetic generator. Their system cost on average 59W of metabolic power to carry without harvesting energy, whereas an additional 5W of metabolic power was required to produce 4.8W of electrical power [7]. Although their harvesting mechanism was extremely efficient, they could have achieved larger efficiency gains if their system's mass was reduced and the device did not alter gait patterns. This begs the question: why are auxiliary mechanical components required for an electromagnetic biomechanical energy harvester?

The system by Donelan et al. was relatively heavy because it required additional componentry for it to work efficiently: electromagnetic generators produce more energy during a single rotation or stroke as velocity increases and are poorly suited for the low induced velocities associated with walking unless augmented mechanically. Thus the viability of harvesting energy from human motions could be improved by utilizing a different energy harvesting technology.

We propose the key characteristics of an ideal technology for harvesting biomechanical energy below:

1. Efficient operation at low biomechanical "walking" speeds.
2. Ability to couple directly to large walking motions.
3. A high energy density.
4. Good mechanical impedance matching to muscle.
5. Cheap to produce and maintain.

The first two criteria eliminate the need for additional mechanisms to condition the mechanical energy when it is transferred to the generator. The first three criteria therefore provide low mass/bulk. By having good impedance matching to muscle, the generator will be comfortable to wear, reducing its effects on the person wearing it. For the mass market, it is essential that the generator is low cost because consumers are unlikely to pay a premium for generators that produce power of approximately 1 Watt.

The research efforts of the authors of this chapter have focused on an energy harvesting technology called dielectric elastomer generators (DEG) which have been identified as a highly promising technology for biomechanical energy harvesting [8-11]. Furthermore, DEG fit the five listed criteria particularly well, so this chapter will focus on our recent developments towards portable and wearable DEG. For the interested reader, Anderson et al. provide a good, broad review of dielectric elastomer transducers [12].

2. The dielectric elastomer generator

DEG are a class of variable capacitance generators that are fabricated from a rubbery dielectric material sandwiched between stretchable electrodes. They have excellent impedance matching to natural muscle [13], can be fabricated from a wide range of low cost materials (commonly acrylic or silicone membranes sandwiched between carbon-based electrodes), have demonstrated extremely high energy densities, can undergo strains in excess of 100% [14], and have the ability to work over a wide frequency range without sacrificing efficiency [15].

Mechanical energy can be converted to electrical energy by cyclically deforming a DEG and placing charges on its flexible electrodes in the deformed state. DEG are typically produced from incompressible polymers so that an area stretch results in a decreased thickness. Relaxation of the charged, deformed DEG forces the opposite charges apart and packs like charges closer together and this transfers the mechanical energy to the electrical charges. The energy flows during an ideal DEG cycle are highlighted by the red arrows in Figure 1. The system receives 1 unit each of mechanical and electrical energy and converts the unit of mechanical energy into electrical, so that it outputs two units of electrical energy.

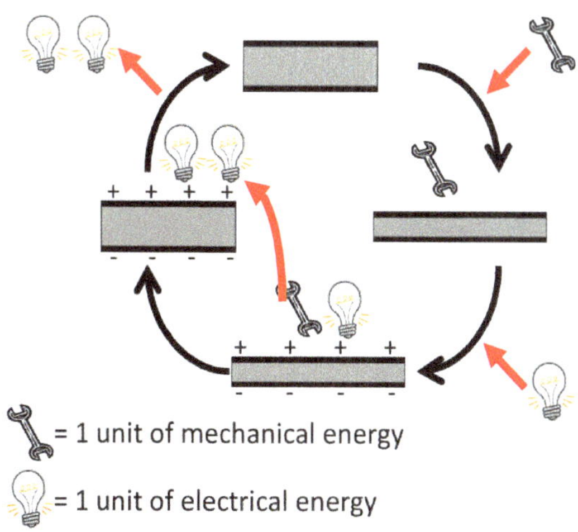

Figure 1. Schematic of the DEG states during a generation cycle. The grey area represents the dielectric and the black area represents the electrodes. From top moving clockwise, mechanical energy is input to the system deforming the DEG; an electrical energy input then charges the stretched DEG; the mechanical energy is then transferred to the charges by separating opposite and compressing like charges together. The electrical energy is then extracted and the cycle repeats (From [16]).

The major advantage that DEG hold for biomechanical energy harvesting is their ability to directly harvest low frequency motions without any gear mechanisms. To illustrate why this is true we will briefly describe their fundamental energy harvesting mechanism. As illustrated in Figure 1, DEG convert mechanical energy to electrical when the deformation of a stretched charged DEG is relaxed. During this relaxation period the thickness of the dielectric increases and the electrode area decreases, both resulting in a reduction of the capacitance. If charge is trapped on the generator during this relaxation phase, the voltage on the DEG will increase and there will be an increase in energy given by equation 1 where C_d and V_d are the capacitance and voltage of the DEG in its deformed state, and C_r and V_r are the relaxed DEG's capacitance and voltage, respectively.

$$E = 0.5C_r V_r^2 - 0.5C_d V_d^2 \tag{1}$$

Since we are considering the case where the charge on the DEG is fixed during the relaxation period, and that the charge on a capacitor can be calculated from its voltage and capacitance (Q=CV), we can relate the voltage of the deformed DEG to the voltage of the generator in its relaxed state using equation 2. We substitute equation 2 into equation 1 to get equation 3.

$$V_R = \frac{C_d}{C_r} V_d \tag{2}$$

$$E = 0.5\left(\frac{C_d^2}{C_r} - C_d\right) V_d^2 \tag{3}$$

These equations emphasise that along with driving voltage, C_r and C_d are the key parameters that influence the energy output of a DEG. The capacitance is dependent on the geometry of the DEG and the driving voltage is controlled by the generator's associated electronics. Neither of these parameters are dependent on the velocity at which the generator is deformed, so the fundamental mechanism of DEG is not dependent on driving velocity.

Although we have highlighted that DEG have a highly suitable mechanism and characteristics for harvesting biomechanical energy, Figure 1 highlights that DEG need an electrical circuit that will control the flow of charge onto and off of the generator at appropriate stages of the energy harvesting cycle. Such circuitry can add weight and is typically composed of stiff and bulky parts. Furthermore the requirement for the electronics to replace charge delivered to the load has traditionally reduced the portability of DEG because they have been either tethered to the grid or used batteries that need periodic replacement [10]. This chapter will focus now on recent developments that eliminate the need for a secondary power source and reduce external DEG circuitry mass and stiffness.

3. Portable dielectric elastomer generator electronics

A passive DEG circuit for controlling the charge state of the DEG appropriately as it is mechanically cycled called the Self-Priming Circuit (SPC) [17, 18] has been developed to eliminate the need for a secondary power source and thus improve DEG portability. The circuit works as a charge pump that provides energy in a higher charge form than the energy supplied to it. The self-priming circuit is configured so that it harvests energy from a DEG and then supplies that energy in a higher charge form to a load or to the DEG when it requires priming. This effectively boosts the charge of the generated energy before it is used, thus a secondary power source is not necessary because the generated energy is used to replace circuit charge losses and charge delivered to the load. A schematic of the simplest form of an SPC is given in Figure 2, showing that an SPC is simply a capacitor bank that has diodes to convert the topology of the circuit between a low capacitance when it is charged (Figure 2c, capacitors in series) and a high capacitance when it discharges (Figure 2d, capacitors in parallel). This toggling of state provides the SPC with a higher output capacitance which converts the energy to a higher charge form. This can be explained using equation 4, which provides two expressions for the energy (E) stored on a capacitor. If the energy is conserved when the capacitance (C) of the SPC increases, then the charge (Q) must increase too, and since the SPC is an adiabatic process the increased charge is accompanied by a decreased voltage (V). Thus the SPC outputs energy in a higher charge, lower voltage form than the input.

$$E = \frac{Q^2}{2C} = \frac{CV^2}{2} \tag{4}$$

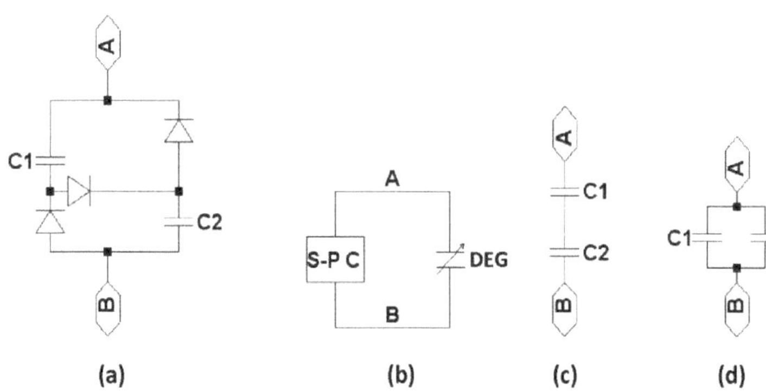

Figure 2. Schematic of a self priming circuit (a) which connects to a DEG in parallel (b), the diodes control current flow so that the SPC capacitor bank takes on the form given in (c) when energy is transferred to it from the DEG, when energy is transferred off the SPC the diodes configure it to take on the higher charge form given in (d).

The SPC has additional benefits for portable DEG. First, the SPC is passive and requires no active switching or control, so it does not require any power to drive it. Second, the SPC accumulates charge from cycle to cycle, this means that the system voltage climbs (see Figure 3). The ability to boost its own voltage is highly desirable because as demonstrated in equation 3, generated energy climbs with priming voltage. This high voltage has traditionally been supplied by a high voltage power supply or converter, so the SPC eliminates the need for these typically high cost components.

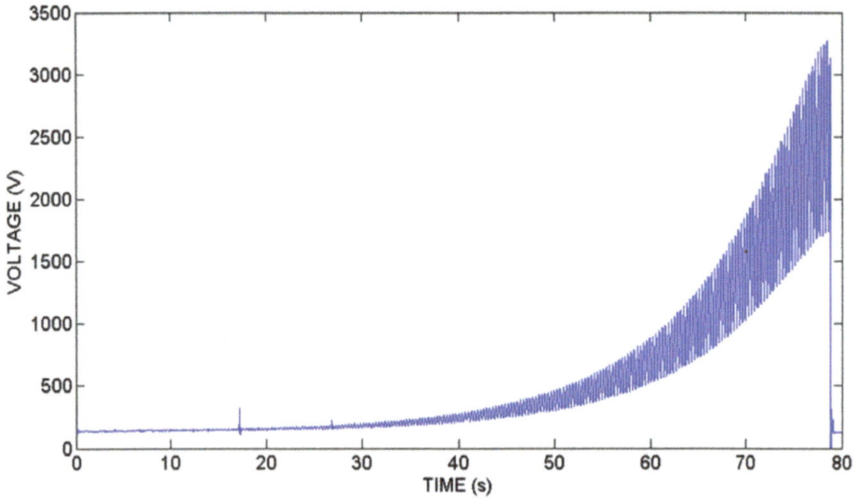

Figure 3. The output voltage waveform of a self-priming generator mechanically deformed at 3 Hz. The voltage climbs from cycle to cycle because the generated energy accumulates in the form of extra charge stored on the generator and priming circuit (From [19]).

Although the SPC is low cost, low complexity, and autonomous it still adds considerable mass and stiffness to the DEG system. The SPC consists of diodes and capacitors. The function of the capacitors is to store priming charges and the diodes control the transfer of charge to and from the DEG, so that an appropriate generation cycle is achieved. We will now discuss how these functions have been integrated onto the DEG membrane to produce a generator that can be fabricated entirely from soft elastomers.

4. Dielectric elastomer generators with integrated soft electronics

The DEG membrane is essentially a soft capacitor and one can take advantage of this, using the DEG to provide the energy storage function of the SPC. This means the generator can be fabricated with its external circuitry consisting solely of a few diodes [16, 20]. One configuration that has been used to integrate the SPC storage function into the DEG membrane is given in Figure 4 where DEG1 is electrically configured as an SPC [16]. Because the generator's elastomer membranes are integrated into the self-priming circuit, this system has been referred to as the integrated self-priming circuit [21]. The generator was configured into two DEG membranes which were deformed so that as one membrane was stretched the other was relaxed. When the voltage of DEG1A and DEG1B connected electrically in parallel exceeded the voltage of DEG2, energy was transferred from the SPC to DEG 2 through the path shown in Figure 4c; when the voltage of DEG2 exceeded that of

DEG1A and DEG1B connected in series, energy was transferred onto the SPC from DEG2 through the path shown in Figure 4d.

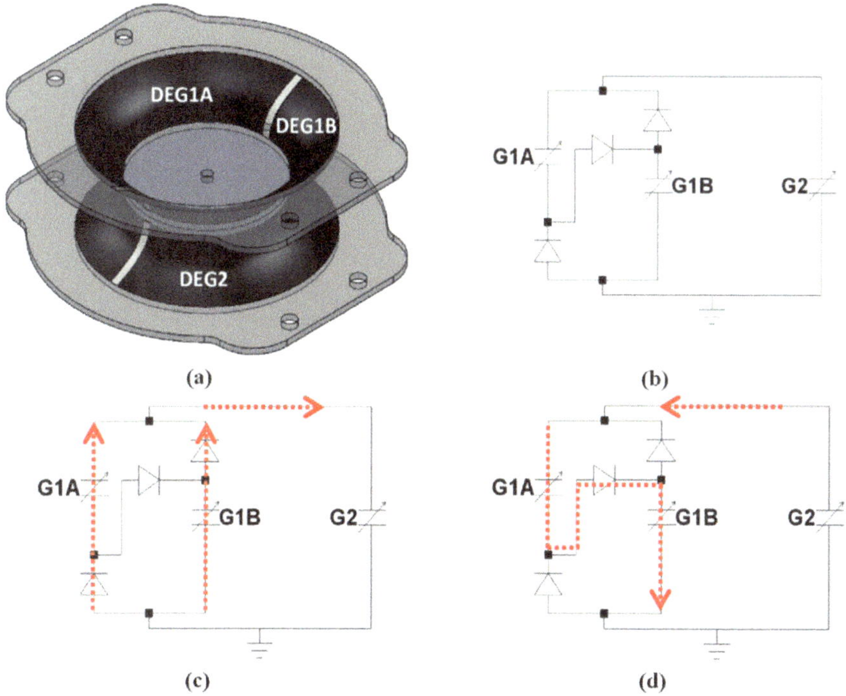

Figure 4. A segmented generator which consists of two membranes that are antagonistically deformed when the inner hub displaces up and down (a). The top membrane DEG1 can be electrically interconnected with diodes to form a self-priming circuit (b). The paths along which current flows off (c) and onto (d) the SPC are also illustrated.

For a small scale portable and potentially wearable generator the diodes of the SPC can represent a significant mass. For instance, the prototype of the generator shown in Figure 4 consisted of DEG membranes with a combined mass of 0.35 grams, whereas the diodes weighed 0.63 grams. If we could remove the diodes, 64% of the total mass (ignoring the mass of the frame) could be eliminated.

The function of the diodes is to control the charge transfer between the DEG and SPC. The diodes simply allow current to flow along one path when one diaphragm is stretched, and along an alternative path when that diaphragm is relaxed. This means that the diodes can be replaced by switches that are toggled at the appropriate time. Since this timing is dependent on the material stretch state, the diodes can be replaced by stretch-sensitive switches coupled to the DEG.

Stretch sensitive electronics called Dielectric Elastomer Switches (DES) have been used to integrate the functionality of the diodes into the DEG membrane. DES consist of piezoresistive electrodes fabricated directly onto a highly stretchable dielectric elastomer membrane. They exhibit very large changes in resistance with stretch. O'Brien et al. first presented the concept [22] and characterized M-shaped DES as illustrated in Figure 5 [23]. These DES had a resistance of several MΩ in their rest state, which increased to several GΩ when they were stretched to approximately 1.4 times their original length (see Figure 5b) [23].

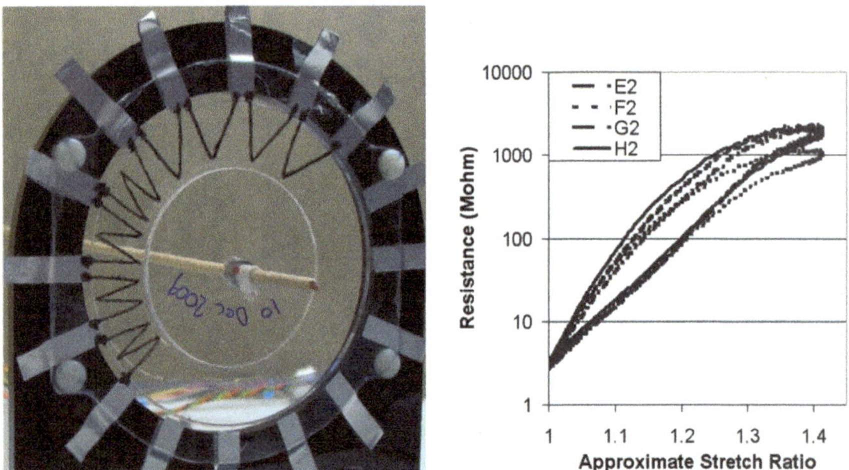

(a) (b)

Figure 5. Carbon powder-based DES applied to a dielectric elastomer diaphragm (black "M" and "V" shaped tracks) for characterisation by O'Brien et al. (a), and a plot of measured DES resistance versus approximate radial stretch ratio showing that the resistance climbed several orders of magnitude when the DES were stretched (b). Images taken from O'Brien [23].

The generator illustrated in Figure 4 was redesigned by placing DES onto the membrane to replace the hard diodes as illustrated in Figures 6 and 7. The SPC capacitors were fabricated onto diaphragm 1. When diaphragm 1 was relaxed the switches Q1 and Q2 were also relaxed and therefore conducted, configuring the SPC capacitors into a parallel topology (high charge form). When diaphragm 1 was stretched so too were Q1 and Q2, so they no longer conducted, but diaphragm 2 was simultaneously relaxed, causing Q3 to conduct, connecting the SPC capacitors in series. This generator can be fabricated entirely from soft elastomers, so we refer to it as the soft generator.

The output voltage of a prototype soft generator is given in Figure 8. When compared to the output of the self-priming circuit given in Figure 3, the soft generator's voltage climbs very rapidly. But the most profound advantages are the ability to produce DEG that maintain the advantages of high flexibility, softness, low volume, low cost, low component count, and low mass at the system level.

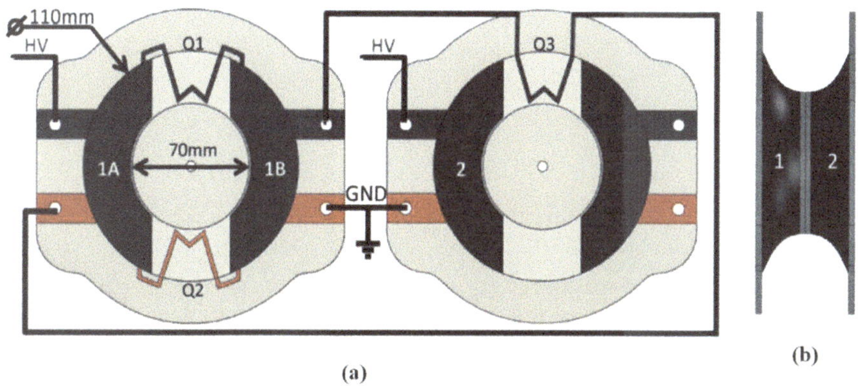

Figure 6. A schematic of the soft generator. The switches Q1-Q3 control the charge flow within the self-priming circuit in a similar manner to the diodes in the integrated self-priming circuit. The two diaphragms in (a) are connected together to form the antagonistic pair shown in (b). (From [24]).

Figure 7. A photograph of a soft generator prototype. The large black regions are the generator electrodes and the thin "M-shaped" electrodes are the piezo-resistive switches.

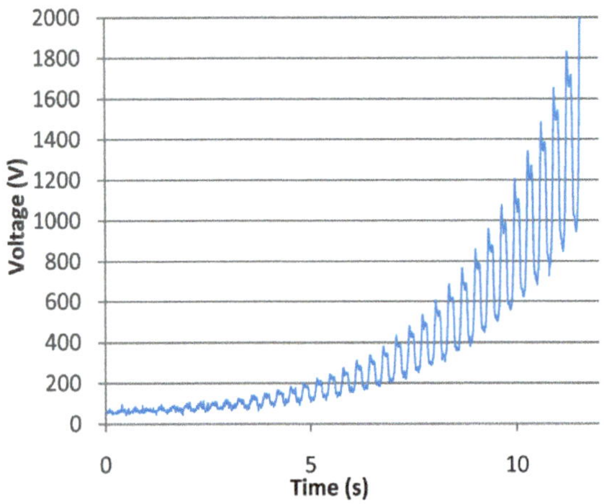

Figure 8. The output voltage waveform of a soft generator initially primed to 10 Volts then mechanically deformed at 3 Hz showing that the generator rapidly boosts its operating voltage through accumulation of generated energy (From [24]).

The total mass of biomechanical energy harvesters has a great influence on the metabolic cost of their use and can be easily compared in a quantitative manner. In Figure 9 we compare the energy density of each DEG system described in this chapter with the mass of their associated electronics included in the calculations (see equation 5).

$$EnergyDensity = \frac{EnergyGenerated}{Mass_{dielectric} + Mass_{ExternalCircuit}} \quad (5)$$

The energy and energy densities produced by similarly sized DEG membranes (~0.3 grams) mechanically cycled at a rate of 3 Hz and operating at 2 kV are compared in Figure 9. The soft generator energy density was superior to both the integrated and external SPC generators because their respective external circuit masses are approximately 0 grams, 0.6 grams, and 13.4 grams. The soft generator's energy density of 9.5 mJ/g is highly competitive with the predicted practical maxima of electromagnetics (4 mJ/g) and piezoelectrics (17.7 mJ/g) at the ~1cm3 scale [25], demonstrating the utility of DEG for small-scale energy harvesters.

The recent developments discussed in this chapter have provided progress towards wearable, soft power generators becoming a reality, but there are still issues that need to be addressed:

1. The switching technology used in the soft generator, DES, are in their infancy. Material and process developments are required to create more reliable DES with resistances that can be tuned to their application.

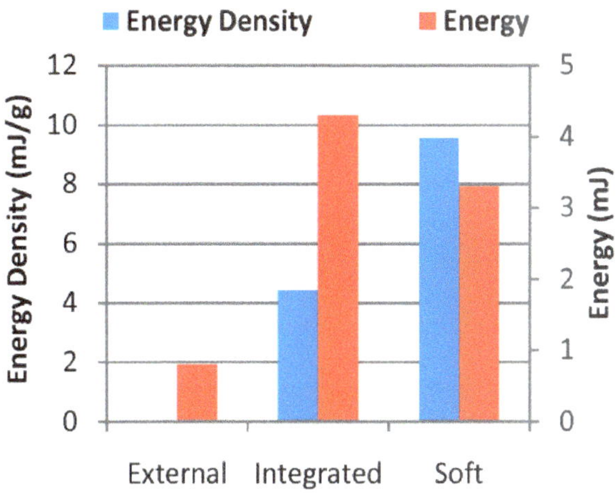

Figure 9. The energy and energy density produced by the external SPC, integrated SPC, and soft generators.

2. The prototypes discussed in this chapter are 1 to 2 orders of magnitude smaller than those required to produce sufficient power to drive the more power hungry portable electronics such as smart phones, which begs the question: How do we expect them to scale? Figure 9 illustrates that the integrated SPC generator produced more energy than the soft generator, so as the generator is scaled up and the diode mass becomes insignificant the integrated generator may outperform the soft generator in terms of energy density. When scaled up to produce an order of a Watt, are the soft generator's advantages sufficient to provide a worthwhile alternative to more complex DEG control strategies that will provide superior energy production [26-29].
3. The soft and integrated generators have an antagonistic pair of membranes. What is the best configuration for harvesting energy from a given biomechanical source?
4. DEG provide the best energy production when operated at high voltages (typically kV range). Compact energy-efficient circuitry is required to reduce this voltage down to consumer electronics levels (~5 Volts) or perhaps the high voltage energy can be used to drive another high voltage device such as a dielectric elastomer actuator.

With these challenges comes an exciting future: the emergence of smart, wearable soft devices. The DEG membranes are multifunctional; they can be operated as actuators or generators and simultaneously sense strain [12]. Furthermore DES have been used to fabricate rubber-based NAND-gates and memory elements, the two primitives required to build a computer [30-32]. Perhaps the future will not only include soft generators to power portable electronic devices, but it may also include a new breed of smart, multifunctional self-powered portable soft devices. These possibilities open the door for smart devices to be integrated directly into clothing. So soft wearable generators will not only revolutionize the use of today's portable devices, but they will be an integral part of a distributed body-area network of sensors and smart devices that will improve future prosthetic devices, sports monitors, and video gaming interfaces.

Author details

Thomas G. McKay and Benjamin M. O'Brien
The Biomimetics Laboratory of the Auckland Bioengineering Institute, The University of Auckland, Level 6, Auckland, New Zealand

Iain A. Anderson
The Biomimetics Laboratory of the Auckland Bioengineering Institute, The University of Auckland, Level 6, Auckland, New Zealand
Department of Engineering Science, School of Engineering, The University of Auckland, Level 3, Auckland, New Zealand

Acknowledgement

This work was supported by a University of Auckland Bioengineering Institute FRDF postdoctoral fellowship and a Rutherford Foundation postdoctoral fellowship from the Royal Society of New Zealand. The authors acknowledge the support of the Auckland Bioengineering Institute, their colleagues in the Biomimetics lab, Kristal Murray, Michelle O'Brien, and Suzie Pilkington.

5. References

[1] *Adidas miCoach.* 2012 [cited 2012]; Available from: www.micoach.com/.

[2] Schiffer, M.B. and C.L. Bell (2003) *Draw the Lightening Down: Benjamin Franklin and Electrical Technology in the Age of Enlightenment.* Berkeley: University of California Press, Ltd.

[3] Jia, D. and J. Liu (2009) *Human power-based energy harvesting strategies for mobile electronic devices.* Frontiers of Energy and Power Engineering in China. 3(1): p. 27-46.

[4] Carroll, A. and G. Heiser (2010) *An Analysis of Power Consumption in a Smartphone. USENIX ATC '10.*

[5] Starner, T. and J.A. Paradiso (2004) *Human Generated Power for Mobile Electronics* in *Low Power Electronics Design* C. Piguet Editor CRC Press. p. 45-1 - 45-35.

[6] Shorten, M.R. (1993) *The energetics of running and running shoes.* Journal of Biomechanics. 26, Supplement 1(0): p. 41-51.

[7] Donelan, J.M., Q. Li, V. Naing, J.A. Hoffer, D.J. Weber, and A.D. Kuo (2008) *Biomechanical Energy Harvesting: Generating Electricity During Walking with Minimal User Effort.* Science. 319(5864): p. 807-810.

[8] Jean-Mistral, C. and S. Basrour (2010) *Scavenging energy from human motion with tubular dielectric polymerElectroactive Polymer Actuators and Devices (EAPAD) 2010* 7642: p. 764209-12.

[9] Jean-Mistral, C., S. Basrour, and J.J. Chaillout (2008) *Dielectric polymer: scavenging energy from human motion.* Proc. SPIE. 6927: p. 692716.

[10] Pelrine, R., R.D. Kornbluh, J. Eckerle, P. Jeuck, S. Oh, Q. Pei, and S. Stanford (2001) *Dielectric elastomers: generator mode fundamentals and applications.* Proc. SPIE. 4329: p. 148.

[11] Prahlad, H., R. Kornbluh, R. Pelrine, S. Stanford, J. Eckerle, and S. Oh 2005 *Polymer Power: Dielectric Elastomers and Their Applications in Distributed Actuation and Power Generation* in *Proc. ISSS*: Bangalore, India. p. 100.

[12] Anderson, I.A., T.A. Gisby, T.G. McKay, B.M. O'Brien, and E.P. Calius (2012) *Multi-functional dielectric elastomer artificial muscles for soft and smart machines.* Journal of Applied Physics. 112(4): p. 041101-20.

[13] Bar-Cohen, Y. (2004) *Electroactive Polymer Actuators as Artificial Muscles: Reality, Potential, and Challenges* SPIE: New York.

[14] Pelrine, R., R. Kornbluh, P. Qibing, and J. Joseph (2000) *High-speed electrically actuated elastomers with strain greater than 100%.* Science. 287(5454): p. 836-9.

[15] Pelrine, R. and H. Prahlad (2008) *Generator Mode: Devices and Applications* in *Dielectric Elastomers as Electromechanical Transducers* F. Carpi, et al. Editors Elsevier: Oxford, UK. p. 146-155.

[16] McKay, T.G., B.M. O'Brien, E.P. Calius, and I.A. Anderson (2011) *Realizing the Potential of Dielectric Elastomer Generators. Proc. SPIE* 7976: p. 7976-10.

[17] McKay, T., O'Brien, B., Calius, E., Anderson, I. (2010) *Self-priming dielectric elastomer generators.* Smart Materials and Structures. 19(5).

[18] Pelrine, R.E., R.D. Kornbluh, J.S. Eckerle, and H. Prahlad 2008 *Circuits for electroactive polymer generators* SRI International.

[19] McKay, T.G., B. O'Brien, E. Calius, and I. Anderson (2010) *Self-priming dielectric elastomer generators.* Smart Mater. Struct. 19(055025): p. 055025.

[20] McKay, T., O'Brien, B., Calius, E., Anderson, I. (2010) *An integrated, self-priming dielectric elastomer generator.* Applied Physics Letters. 97(6).

[21] McKay, T., B. O'Brien, E. Calius, and I. Anderson (2010) *An integrated, self-priming dielectric elastomer generator.* Applied Physics Letters. 97(062911): p. 062911.

[22] O'Brien, B.M., E.P. Calius, T. Inamura, S.Q. Xie, and I.A. Anderson (2010) *Dielectric elastomer switches for smart artificial muscles.* Appl. Phys. A.: p. 1-5.

[23] O'Brien, B. Simulation, Fabrication, and Control of Biomimetic Actuator Arrays PhD Auckland Bioengineering Institute 2010 The University of Auckland.

[24] McKay, T.G., B.M. O'Brien, E.P. Calius, and I.A. Anderson (2011) *Soft generators using dielectric elastomers.* Applied Physics Letters. 98(14): p. 142903-3.

[25] Roundy, S., P.K. Wright, and J.M. Rabaey (2004) *Energy Scavenging For Wireless Sensor Networks*Boston, MA: Kluwer Academic Publishers. 212.

[26] Due, J., S. Munk-Nielsen, and R.O. Nielsen (2010) *Energy harvesting with Di-Electro Active Polymers.* IET Conf. Pub. 2010(CP563): p. WE244-WE244.

[27] Graf, C., J. Maas, and D. Schapeler (2010) *Energy harvesting cycles based on electro active polymers. Proc. SPIE* 7642: p. 764217-12.

[28] Kaltseis, R., C. Keplinger, R. Baumgartner, M. Kaltenbrunner, T. Li, P. Machler, R. Schwdiauer, Z. Suo, and S. Bauer (2011) *Method for measuring energy generation and efficiency of dielectric elastomer generators.* Appl. Phys. Lett. 99(16): p. 162904.

[29] Koh, S.J.A., X. Zhao, and Z. Suo (2009) *Maximal energy that can be converted by a dielectric elastomer generator.* Appl. Phys. Lett. 94(26): p. 262902.

[30] O'Brien, B.M. and I. Anderson (2011) *An Artificial Muscle Ring Oscillator.* Mechatronics, IEEE/ASME Transactions on. PP(99): p. 1-4.

[31] O'Brien, B.M., E.P. Calius, T. Inamura, S.Q. Xie, and I.A. Anderson (2010) *Dielectric Elastomer Switches for Smart Artificial Muscles.* Applied Physics A: Materials Science & Processing. 100: p. 385-389.

[32] O'Brien, B.M., T.G. McKay, S.Q. Xie, E.P. Calius, and I.A. Anderson (2011) *Dielectric Elastomer memory.* Proc. SPIE. 7976: p 797621-7.

Chapter 8

Ionic Polymer Metal Composite Transducer and Self-Sensing Ability

Doan Ngoc Chi Nam and Ahn Kyoung Kwan

Additional information is available at the end of the chapter

1. Introduction

During the recent decade, ionic polymer metal composite (IPMC), as a smart material, is gaining great importance in the use for both sensors and actuators. An ion polymer metal composite (IPMC) is an Electro-Active Polymer (EAP) that bends in response to a small applied electrical field as a result of mobility of cations in the polymer network [1] and vice versa. A typical IPMC sheet is constructed with a thin (200 μm) ionic polymer membrane like Nafion or Flemion and two metalic electrode layers (10 μm thick) outside such as platinum or gold. When a low voltage electrical field (1 – 5V) is applied, the transport of hydrated cations within the solvated IPMC and the associated electrostatic interactions lead to bending motions of the IPMC sheet. Thus, an IPMC can work as a small size actuator. Figure 1 illustrates the mechanism of the IPMC as actuator. On the other hand, when an IPMC is mechanically bent, it will generate a low voltage (order of milivolts) between the two electrodes. The generated voltage is due to the non-uniform concentration of ions in the IPMC membrane. Figure 2 illustrates the behavior of the IPMC as sensor.

Since firstly introduced, the IPMC has been applied in many research fields such as biomedical system, biomimetic robot, and MEMS devices. In these applications, the IPMC may give advantages thank to the low driven voltage, flexible operation, and simple actuating structure. Some applications using IPMC material have been a snake-like robot with IPMC actuator [7], a micro pump [8], a scale biped walking robot [9], an underwater micro robot [10], etc. Because of the increasing importance of IPMC in the smart material field, it is important to clearly understand the behavior inside the IPMC actuator.

Recently, it has been found that the IPMC possesses an attractive characteristic which is so called the "self – sensing" behavior. The self – sensing ability originates from the variation of surface resistance of the two electrode layers during operating period [11 – 12]. If one can capture these variations of surface resistance, the bending curvature can be figured out. The self – sensing ability has been one of the most interesting features of the IPMC transducer.

By employing self – sensing feature, the sensing problem will be easier for the field of biomedical devices, biomimetic and micro robots.

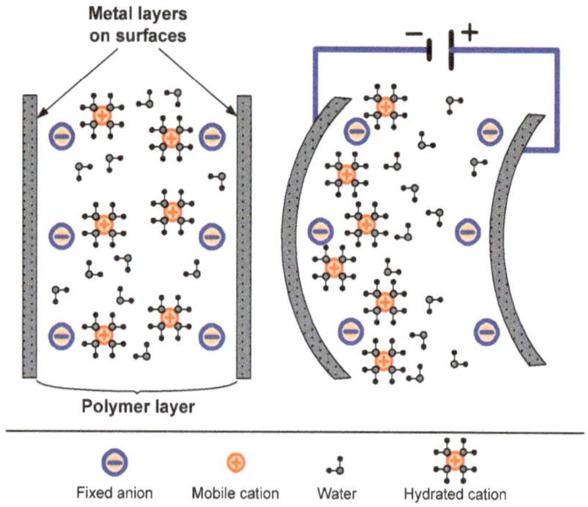

Figure 1. Operating principle of IPMC as actuator

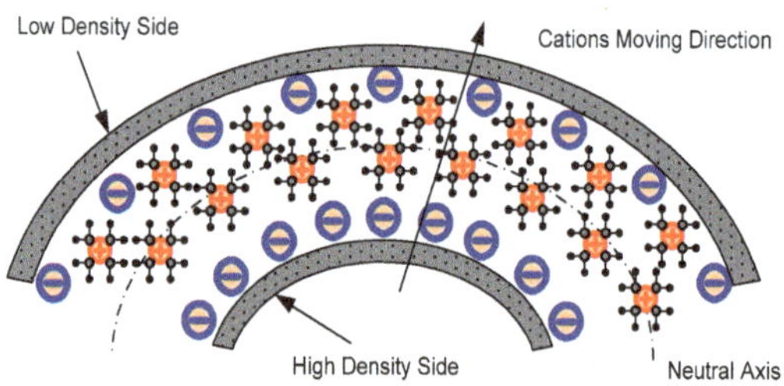

Figure 2. Fundamental of IPMC as a motion sensor

In order to investigate the behavior of IPMC and its self- sensing ability, this chapter will cover physical analysis on an IPMC actuator and its self – sensing model. Firstly, the mechanism inside the IPMC actuator under an applied input voltage signal is carried out. The analysis starts from the governing partial differential equation (PDE) for charge distribution in an IPMC, which was introduced firstly by Nasser et al. in [13] and then used in [14] and [15] for investigating the actuation and sensing response. The effect of distributed surface resistance is also considered to obtain accurate boundary conditions for the PDE. Solving the PDE in asymmetric boundary conditions, the charge density inside the IPMC is obtained. Consequently, the induced stress of the IPMC is obtained. Using this induced stress model, a visual simulation technique is proposed for investigating the actuation capabilities of IPMC actuator via ANSYS environment [18].

Next, the self – sensing ability of the IPMC actuator is analyzed. The self sensing technique is inspired by the variation of surface resistance along the IPMC. An equivalent circuit of

IPMC actuator is present and analyzed. Consequently, a simple technique to implement the self – sensing behavior of IPMC is carried out. In this technique, the variation of surface resistance along the IPMC is figured out using half clamped configuration. Four voltage feedback signals are then employed to estimate the bending behavior. Experimental results show that this technique has a strong ability to apply the self – sensing technique into real application, i.e. micro robot, biomimetic robot.

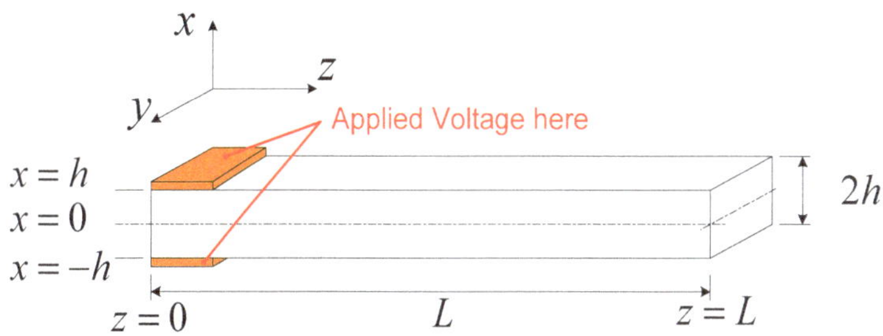

Figure 3. Geometric description of IPMC transducer

2. Electromechanical model of IPMC actuator

The electromechanical model developed for the Nafion based IPMC relies on field and continuity equations. Let D, E, ϕ , and ρ denote the electric displacement, the electric field, the electric potential, and the charge density, respectively, the field equations and continuity equation using in this analysis are as follows:

$$E = \frac{D}{\kappa_e} = -\nabla\phi \tag{1}$$

$$\nabla \cdot D = \rho = F\left(C^+ - C^-\right) \tag{2}$$

$$\nabla \cdot J = -\frac{\partial C^+}{\partial t} \tag{3}$$

where, κ_e is the effective dielectric constant of the polymer, F is Faraday's constant, J is the ion flux vector, and C^+ and C^- are the cation and anion concentrations, respectively. The ion flux is considered to be composed of three primary terms: the diffusion, migration and convection terms:

$$J = -d\left(\nabla C^+ + \frac{C^+F}{RT}\nabla\phi + \frac{C^+\Delta V}{RT}\nabla p\right) + C^+ v \tag{4}$$

where, d is the ionic diffusivity, R is the gas constant, T is the absolute temperature, p is the fluid pressure, v is the free solvent velocity field, and ΔV is the volumetric change.

Considering the geometric description of IPMC as in figure 3, since the thickness of an IPMC is much smaller than its length or width, the first assumption is claimed as: inside the polymer, D, E, and J are all restricted to the thickness direction (x-direction). This assumption enables to drop the boldface notation for these variables. Rewrite equation (2) as:

$$C^+ = \frac{1}{F}\rho + C^- \tag{5}$$

Because the anions are constrained to the polymer backbone, the concentration of anions inside the IPMC is considered to be spatially and temporally constant. Therefore, the cation gradient can be rewritten specifically in terms of the electric field E :

$$\nabla C^+ = \frac{\kappa_e}{F}\nabla^2 E \tag{6}$$

Employing a modified version of Darcy's Law [29], the free solvent velocity field can be calculated as:

$$v = k'\left[C^- FE - \nabla p\right] \tag{7}$$

where k' represents the hydraulic permeability. The second assumption is that the convection term can be neglected: $v \to 0$. Then, equation (7) becomes:

$$\nabla p = C^- FE \tag{8}$$

By employing equations (5), (6) and (8), the original flux equation (4) can be rewritten as:

$$J = -d\left(\frac{\kappa_e}{F}\nabla^2 E - \frac{\left(1 - C^-\Delta V\right)}{RT}E\left(\kappa_e \nabla E - FC^-\right)\right) \tag{9}$$

By taking the gradient of (9) and employing (3), the nonlinear PDE in term of electrical field is obtained as:

$$-\frac{\kappa_e}{F}\frac{\partial(\nabla E)}{\partial t} = -d\left(\frac{\kappa_e}{F}\nabla^3 E - \frac{\kappa_e\left(1 - C^-\Delta V\right)}{RT}\left(\nabla^2 E.E + (\nabla E)^2\right) - \frac{FC^-\left(1 - C^-\Delta V\right)}{RT}\nabla E\right) \tag{10}$$

Consider the properties of the IPMC actuator as shown in table 1. It can be seen that the coefficient ratio of the linear term and the nonlinear term:

$$ratio = \frac{FC^-}{\kappa_e} \tag{11}$$

is around the order of 5×10^{10}. Hence, it can be concluded that:

$$\frac{\kappa_e\left(1-C^{-}\Delta V\right)}{RT}\left(\nabla^2 E.E+\left(\nabla E\right)^2\right) << \frac{FC^{-}\left(1-C^{-}\Delta V\right)}{RT}\nabla E \tag{12}$$

Neglecting small nonlinear term, the PDE for the electric field can be obtained as:

$$\frac{\partial}{\partial x}\left[\frac{\partial(\kappa E)}{\partial t}-d\frac{\partial^2(\kappa E)}{\partial x^2}-\frac{F^2 dC^{-}}{\kappa RT}\left(1-C^{-}\Delta V\right)(\kappa E)\right]=0 \tag{13}$$

or in the form of charge density ρ:

$$\frac{\partial \rho}{\partial t}-d\frac{\partial^2(\rho)}{\partial x^2}-\frac{F^2 dC^{-}}{\kappa RT}\left(1-C^{-}\Delta V\right)\rho=0 \tag{14}$$

where, C^{-} is the anion concentration and κ is the effective dielectric constant of the polymer.

$F=96487\ C/mol$	$R=8.3143\ J/mol\cdot K$	$T=300K$	$w=0.02mm$
$h=100\,\mu m$	$d=1.030\times10^{-10}m^2/s$	$C^{-}=1200mol/m^3$	$\kappa=1.88\mu F/m$

Table 1. Parameters of IPMC charging model

The solutions of (13) and (14) are analyzed as follows:

It is noted that the IPMC transducer consists of two noble metallic layers outside as electrodes and one ion-exchange polymer membrane with fixed anions at both sides and movable cations. Because of the fixed anion layers, an applied step voltage will lead to an asymmetry in the ion redistribution [16 – 17].

Considering the steady state where $J=0$, the PDE (13) becomes:

$$\frac{\partial^2 E}{\partial x^2}-KE=0 \tag{15}$$

where:

$$K \triangleq \frac{F^2 C^{-}}{\kappa RT}\left(1-C^{-}\Delta V\right) \tag{16}$$

Since fixed anions locate at both sides of the polymer membrane (see figure 1), the charge density in the anode boundary layer keeps a constant value, given by:

$$\rho_{(1)}=-C^{-}F \tag{17}$$

Whereas, in the remaining part, the charge density is given as:

$$\rho_{(2)}=\left(C^{+}-C^{-}\right)F \tag{18}$$

Thus, the charge density ρ, electric field E, and electric potential field ϕ in the anode boundary layer are then given by:

$$\begin{cases} \rho_{(1)}(x) = -C^{-}F \\ E_{(1)}(x) = \dfrac{1}{\kappa}\left(-C^{-}Fx + E_0\right) \qquad \text{for}\left(-h \le x \le -h + w\right) \\ \phi_{(1)}(x) = \dfrac{1}{\kappa}\left(\dfrac{C^{-}Fx^2}{2} - E_0 x\right) + A_0 \end{cases} \tag{19}$$

where, w is the thickness of the anode boundary layer, E_0, and A_0 are an integration constants. In the remaining region, solution of the ODE (5) can be obtained as:

$$\begin{cases} E_{(2)}(x) = C_1 \exp(\sqrt{K}x) + C_2 \exp(-\sqrt{K}x) \\ \phi_{(2)}(x) = \dfrac{1}{\sqrt{K}}\left(-C_1 \exp(\sqrt{K}x) + C_2 \exp(-\sqrt{K}x)\right) + C_3 \quad \text{for}\left(-h + w \le x \le h\right) \\ \rho_{(2)}(x) = \kappa\sqrt{K}\left(C_1 \exp(\sqrt{K}x) - C_2 \exp(-\sqrt{K}x)\right) \end{cases} \tag{20}$$

where, C_1, C_2, and C_3 are integration constants. Consequently, the function of charge density in steady state is obtained for an IPMC actuator under a step voltage input signal:

$$\rho^{*}(x) = \begin{cases} -C^{-}F & \text{for}\left(-h \le x \le -h + w\right) \\ \kappa\sqrt{K}\left(C_1 \exp(\sqrt{K}x) - C_2 \exp(-\sqrt{K}x)\right) & \text{for}\left(-h + w \le x \le h\right) \end{cases} \tag{21}$$

3. Effect of distributed surface resistance in an IPMC actuator

In order to find correct value for the charge density ρ, electric field E, and electric potential field ϕ, a sufficient set of boundary conditions must be considered. In this section, the effect of distributed surface resistance is investigated to serve as a boundary condition for solving the distribution of charge density, electric field, and electric potential along the z axis. Since the chemical reduction of salt is typically used to build the electrode layers, the surfaces of an IPMC typically consist of aggregated nano particles of noble metal. Thus, the affect of surface resistance is non-negligible in the behavior of IPMC during sensing and actuating process.

Considering the effect of distributed surface resistance, the equivalent circuit of the IPMC is shown as in figure 4. Let the electrode resistance per unit length be r_1 in z direction and r_2 in x direction. As shown in figure 4, $i_p(z,t)$ is the distributed current per unit length going through the polymer due to the ion movement, $i_k(z,t)$ represents the leaking current per unit length, and $i_s(z,t)$ is the surface current on the electrodes. R_{dc} denotes the leakage resistance per unit length. The surface current $i_s(0,t)$ at $z = 0$ is the total actuation current. The following equations hold for the impedance model:

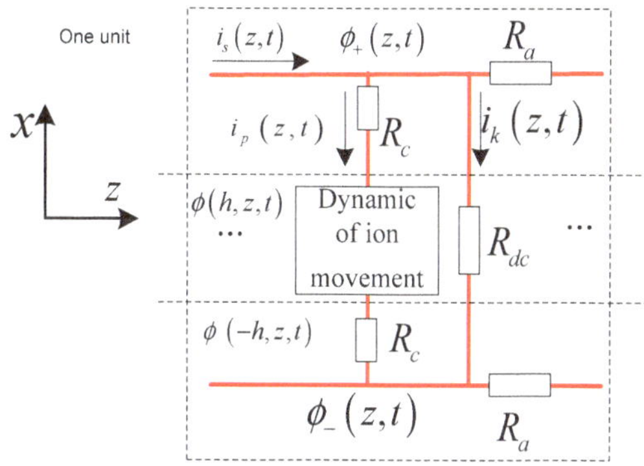

Figure 4. Equivalent circuit of IPMC in a unit length

$$\frac{\partial \phi_{\pm}(z,t)}{\partial z} = \mp R_a i_s(z,t) \tag{22}$$

$$\frac{\partial i_s(z,t)}{\partial z} = -\left(i_p(z,t) + i_s(z,t)\right) \tag{23}$$

Under the effect of distributed surface resistance, the first two boundary conditions for the PDE are:

$$\frac{\partial \phi(h,z,t)}{\partial z} = \int_0^z R_a i_s(\tau,t)d\tau - R_c i_p(z,t) \tag{24}$$

$$\frac{\partial \phi(-h,z,t)}{\partial z} = \phi_0 - \int_0^z R_a i_s(\tau,t)d\tau - R_c i_p(z,t) \tag{25}$$

Other boundary conditions are described as:

$$C^+(-h+\mathrm{w}) = 0 \tag{26}$$

$$E_{(1)}(-h+\mathrm{w}) = E_{(2)}(-h+\mathrm{w}) \tag{27}$$

$$\phi_{(1)}(-h+\mathrm{w}) = \phi_{(2)}(-h+\mathrm{w}) \tag{28}$$

These constraints are continuity, and the overall charge-balance conditions. By employing the above constraints, one can solve for E_0, A_0, C_1, C_2, and C_3 to obtain the distribution of charge density, electric field, and voltage potential along the z axis.

Approximately, the dynamic of charging model can be express as:

$$\rho(x,t)=\left(1-\exp\left(\frac{-t}{\tau}\right)\right)\rho^{*}(x),\quad \tau=\frac{\kappa RT}{dC^{-}F} \tag{29}$$

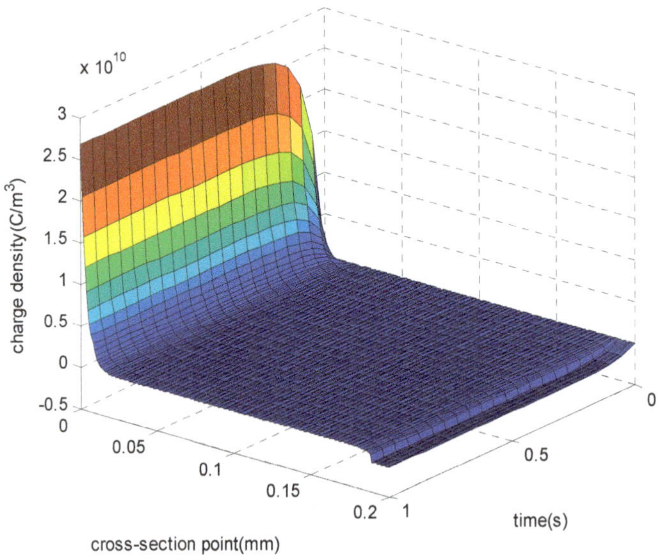

Figure 5. Simulation result for the charge density profile

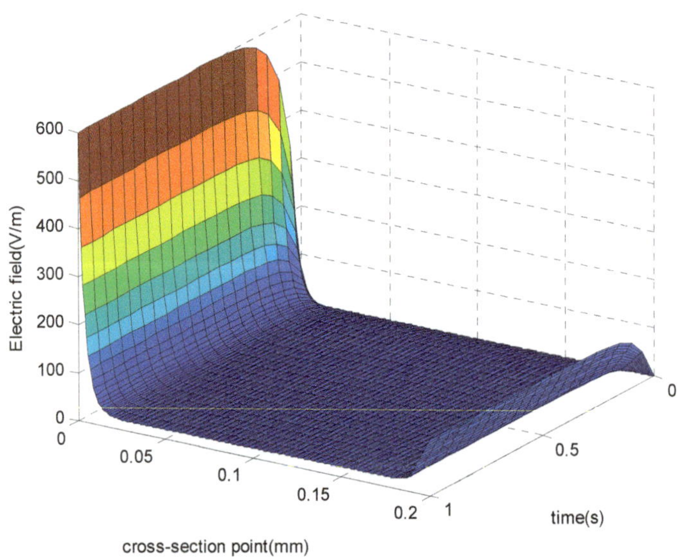

Figure 6. Simulation result for the distributed electric field

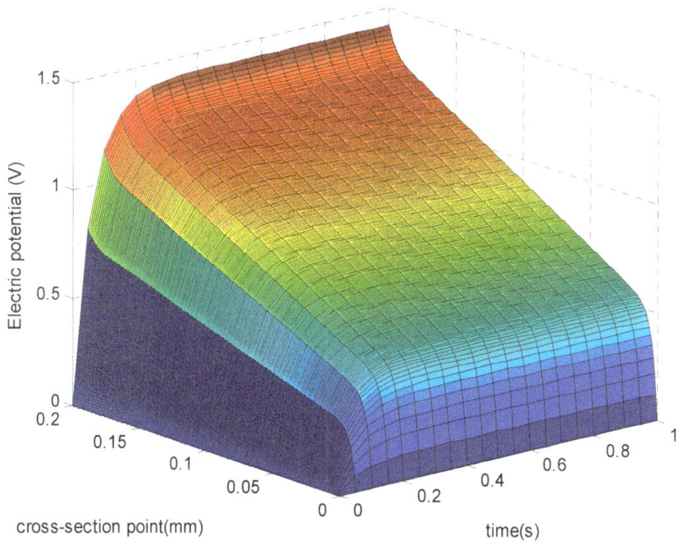

Figure 7. Simulation result for the distributed voltage potential

Here, τ is the natural time scale which characterizes relative speed of cation redistribution. Applying the above analysis, the charging behavior, the distributed electric field, and the distributed electric potential can be trivially obtained. Figure 5, figure 6, and figure 7 are the step response of the charge density, electric field, and electric potential when a 1.5V step input voltage signal is applied to the IPMC actuator.

4. Visual analysis of IPMC actuator

Since the charge distribution is obtained, the induced stress inside the IPMC actuator is available. The key assumption for analysis the electromechanical behavior inside the IPMC transducer is that the induced stress inside the IPMC σ is proportional to the electric charge density ρ :

$$\sigma(x,y,z,t)=\alpha_0\rho(x,y,z,t) \tag{30}$$

where, α_0 is the coupling constant to be determined by experiment. The internal stress σ produces the deformation of the IPMC actuator. Generally, the ideal linear beam theory is employed to analyze the bending behavior of a beam configured IPMC actuator. The bending moment is obtained by integration of induced stress along the x axis. However, this is a complicated work since the induced stress σ is not in simple form. Moreover, if the configuration of IPMC is complex (i.e. round shape, triangular shape), the above technique is unavailable.

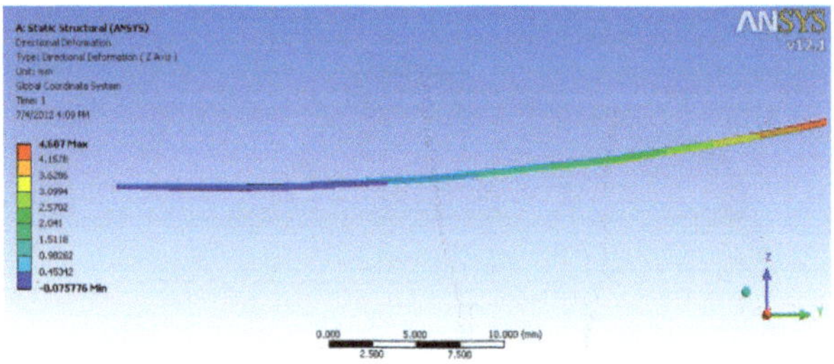

(a) Displacement of cantilever IPMC actuator (iso view)

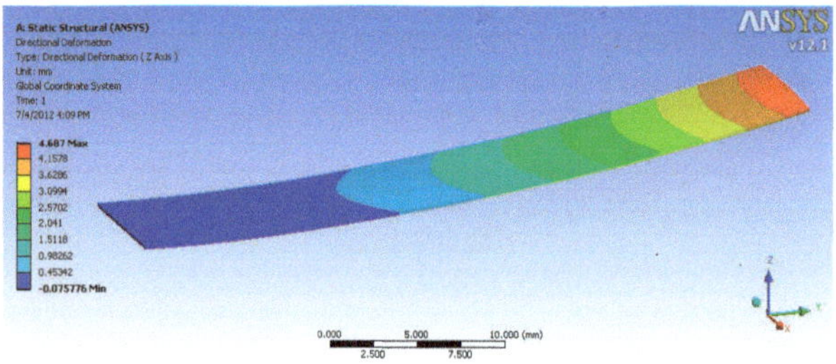

(b) Displacement of cantilever IPMC actuator (front view)

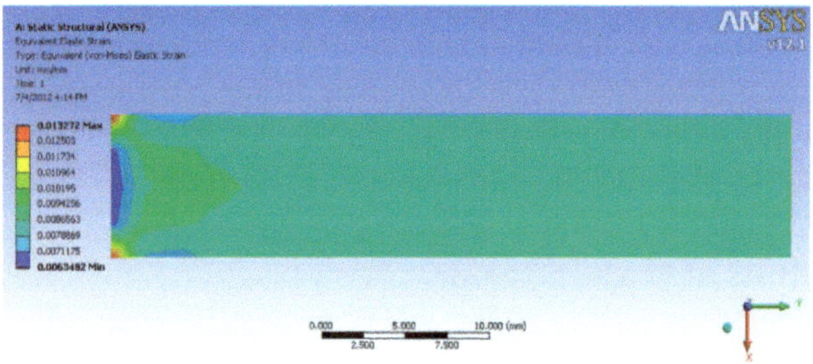

(c) Strain of cantilever IPMC actuator (top view)

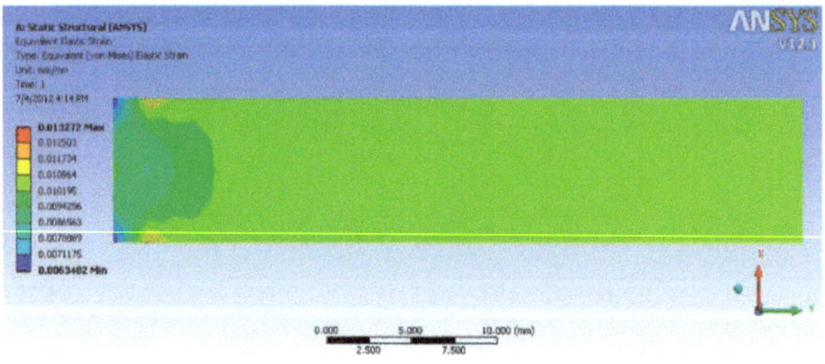

(d) Strain of cantilever IPMC actuator (bottom view)

Figure 8. Simulation of a cantilever IPMC actuator

For ease of analysis, this chapter proposes a novel visual simulation technique for the IPMC actuator using a commercial mechanical analysis tool, ANSYS 12.1 (2009) [18]. In this technique, a finite element model (FEM) is created to model an IPMC actuator. By using the command "INISTATE", the induced stress, obtained from equation (30), is added to the FEM as the deformation mechanism. Next, the constraints at the clamped edges of the IPMC actuator are considered as boundary conditions (fixed support to the clamped edges). Displacement constraints are employed as boundary conditions for neutral plane inside the IPMC actuator.

An example of visual simulation is shown as figure 8 for the cantilever IPMC actuator with a constant 3V input signal. The material properties of the IPMC actuator, employed for the FEM, are shown as in Table 2. A FEM with total 400 elements was created. As shown in figure 8a and figure 8b, the deformation of the IPMC can reach 4.69 mm at maximum under 3V input signal. Figure 8c and figure 8d show the internal strain ratio of the IPMC actuator under operation (maximum strain ratio is 0.013). Figure 9 shows the response of the IPMC beam under a step input voltage signal, and figure 10 shows the displacement versus input voltage graph at the steady state. Modeling results show that the visual simulation technique possesses a strong ability to be applied in analyzing operation of IPMC. It is obvious that the proposed visual simulation technique helps to analyze the behaviors of IPMC actuators more conveniently. In high input voltage signal range, the experimental displacements of IPMC are smaller than simulations. The reasons may be caused by the incompetence of the key assumption (see equation 30), the neglecting of nonlinear term in the PDE, or the variation of surface resistance along the z axis when the IPMC deforms.

	Elastic modulus (Pa)	Poisson's ratio	Size of IPMC actuator (mm)	Diaphragm radius (mm)	Thickness (mm)
Specification	5.71×10^{8}	0.487	$30\times6\times0.2$	5	0.2

Table 2. Material properties of IPMC diaphragm

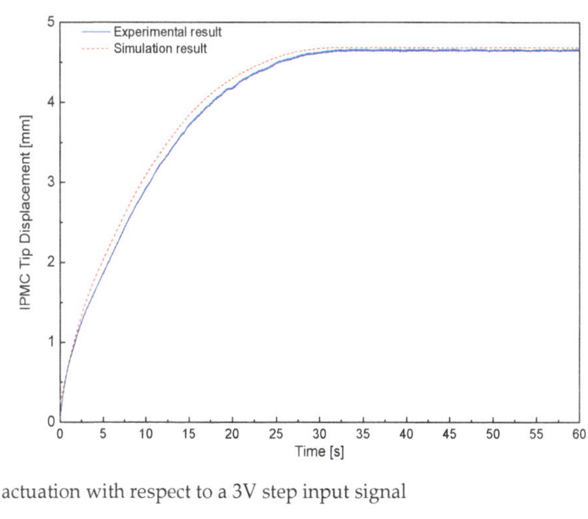

Figure 9. IPMC actuation with respect to a 3V step input signal

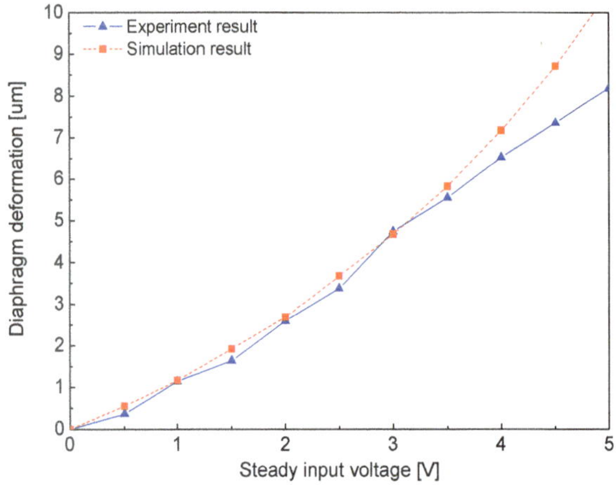

Figure 10. Displacement vs. input voltage

5. Variation of surface resistance and self sensing ability

5.1. Physical model of self sensing ability

During the bending operation, surface resistances of the two electrode layers may change due to the non – uniform material distribution between the two electrode layers as reported in [11 – 12]. During the deforming process, the surface resistance on the stretched side of IPMC increases corresponding to the reduction of conductive material density. Meanwhile, the surface resistance on the compressed side of IPMC decreases because the density of conductive material on the surface layer increases. In order to investigate the bending curvature - surface resistance correlation, a simple model of an IPMC sheet is derived out as in figure 11. Here, half of the IPMC sheet is fixed by a clamp while the other half sheet is free

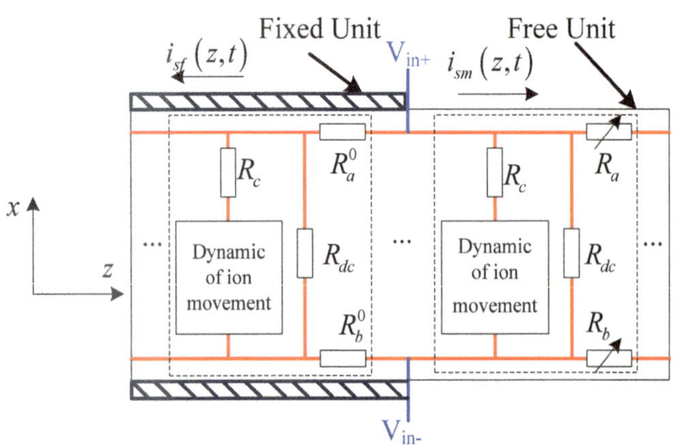

Figure 11. Equivalent circuit of an IPMC actuator

for moving. The actuator is then represented by the equivalent circuit in which affecting factors are defined as follows:

- R_a^0, R_b^0 : the fixed surface resistances at both sides of the clamped part.
- R_a, R_b : the variable surface resistances at both sides of the movable part.
- $i_{sf}(z,t)$: the surface current of the movable part.
- $i_{sm}(z,t)$: the surface current of the movable part.

The analysis of self – sensing behavior deals with the following assumptions:

- The properties of the material are uniform. Consequently, surface resistances on both sides of the clamped part are equal:

$$R_a^0 = R_b^0 \tag{31}$$

- The variation of surface resistances is proportional to the unit bending displacement (the bending behavior corresponding to a unit length along the z axis). Given the unit bending displacement $\psi(z,t)$, the surface resistances on the movable part are calculated as follows:

$$\begin{cases} R_a(z,t) = \beta_a \psi(z,t) + R_a^0 \\ R_b(z,t) = \beta_b \psi(z,t) + R_b^0 \end{cases} \tag{32}$$

where, and β_a, β_b are coupling constants to be determined by experiment.

- The neutral axis of the resistance variation is the center line of the IPMC along the z axis. This assumption allows one to treat β_a, β_b at same value:

$$\beta_a = -\beta_b = \beta \tag{33}$$

With the above assumption, the self – sensing ability of IPMC is discussed as follows. Define:

$$\begin{cases} \Delta R_a(z,t) = \beta_a \psi(z,t) \\ \Delta R_b(z,t) = \beta_b \psi(z,t) \end{cases} \tag{34}$$

Employing equation (33), the variant of surface resistance in equation (34) is rewritten as:

$$\Delta R_a(z,t) = -\Delta R_b(z,t) = \beta \psi(z,t) \tag{35}$$

A simplified circuit for the self – sensing actuator is carried out (figure 12)

Based on the equation (35), the following statements hold:

$$Z_F = Z_M = Z_{eq} \tag{36}$$

$$i_{sf}(z,t) = i_{sm}(z,t) = i_s(z,t) \tag{37}$$

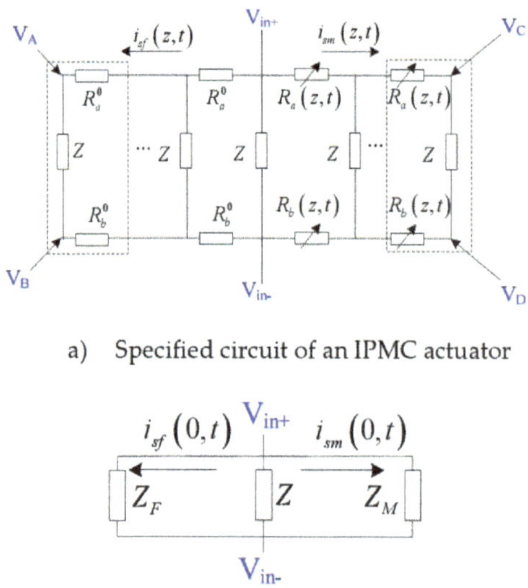

Figure 12. Simplified equivalent circuit of an IPMC actuator

where, Z_F, Z_M are equivalent impedance of fix part and movable part, respectively. The voltage signal at 4 points *A, B, C,* and *D* are in the forms:

$$V_A(t) = V_{in+}(t) - \int_0^{L/2} R_a^0 i_s(\tau,t)d\tau \tag{38}$$

$$V_B(t) = V_{in-}(t) + \int_0^{L/2} R_b^0 i_{sf}(\tau,t)d\tau \tag{39}$$

$$V_C(t) = V_{in+}(t) - \int_0^{L/2} R_a(\tau,t) i_{sm}(\tau,t)d\tau \tag{40}$$

$$V_D(t) = V_{in-}(t) + \int_0^{L/2} R_b(\tau,t) i_{sm}(\tau,t)d\tau \tag{41}$$

The following analyses are carried out:

$$\begin{aligned} V_{AC}(t) = V_A(t) - V_C(t) &= \int_0^{L/2} R_a(\tau,t) i_s(\tau,t)d\tau - \int_0^{L/2} R_a^0 i_s(\tau,t)d\tau \\ &= \int_0^{L/2} \left(R_a^0 + \Delta R_a(\tau,t)\right) i_s(\tau,t)d\tau - \int_0^{L/2} R_a^0 i_s(\tau,t)d\tau \\ &= \int_0^{L/2} \Delta R_a(\tau,t) i_s(\tau,t)d\tau \\ &= \beta \int_0^{L/2} \psi(\tau,t) i_s(\tau,t)d\tau \end{aligned} \tag{42}$$

Denote $V_{in+A}(t) = V_{in+}(t) - V_A(t)$, the derivatives of equation (38) and equation (42) give:

$$i_s(z,t) = \frac{1}{R_a^0}\frac{\partial V_{in+A}(t)}{\partial z} \tag{43}$$

$$\psi(z,t) = \frac{1}{\beta i_s(z,t)}\frac{\partial V_{AC}(t)}{\partial z} \tag{44}$$

$$\psi(z,t) = \frac{R_a^0}{\beta}\frac{\frac{\partial V_{AC}(t)}{\partial z}}{\frac{\partial V_{in+A}(t)}{\partial z}} \tag{45}$$

Note that R_a^0 is a constant. Equations (43), (44) and (45) imply that if one can observe the change of voltage signal along the length of IPMC actuator, the unit bending displacement can be obtained accurately. Consequently, the tip displacement of the IPMC can be obtained as follows:

$$d(t) = \int_0^{\frac{L}{2}} \psi(z,t)dz \tag{46}$$

In practice, the bending curvature can be estimated by using the feedback voltage signals at several points along the z axis. The following section presents the bending estimation of IPMC actuator using four points feedback configuration.

5.2. Experimental results for self sensing of an IPMC actuator

The self – sensing test rig using an IPMC actuator is shown in figure 13. Here, the actuator is a sheet of IPMC clamped by two plastic plates to which electrodes (V_{in+}, V_{in-}, A, B, C, and D) are attached and also contacted with two sides of the IPMC. The IPMC sheet in size of 30 x 6 x 0.2 mm manufactured by Environmental Robots Inc. can operate in both wet and dry environments. A processing system is built on a personal computer (Intel ® CoreTM2 Duo 1.8 GHz) within Simulink environment combined with Real-time Windows Target Toolbox of MATLAB. Two multi-function data acquisition Advantech cards, A/D 1711 and D/A 1720, are installed on PCI slots of the PC to perform the peripheral buses. In addition, a CCD laser displacement sensor, LK-081, from Keyence Corp. is installed on the rig base to measure the IPMC tip displacement.

Employing the technique described in section 5.1, the voltage signals at four points A, B, C, D are collected. In ideal case when the assumption in equations (31), (32), and (33) hold, the estimation can be done with only two feedback voltage signals at A and C. However, for more accurate result, the two more voltage signals at B and D are obtained. Since this implementation uses only four feedback signals, the equation (45) is rewritten as:

$$\psi(z,t) = \frac{R_a^0}{2\beta}\left(\frac{V_{AC}(t)}{V_{in+A}(t)} + \frac{V_{BD}(t)}{V_{in-}(t) - V_B(t)}\right) \tag{47}$$

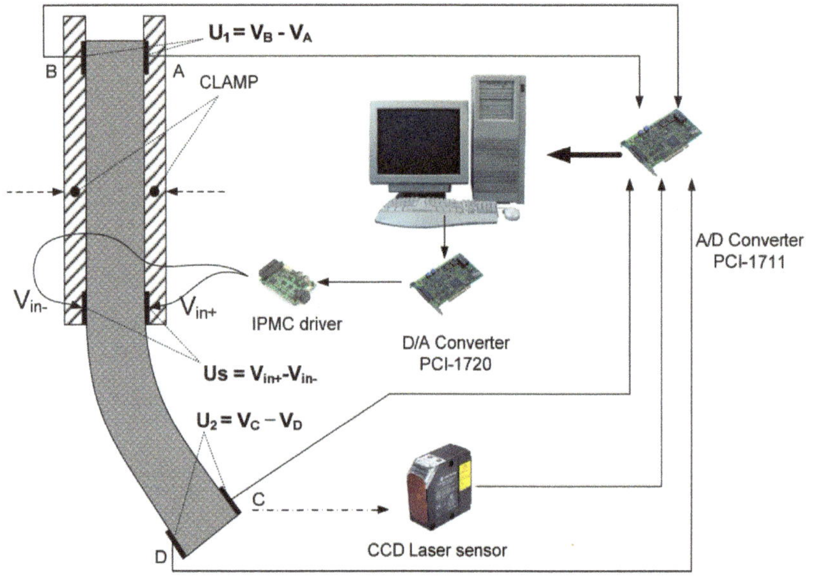

a) Configuration of IPMC in self – sensing application

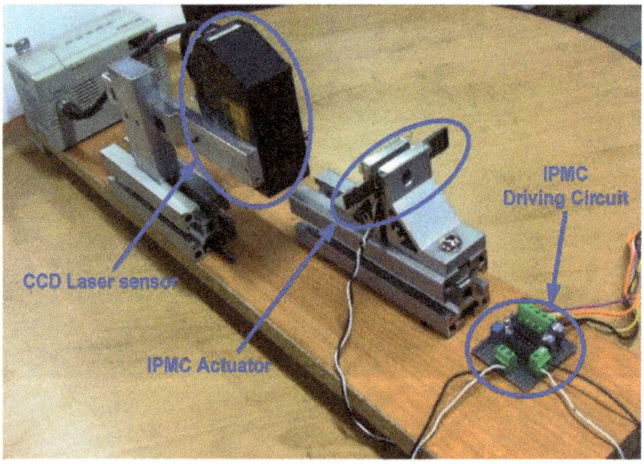

b) Experimental self – sensing IPMC actuator

Figure 13. Configuration of IPMC in self – sensing application

By applying equations (46) and (47), the self sensing has been carried out corresponding to four cases: 2V, 4V, 4.5V and 5V driving square input voltages signal, 0.05 Hz in frequency. Consequently, the experimental results are then shown in figure 14. Experimental results prove that the developed self – sensing technique has the ability to measure the bending behavior of the IPMC actuator. In order to increase the accuracy of the self sensing performance, one can increase the number of captured voltage signals. However, the accuracy requirements depend on the practical application.

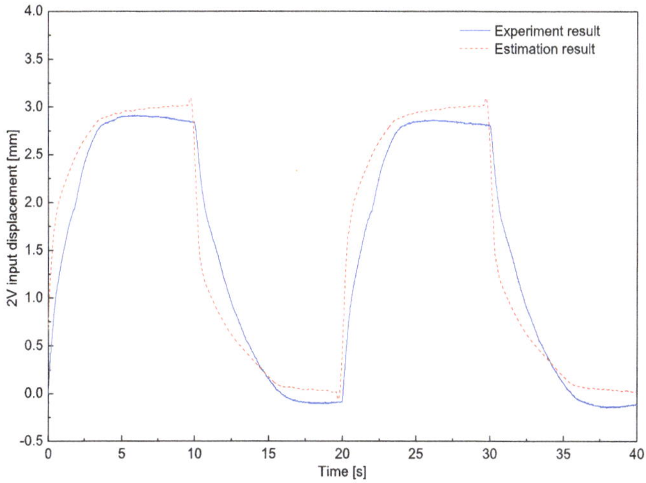

a) Estimation result – 2V square input signal

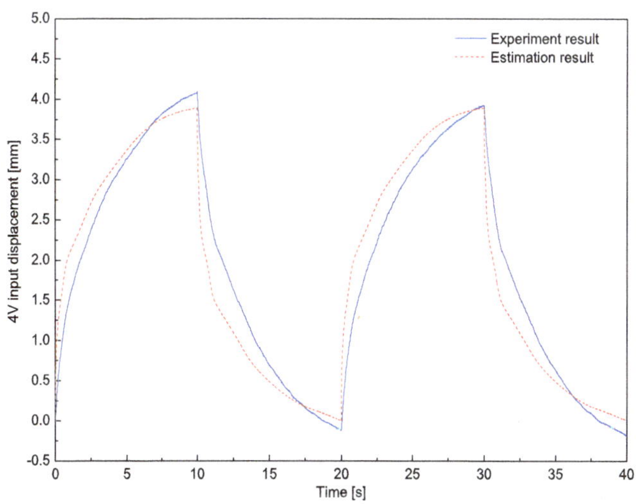

b) Estimation result – 4V square input signal

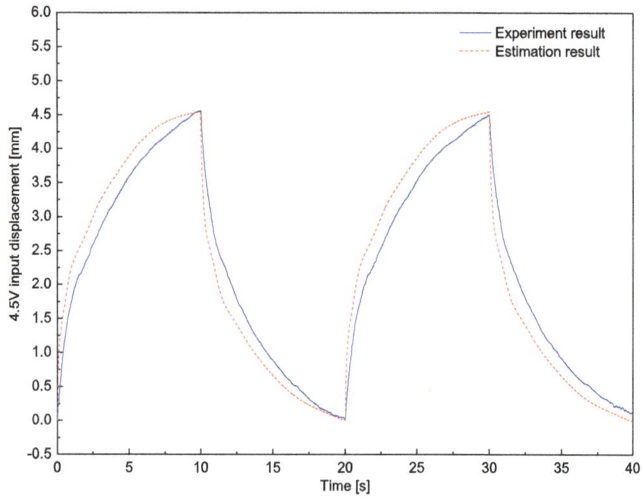

c) Estimation result – 4.5V square input signal

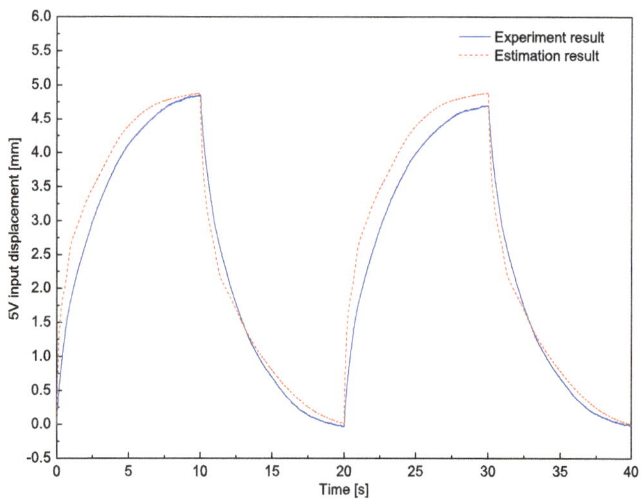

d) Estimation result – 5V square input signal

Figure 14. Self - sensing behavior of an IPMC actuator

6. Conclusion

This chapter gives a physical view in modeling the IPMC actuator and its self – sensing ability. The modeling technique provides a powerful tool to investigate the characteristic of

IPMC actuator via incorporation of physical induced stress model and visual ANSYS environment. The self – sensing model gives a powerful methodology to measure the displacement of IPMC actuator. This is a great advantage for biomedical, biomimetic, micro or mobile robotic applications where external sensing systems are usually unavailable.

Author details

Ahn Kyoung Kwan*
School of Mechanical and Automotive Engineering, University of Ulsan, Korea

Doan Ngoc Chi Nam
Graduate School of Mechanical and Automotive Engineering, University of Ulsan, Korea

Acknowledgement

This work was supported by the Korea Science and Engineering Foundation (KOSEF) grant funded by the Korea government (MEST) (No. 2009-0080924)

7. References

[1] Ba-Cohen Y. (2001) Electro-Active polymer (EAP) actuators as artificial muscles - Reality, potential, and challenges, Spie press .
[2] Abadi S., Dehghani A., Nelson E.A. (2008) A soft sensing method for monitoring ambulatory activities of patients with venous ulceration, Proc. of the 4th IET Int. Conf. on Adv. in Medical, Sig. and Inf. Processing, MEDSIP. pp.1 – 4.
[3] Paola B., Fortuna L., Giannone P., Graziani S., Strazzeri S. (2008) IPMCs as Vibration Sensors, Proc. of IEEE Int. Instrumentation and Measurement Tech. Conf. pp.2065 – 2069.
[4] Bonomo C., Fortuna L., Giannone P., Graziani S. (2004) A sensor-actuator integrated system based on IPMCs [ionic polymer metal composites], Sensors, Proc. of IEEE conf. pp. 489 – 492.
[5] Fang B.K., Lin C.-C.K., Ju M.-S. (2010) Development of Sensing/Actuating Ionic Polymer Metal-Composite (IPMC) for Active Guide-Wire System, Sensors & Actuators: A. Physical, v 158, pp. 1-9.
[6] Chen Z., Kwon K.Y., Xiaobo T. (2008) Integrated IPMC/PVDF sensory actuator and its validation in feedback control, Sen. and Act. A: Physical 144(2), pp. 231-241.
[7] Yamakita M., Kamamichi N., Kozuki T., Asaka K., Zhi-Wei L. (2005) A snake-like swimming robot using IPMC actuator and verification of doping effect, Proc. of IEEE/RSJ Int. Conf. pp.2035 – 2040
[8] Tung N.T., Khanh N.V., Youngtai Y., Goo N.S. (2006) A Novel Polymeric Micropump based on a Multilayered Ionic Polymer-Metal Composite, Proc. of IECON - 32nd Annual Conference, pp.4888 – 4893.

* Corresponding Author

[9] Yamakita M., Kamamichi N., Kozuki T., Asaka K., Zhi-Wei L. (2005) Control of Biped Walking Robot with IPMC Linear Actuator, Proc. of IEEE/ASME International Conference, Sep. pp.48 – 53.

[10] Guo S., Shi L., Asaka K. (2008) IPMC actuator-sensor based a biomimetic underwater microrobot with 8 Legs, Proc. of ICAL. IEEE International Conference, Sep. 2008, pp.2495 – 2500.

[11] Punning A., Kruusmaa M. and Aabloo A. (2009) Surface resistance experiments with IPMC sensors and actuators Sen. and Act. A: Physical 133(1). Pp. 200-209.

[12] Punning A., Kruusmaa M. and Aabloo A. (2007) A self-sensing ion conducting polymer metal composite (IPMC) actuator Sen. and Act. A: Physical 136(2). Pp. 656-664.

[13] Nemat-Nasser S. and Li J. (2000) Electromechanical response of ionic polymer-metal composites, Journal of Applied Physics, 87 (7), pp. 3321–3331.

[14] Farinholt K. M. (2005) Modeling and characterization of ionic polymer transducers for sensing and actuation, Ph.D. dissertation, Virginia Polytechnic Institute and State University.

[15] Chen Z. and Tan X. (2008) A control-oriented and physics-based model for ionic polymer –metal composite actuators, IEEE/ASME Trans. Mechatronics. 13, pp. 519–529.

[16] Chen Z., Hedgepeth D. R. and Tan X. (2009) A nonlinear, control-oriented model for ionic polymer–metal composite actuators, Smart Mater. Struct. 18055008 (9pp).

[17] Nemat-Nasser S. (2002) Micromechanics of Actuation of Ionic Polymer-metal Composites, Journal of Applied Physics, 92 (5) (21pp).

[18] http://www.ansys.com/Support/Documentation

Chapter 9

Large Scale Capacitive Skin for Robots

P. Maiolino, A. Ascia, M. Maggiali, L. Natale, G. Cannata and G. Metta

Additional information is available at the end of the chapter

1. Introduction

In order to allow robots to share our space and chores, tactile sensing is crucial. Indeed it allows safe interaction of robots with people and objects, because it provides the most direct feedback to control contact forces both in voluntary and involuntary interactions. Furthermore, it allows improving performance in tasks that require controlled physical interactions in uncontrolled environments where the location and the characteristics of contact cannot be predicted or modeled in advance and more complex forms of interactions are required. Therefore, a tactile sensor system capable of measuring contact forces over large areas is needed. Tactile sensing in robotics has been widely investigated in the past 30 years and many examples of engineering solutions to tactile sensing have been presented in the literature [1]. Research in this field has focused largely on transduction principles and transduction technologies [2]; however, various technical issues have limited the transition from a single tactile element (or a small matrix prototype) to a large scale integrated solution: it is easy to understand that a sensitive robot skin cannot be achieved by simply aggregating a large number of single sensors. In fact, the concept of robot skin entails a number of system level problems that simply do not appear when focusing on small tactile sensors or small area arrays:

- Modularity and Conformability: the system should be modular and it should be simple to tailor it to the shape of the surface of the robot.
- Infrastructure and Networking: embedded electronics and distributed computation are necessary to facilitate the integration in the robot, otherwise a large number of wires would impede mobility and dexterity of the robot.
- Coverage: The sensors must cover the largest possible area of the robot without increasing the number of wires or the size of the ancillary electronics.
- Compliance: The sensors must be covered by a soft layer that allows safe interaction with humans and the environment and that forms a protective layer for the sensors themselves.

- Dynamic range and sensitivity: The system should be able to detect a wide range of contact pressures.
- Total Weight and installation space: In order to cover the whole robot body surface the weight of the sensors must be small as well as the required space for installation.
- Costs: The system must be as cheap as possible and off-the-shelf components should possibly be used to decrease the manufacturing cost.
- Manufacturing and deployment: the ease and speed of production have to be taken into account. The deployment procedures must be reproducible on robots with different characteristics.

Several tactile systems have been integrated into robots and described in the literature. Some of them are modular and include hierarchical data processing. Yet, the modules are usually big and cannot be installed on small robot parts. In many cases the spatial resolution is low; in others the sensory modules connectivity is cumbersome. This is clearly in contrast with the requirements above. Small sensor modules are typically necessary in order to allow the communication between them; size constraints are then very critical as not enough space typically can be found on the sensor modules to the local integration of microcontrollers or other networking electronics. One of the first examples of full-body skin for robots was proposed by Inaba et al. [3] that describes a tactile sensor system composed of a layered structure of electrically conductive fabric implementing a matrix of pressure sensitive switches. However, the complexity of the manufacturing process and the limited sensing capabilities represent the most significant limits of this approach. Force detectable surface covers have been introduced by Iwata et al. [4]. Information originating from both resistive and force sensors to correlate pressure information and exerted force is exploited. Although the presented system can be used to cover large surfaces, the actual design is strictly dependent on the actual robot surface. However, a principled discussion about design issues is not carried out. Ohmura et al. [5] proposed a conformable and truly scalable robot skin system formed by self-contained modules supported by a flexible printed circuit boards (PCBs) that can be interconnected. Each module contains 32 tactile elements consisting of a photo-reflector covered by urethane foam. In order to adjust the distance between each tactile sensor element, a band-like bendable substrate that can be easily folded (or even cut) is adopted. The system integrates a microcontroller based architecture for data acquisition and networking, however, the dimension of the system are too big for small robot parts and the spatial resolution featured is quite low [6]. The tactile system that has been developed for the robot RI-MAN also uses flexible PCBs with a tree-like shape to conform to curved surfaces [7]. The tactile elements are commercially available piezoresistive semiconductor pressure sensors, and the measurements have less hysteresis than the one shown by Ohmura et al. [5]; to reduce the number of wires, the sensor modules include multiplexers, but this approach requires to connect each module individually to a controller board. The robot ARMAR-III [8] uses skin patches based on piezoresistive sensor matrices with embedded multiplexers. The patches have a flat or a cylindrical shape and are specifically designed for the different parts of the robot; smaller patches are used for the fingers. In Kotaro [9], tactile sensing is achieved by using flexible bandages formed by two flexible PCBs with an intermediate layer of pressure sensitive conductive rubber. Each bandage has 64 taxels, but no integrated data acquisition electronics is mentioned. Piezoelectric transducers are used for the humanoid robots Robovie-IIS [10] and CB2 [11]. The transducers were placed individually on the robots and the sensitive skin has a low spatial resolution.

In Yoshikai et al. [12] the construction method for soft stretchable enclosing type tactile sensing exterior for humanoid robots is presented. A cardigan knit sensor which encloses an upper body of a small humanoid has been developed using the proposed method. The overall sensor is constituted by layers of conductive fabrics that are knitted for improved stretchability. The presented cardigan knit has, however, only 75 electrodes.

Shimojo et al. [13] have developed a mesh of tactile sensors that satisfy the requirements of a high speed and reduced number of wires and that is able to cover surfaces of arbitrary curvatures. The tactile sensors mesh is arranged as a net, where only nearby taxels are connected through wires. Since the shape of a patch must be specifically designed, modularity issues are not properly addressed, but scalability is possible thanks to the smart wiring infrastructure: only 4 wires actually exit from the patch.

Mittendorfer et al. [14] have presented a tactile sensor system made by small hexagonal PCB modules equipped with multiple discrete off-the-shelf sensors for temperature, acceleration and proximity. Each module contains a local controller that pre-processes the sensory signals and actively routes data through a network of modules towards the closest PC connection. The sensory system is embedded into a rapid prototyped elastomer skin material and redundantly connected to neighboring modules by just four ports. The functionality of some modules on a KUKA robot arm has been demonstrated, however a complete integration has not been shown.

The ROBOSKIN tactile system (Patent No. I0128764) is our proposal for a skin that is able to cover large area of a robot body. It incorporates a distributed pressure sensor based on capacitive technology. The transducer consists of a soft dielectric sandwiched by electrodes. When pressure is applied to the sensor, the distance between the electrodes changes, and the capacitance changes accordingly (capacitance is a function of distance). The ROBOSKIN system is made up of a number of tactile elements (taxels) geometrically organized in interconnected modules of triangular shape. Above the flexible PCB, there is a layer of silicone foam (Soma Foama 15 from Smooth-On) that covers the 12 taxels and acts as a deformable dielectric. On top of the silicone foam is placed a deformable conductive layer made of electrically conductive lycra-like fabric. This layer is connected to ground and enables the sensor to respond to contacts with objects of any material.

As we mentioned earlier, manufacturing and deployment processes constitute an important aspect of the robot skin design since the skin has in fact to be integrated on robots with different characteristics. Flexibility in this sense is a good property for a robot skin system. In addition, it is desirable that the skin production process is reproducible so to guarantee resilience against impact and forces and the general wear and tear of the material. In particular, the ROBOSKIN tactile system has been integrated on three different types of robots: iCub [15], KASPAR [16] and NAO [17]. The three robots have different sizes and shapes, and the tactile feedback has been used for different purposes. Nevertheless, the methods that have been used to implement the skin were nearly the same, which demonstrate the portability of the sensor system technology. This chapter is organized as follows: firstly, the ROBOSKIN tactile system is presented in details in section 2; section 3 presents the methods and procedures for ROBOSKIN manufacturing and integration; in particular, section 3.1 is related to the iCub platform; section 3.1.1 presents a new implementation of iCub fingertip with experimental data related to their behavior characterization; section 3.2 is related to KASPAR and section 3.3 presents the ROBOSKIN integration on robot NAO; finally section 4 is dedicated to conclusion.

2. ROBOSKIN technology

As we told before, ROBOSKIN tactile system, incorporates a distributed pressure sensor based on capacitive technology. The transducer consists of a soft dielectric sandwiched by electrodes. When pressure is applied to the sensor, the distance between the electrodes above and below the dielectric changes, and the capacitance changes accordingly (capacitance is a function of distance).

The basis of the sensor is a flexible printed circuit board (PCB). It includes the electronics to obtain 12 measurements of capacitance and send them over a serial bus. In particular, each PCB includes 12 round pads, one for each taxel, and a capacitance to digital converter (CDC) (AD7147 from Analog Devices).

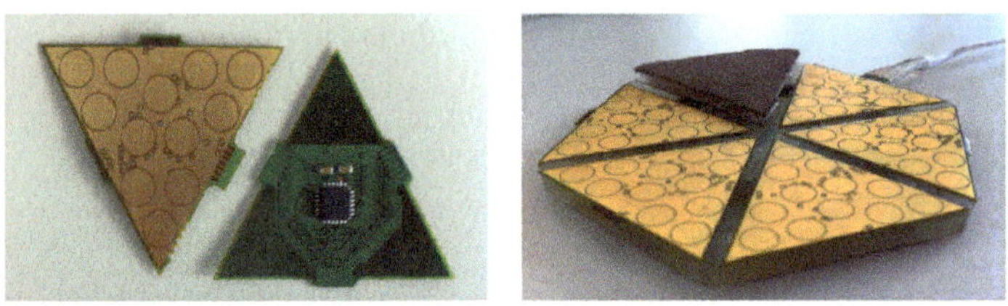

(a) (b)

Figure 1. The triangular module: (a) in the front view it is possible to see the 12 round taxels, in the back view the CDC chip from Analog Devices is shown; (b) An hexagonal patch glued on a cover with the foam elastomer and electrically conductive lycra layers.

The chip can measure the capacitance of all taxels with 16 bit resolution, but we use only 8 bit measurements, as any higher resolution is covered by noise. As a result, one measurement unit corresponds to 2.88 fF. The CDC has an I^2C serial interface and each chip can be assigned with a 2 bit address; therefore up to 4 chips can communicate over the same serial bus. The shape of the PCB is in most cases a triangle (only for the fingertips of the robot iCub we used a unique solution, which we will discuss in section 3.1.1), see figure 1(a). The triangular shape was chosen in analogy to polygonal modeling in 3D computer graphics, which uses triangles to describe the shapes of objects. The triangles can conform up to a certain degree to generic smooth curved surfaces (see figure 2(a) and figure 2(b)).

The triangular PCBs also include the electronics to communicate between themselves: three communications ports placed along the sides of the triangle (one for input and two for output) relay the signals from one triangle to the adjacent one. Up to 16 triangles can be connected in this way (4 serial buses with 4 different addresses each) and only one of them needs to be connected to a microcontroller board (MTB, see figure 3). This is a critical advantage since it reduces the amount of wires and electrical connections that are required. The MTB can also

(a) (b)

Figure 2. (a) A ROBOSKIN patch; (b) ROBOSKIN patch conformed to a curved body parts.

be used to program each CDC to either measure its 12 taxels independently at 50 Hz or an average of them at about 500 Hz.

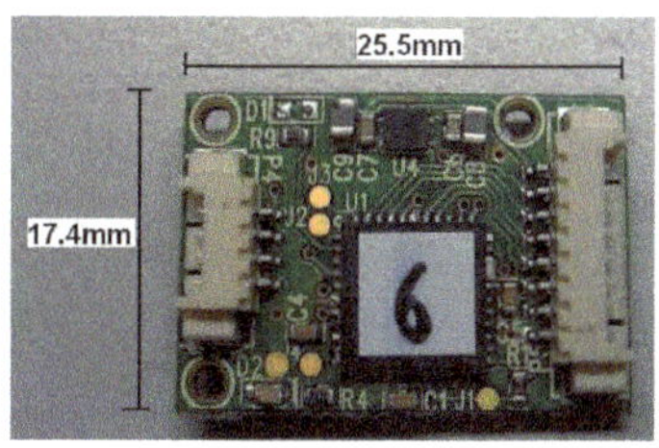

Figure 3. The microcontroller board (MTB). We designed a small microcontroller board, which collects the measurements from up to 16 CDC chips and sends the measurements over a CAN bus.

Above the flexible PCB there is a layer of silicone foam (Soma Foama 15 from Smooth-On). It is 2 mm thick for the hands of the robot iCub, and 3 mm in all other cases. It covers the 12 pads and acts as a deformable dielectric for the capacitive pressure sensor. The foam also makes the skin compliant. On top of the silicone foam there is a second conductive layer made by electrically conductive Lycra. This layer is connected to ground and enables the sensor to respond to objects irrespective of their electrical properties. It serves as the common electrode above the silicone foam for all the taxels, see figure 1(b). When pressure is applied to the sensor, this layer gets closer to the round pads on the PCB, and the sensor measures the distance. This layer also reduces the electronic noise coming from the environment, in particular the stray capacity, which can be a problem for capacitive pressure sensor systems [2]. For the KASPAR and NAO, we didn't use the conductive layer on top of the silicone foam, as in these cases the robot is intended to interact only with humans, in which case the human constitutes the ground plane (like in many consumer products, which are responsive to humans, but are not responsive to insulators, for example).

3. Integration process

In order to integrate the ROBOSKIN sensor on a robot, this general procedure must be followed:

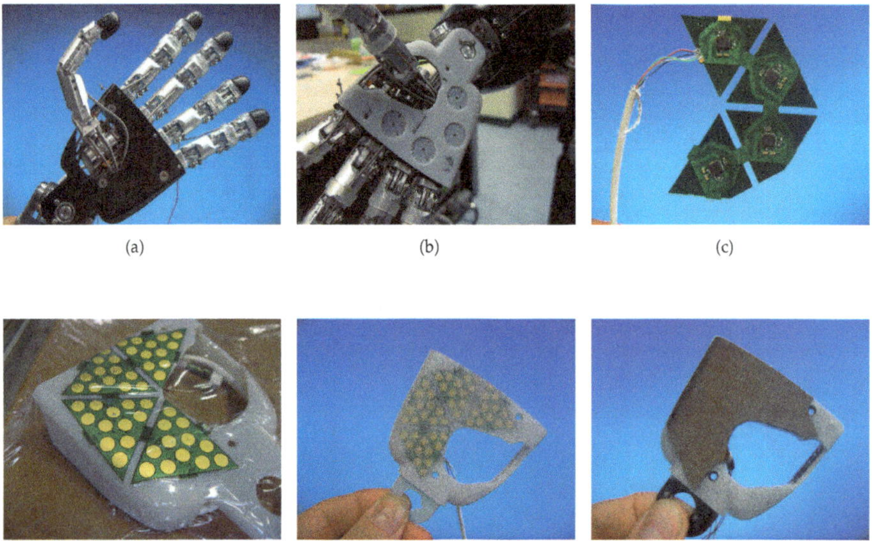

(a) (b) (c) (d) (e) (f)

Figure 4. The production steps for the palm of iCub. For a description of each step, please refer to the text.

- Identification of the part to be covered, see Fig 4(a). If no CAD model is available, obtain the shape with a 3D laser scanner (as for example for the hands of KASPAR).
- Manufacturing of the part (or of a cover) with a 3D printer (Eden 3D printer from *Objet*). The resulting parts look for example like in Fig. 4(b). Round holes provide space for the CDC chip and the other electronic components which are soldered on the PCB.
- Identification and wiring of the mesh of flexible PCBs that is needed to cover the part, see Fig. 4(c).
- Bonding of the PCBs on the part with bi-component glue and the help of a vacuum system, see Fig. 4(d).
- Covering the PCBs with silicone foam, see Fig. 4(e). To this aim we employ a specific purpose-built mold for each part.
- Covering of the silicone foam with a conductive lycra as ground plane, as shown in Fig. 4(f).

3.1. Implementation on iCub robot

The integration of the ROBOSKIN tactile sensor on iCub, has involved forearms, upper arms, torso and hands (palms and fingertips). The current implementation allows obtaining a number of distinguished taxels equal to 2400. In figure 5 the final result is shown.

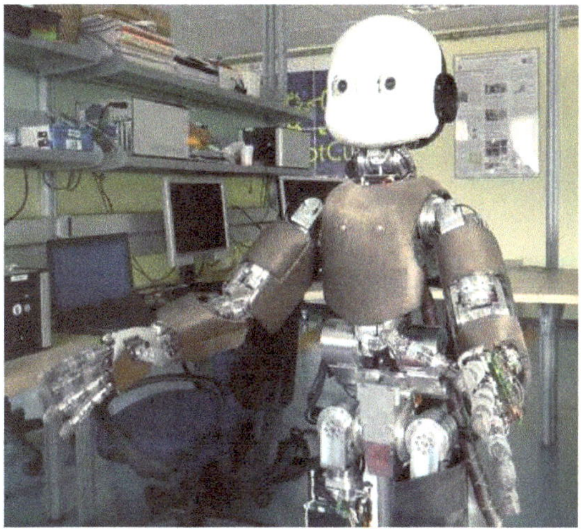

Figure 5. The iCub covered with artificial skin

3.1.1. Implementation on the iCub hands

As it is possible to see in figure 4(a), the palm of iCub is made from carbon fiber; since this is a structural part, we decided not to modify it, but instead, we added another cover above the carbon fiber part as a basis for the sensor: it has a thickness of 1.2 *mm* and provides space for the CDC chip and the other electronic components which are soldered on the PCB. The implementation steps are reported in Fig. 4. While for all the iCub parts the standard ROBOSKIN solution for the PCB has been used, for the iCub fingertips it has been necessary to design a specific solution (see figures 7(a), 7(b), 7(c)), since they have small size and round shape (each fingertip is 14.5 mm long and 13 mm wide and high and has a round shape that resembles a human fingertip).

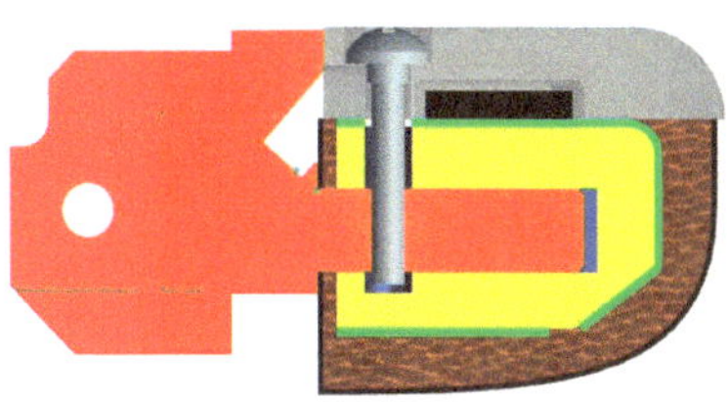

Figure 6. Crosssection of the fingertip.

The structure of the fingertip is illustrated schematically in Fig. 6: the inner support of the fingertip is shown in yellow, and the flexible PCB, that is wrapped around, is depicted in green. To mechanically attach the fingertip to the hand, the last phalanx of each digit (shown in red) has a stick that fits inside a hole in the inner support. A screw is used to secure the fingertip and, in addition, the screw fixes a fingernail on top of the fingertip that covers the

PCB. The dielectric, made of silicone rubber foam, is depicted in brown; around the foam there is the conductive lycra layer shown in black as well as the AD7147 chip.

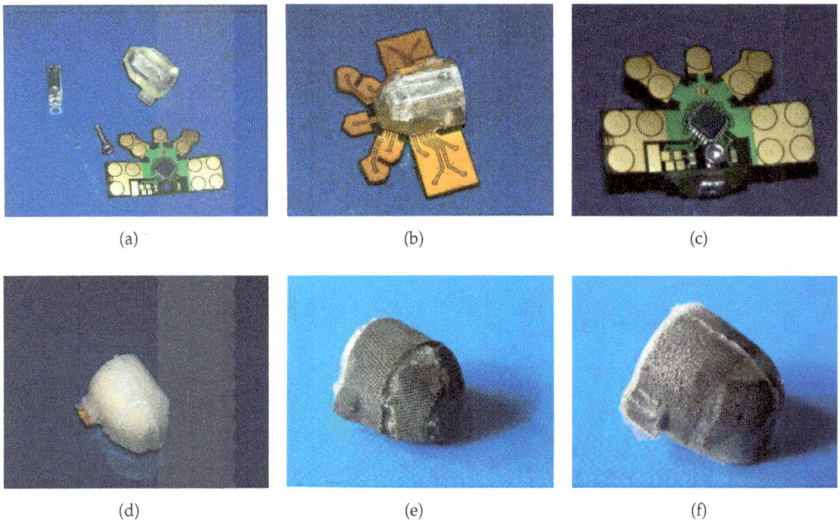

Figure 7. The production steps for the fingertip iCub

The fingertip production protocol involves the following steps:

- The flexible PCB is wrapped around an inner support that was printed with a 3D printer. (see figures. 7(a), 7(b), 7(c)). As we are using an I^2C serial bus, only 4 wires have to be connected to the PCB (Vcc, ground, serial data line and serial clock). They travel along the side of the fingers to small boards at the back of the hand. These boards relay the data from all five fingertips (and the four triangular modules in the palm) to one microcontroller board, which is located in the forearm of iCub.
- A first layer, made with a foam elastomer (Smooth-on Soma Foama 15), is deposited over the PCB; mechanical deformation of this soft dielectric material leads to capacitance variations; therefore, it is possible to detect pressures applied on the fingertip, see figure 7(d).
- The second layer, that is made with conductive lycra, is glued over the silicone foam substrate, allowing the development of a single ground plane above all electrodes placed on the PCB thus enabling the detection of each type of object (conductive and non conductive) within noise reduction. This layer is connected to the digital ground of the CDC by one flat pad on the PCB.
- Finally the third protective layer, made again with Smooth-on Soma Foama 15, is deposited; it is used such as a protection, thus intrinsically increases the lifetime of the fingertip sensor.

Figure 8. Fingertip sensor placed on the support used for measurements together with schematic view of the line where measurements have been taken

With respect to the first implementation of the fingertips [18], where it was used a self-made mixture of silicone (CAF4 from *Rhodia-Silicones*) and carbon-black particles (Vulcan XC72 from *Cabot*) as a ground plane, and, as protective layer, silicone glue (Sil-Poxy from *Smooth-On*) sprayed above it, the new implementation made with conductive lycra, has been chosen because it increases the durability of the fingertip to usury due to friction forces appearing during grasp tasks. In order to characterized the behavior of the new fingertip several experiments have been performed.

The characterization setup consists of a cartesian robot (TT-C3-2020 from IAI) which moves one non-conductive probe against the fingertip. The non-conductive probe is fixed at the top of an off-center load cell (AS1 form Laumas) which measures independently the force applied to the fingertip by the probe during the test. A microcontroller records the CDC output and the load cell circuit output. Therefore the Data are stored in a computer by a dedicated graphic user interface made in Matlab.

The characterization protocol was the following:

- One non-conductive probe of 4 mm diameter has been used. By the use of the Cartesian robot, the non-conductive probe applied different pressures to 13 positions, on the fingertip, separated by 0.5 mm along a 6 mm long straight line of two capacitors, indicated with C1 and C2(see figure 8).
- In each position, different pressures have been applied to the fingertip by vertical displacement of the non-conductive probe using the cartesian robot.
- Each pressure has been applied for 2 seconds, with intervals of 2 seconds. In order to investigate sensor repeatability, each pressure has been repeated fifty times. It must be observed that, along the straight line the probe was perpendicular to the fingertip.

Only the steady-state responses at external pressures of the sensor have been taken into account during post elaboration. Mean values of the fifty steady-state responses of C1 and C2, at different positions of the probe, are reported from figure 9(a) to figure 9(e). It is possible to observe a non linear trend of the capacitor variation as a function of the applied pressure. Therefore, the least squares method has been used to fit a polynomial model on each steady state response at each position. Figure 10(a) shows an example of the choice of the order of the polynomial function for CDC-C1 in the case of the position 0 mm, while, in figure 10(b), it is shown the norm of residual with respect to the polynomial order, on the basis of which,

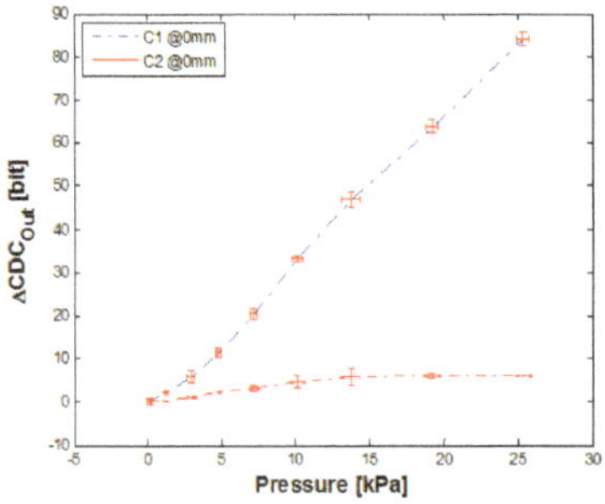

(a)

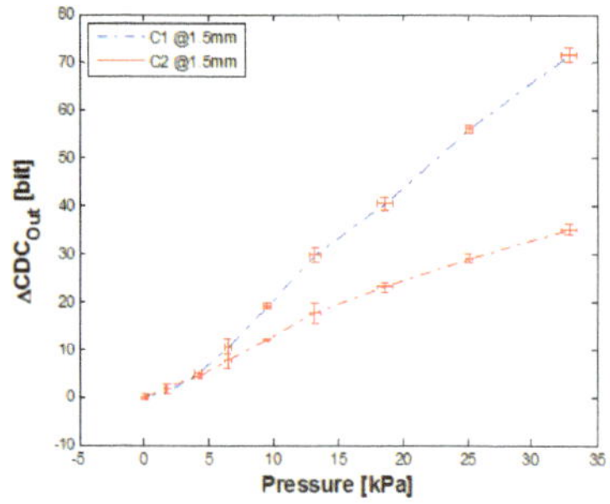

(b)

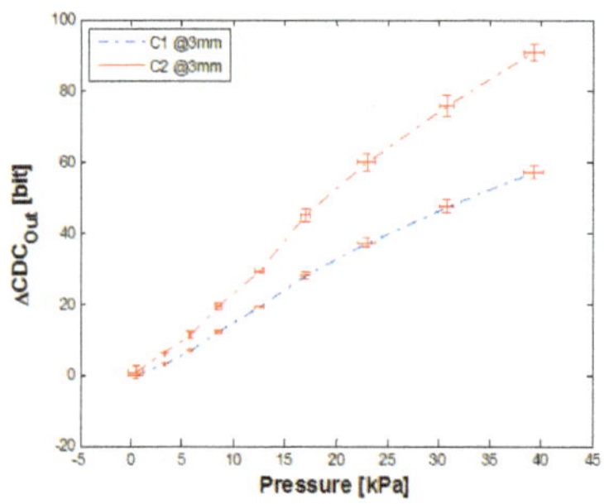

(c)

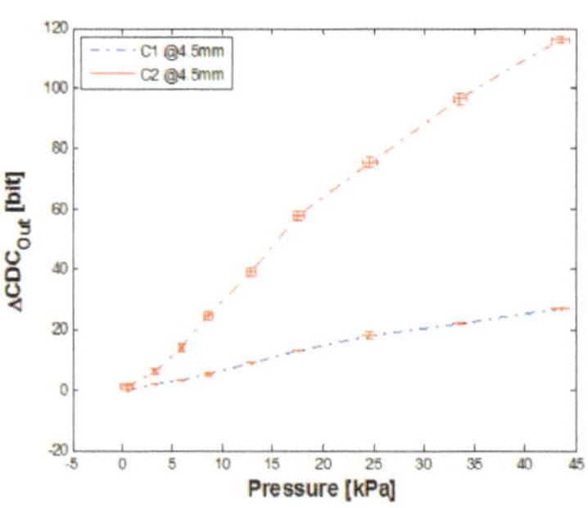

(d)

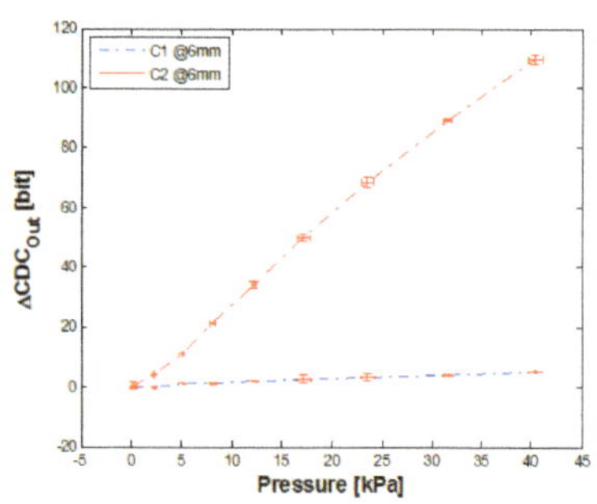

(e)

Figure 9. steady-state response at 0 mm (a), 1.5 mm (b), 3 mm (c), 4.5 (d) mm and 6 mm (e) position point of the straight line of measurement

a 4th polynomial model order has been chosen. Polynomial functions thus obtained, were used to calculate the variation of capacitors C1 and C2 due to a fixed pressure as a function of position along the measurement line. Results are presented from figure 11(a) and figure 11(e), which report the variation of capacitors C1 and C2 as a function of the probe position for five different pressure values from 5kPa to 25 kPa.

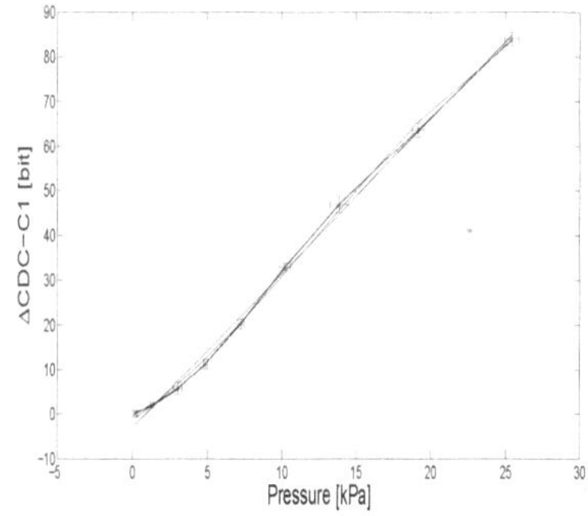

(a)

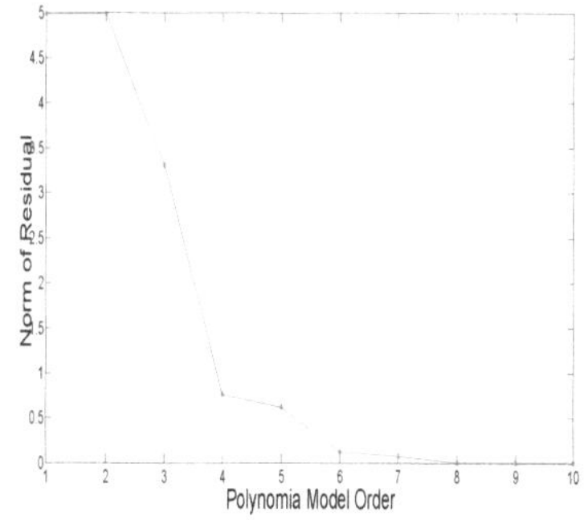

(b)

Figure 10. a) An example of the choice of the order of the polynomial function for CDC-C1 in the case of the position 0 mm; b) Norm of the residual with respect to polynomial order

3.1.2. Implementation on the iCub arms

For the iCub arms, the skin was integrate directly into the covers, so new covers were designed. The production steps are the same as for the palm and the other iCub parts and are illustrated in fig. 12, for the forearm, and fig. 13, for the upper arm. On the two iCub arms we have 1464 contact points: 6 patches (61 triangular module) and 7 MTBs for each arm.

3.1.3. Implementation on the Torso of iCub

For covering the iCub Torso, five patches has been used, obtaining 528 contact points. In figure 15 it is possible to see the front of the torso covered with the glued patches and the silicone foam substrate (figure15(a)), the torso with the conductive lycra substrate on top (figure 15(b)) and finally the back of the torso cover with the 4 MTBs that are needed for the five patches (figure 15(c)).

3.2. Implementation of artificial skin on KASPAR

Despite the other two robot, that have an iconic appearance, KASPAR has a more human-like appearance, in fact in order to have a "natural" shape a child-sized mannequin was used as a basis and the legs, torso and the hands were kept. In order to invite children to touch the hands (which is more like touching a doll), the maintenance of human-like appearance has been an important parameter that has been kept under consideration in the covering process. For this purpose the hands have been covered with coloured foam silicone rubber (it has been

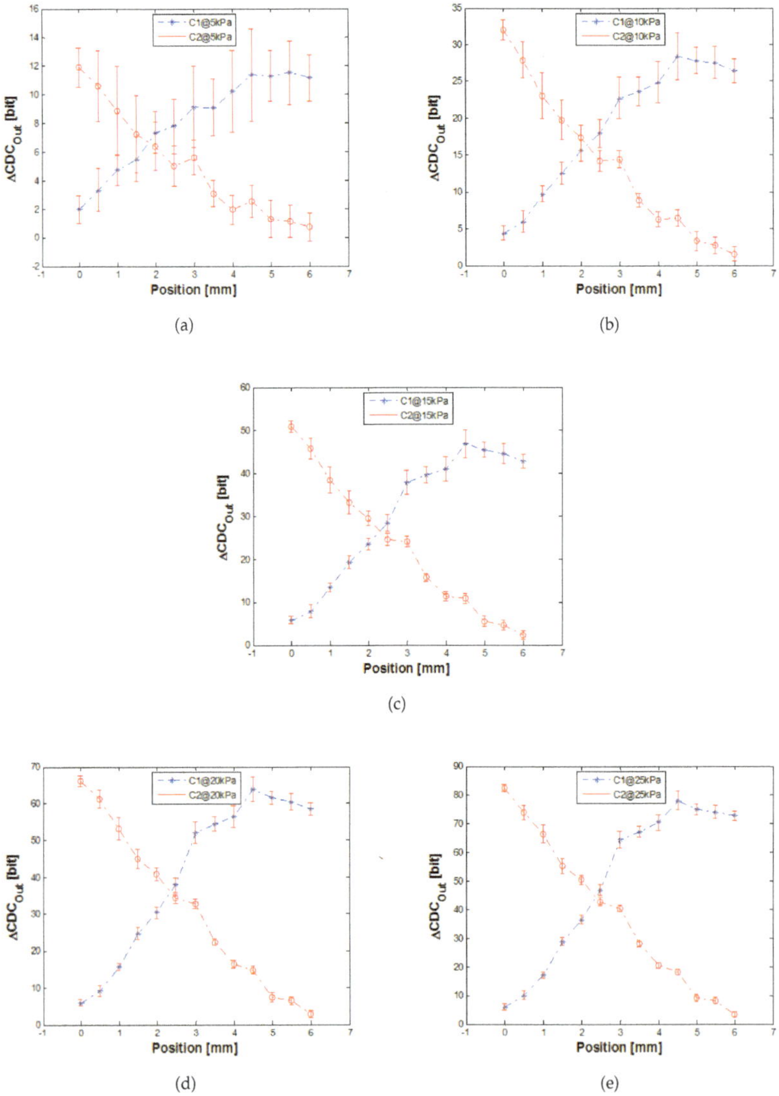

Figure 11. Calculated variations of C1 and C2 due to a fixed pressure applied by the probe along different positions of the straight line of measurement

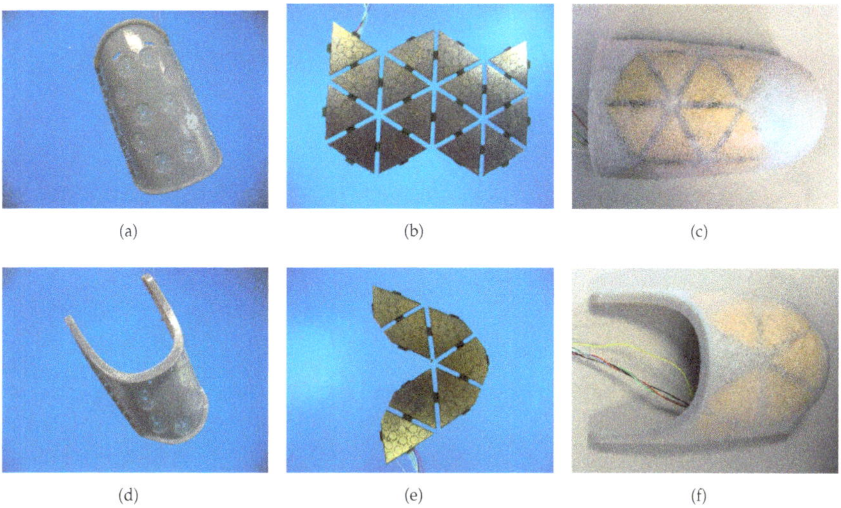

Figure 12. The process steps for iCub forearm: two covers are needed which are mounted together on the robot forearm.

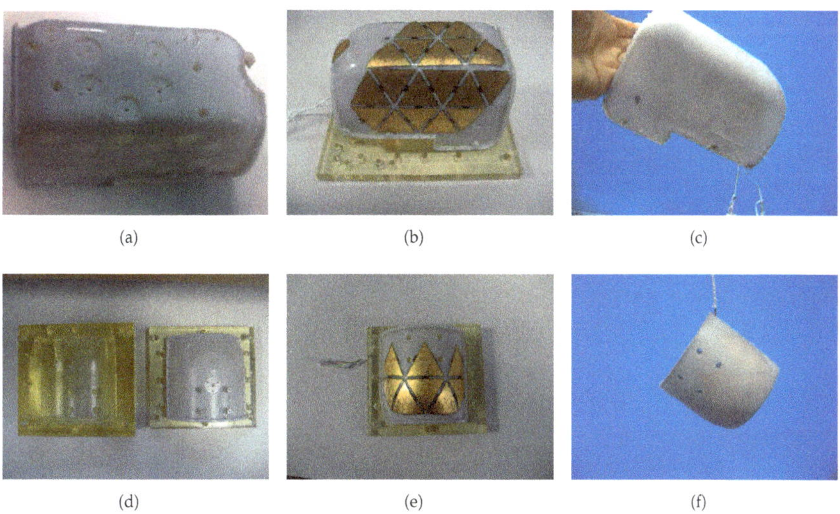

Figure 13. The process steps for iCub upper arm: two covers are needed which are mounted together on the robot upper arm.

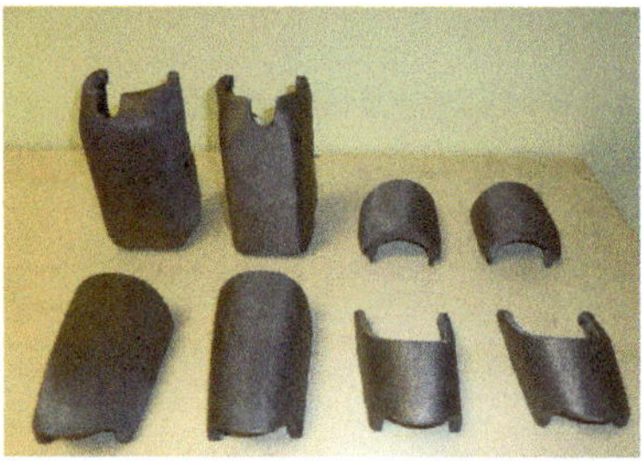

Figure 14. The covers of iCub arms covered with conductive lycra

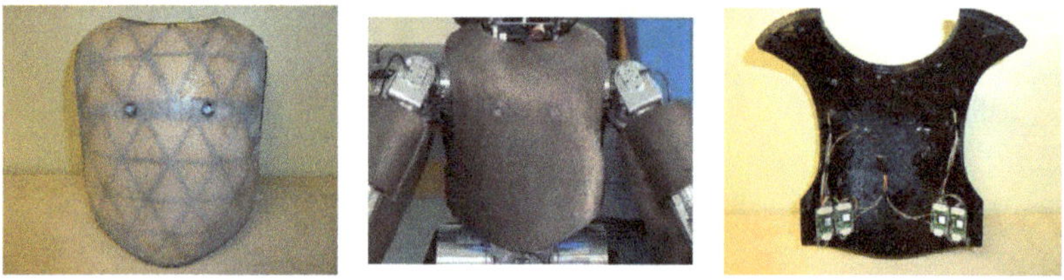

(a) (b) (c)

Figure 15. The process steps for iCub Torso

obtained adding color pigments to the rubber mixture). Furthermore it has been decided to not cover the silicone rubber substrate with ground plane (the KASPAR parts come in contact only with human being that constitute the ground plane during tactile contact). The hands have been scanned with a laser scanner and completely rebuild by a 3D printer in order to have the housing for the electronics (see figures 16(a) and 16(b)); The procedure used was the same explained before for the iCub parts. In figure 16(d) the final result for KASPAR hands is shown. For the other parts (cheeks, torso, upperarms and feet) we didn't take in consideration the human-like appearance because this parts, in the final setup of KASPAR, are covered by clothes. For the upper arms a cover with the housing for the electronics was build by a 3D printer and the patch was covered by a neoprene substrate. For the cheeks, torso and the feet it has not been possible to rebuild the part with the 3D printer (more expensive), so Artificial Skin patches have been applied on the puppet surfaces by glue. In total on KASPAR there are 12 MTB boards and 68 triangles that corresponds to 1128 contact point.

3.3. Implementation on NAO

For NAO the goal was to sensorize the forearms and the upper arms. Therefore, we designed 6 covers (2 for each forearm and 1 for each upper arm) to mount over the robot structural covers as for the iCub palm.

Fig. 18 shows the steps to cover NAO's forearm with skin. In total NAO has 18 triangles that correspond to 216 contact points for each lower arm and 9 triangles that correspond to 108 contact points for each upper arm.

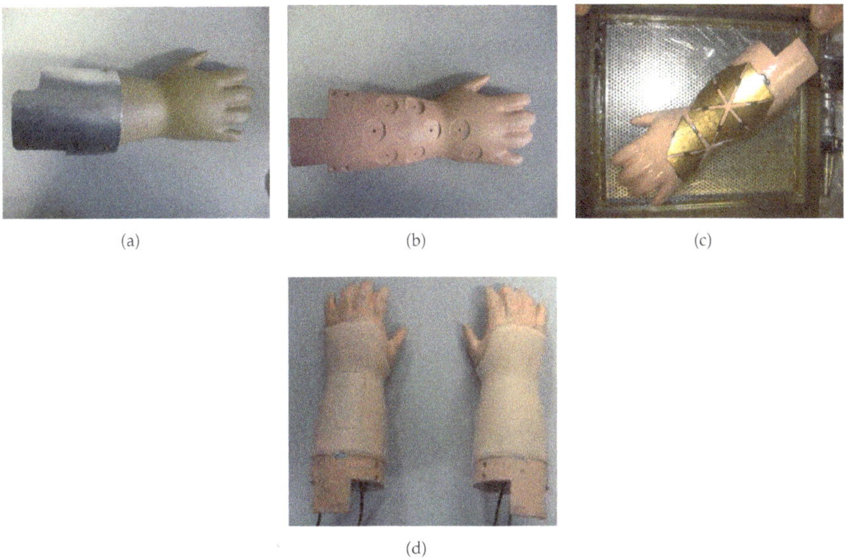

Figure 16. The process steps for KASPAR hands

Figure 17. NAO with his forearms covered with the artificial skin.

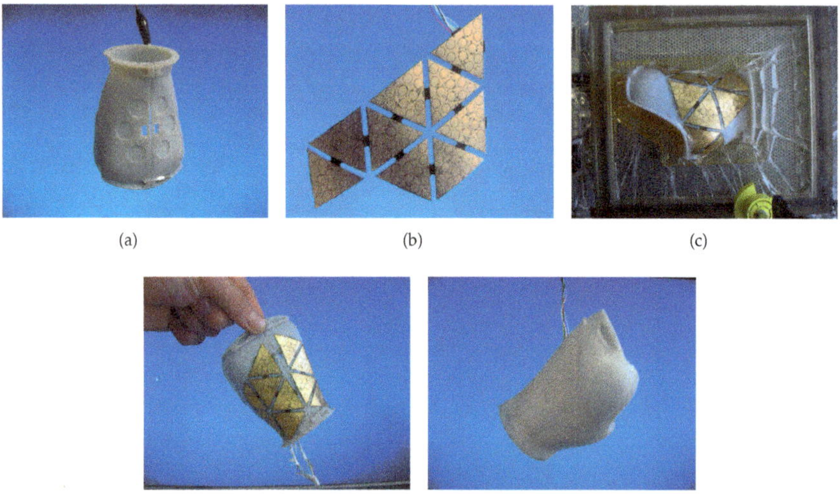

(a) (b) (c)

(d) (e)

Figure 18. The steps for implementing the skin on NAO's hands. (a) The two parts of the cover for NAO's forearm. (b) The triangles for one half of the forearm. (c) The PCBs are glued to the cover. (d) The cover with the PCBs. (e) The silicone foam has been molded onto the forearm. (e) The final result.

4. Conclusion

In the chapter the ROBOSKIN tactile system has been presented; in particular the procedures and methods for its integration in three different robotic platforms have been shown, proving its portability and its capability to cover large area of the robots body. Indeed the three robots have very different characteristics, but the integration steps were the same. A new implementation for the iCub fingertips has been shown and behavior characterization data have been presented. The new implementation allows to have a better robustness increasing the durability of the fingertip during grasp operations.

Acknowledgements

The research leading to these results has received funding from the European Community Seventh Framework Programme(FP7/2007-2013) under Grant Agreement no. 231500 (Project ROBOSKIN).

Author details

P.Maiolino and G.Cannata
Department of Communication, Computer and System Sciences (DIST), University of Genova, Genova, Italy

A.Ascia, M.Maggiali, L.Natale and G.Metta
Fondazione Istituto Italiano di Tecnologia (Italian Institute of Technology), Genova, Italy

5. References

[1] Webster J (1988) Tactile Sensors for Robotics and Medicine. Wiley, NewYork.

[2] Lee M H, Nicholls H R (1999) Tactile sensing for mechatronics: a state of the art survey. Mechatronics. 9:1Ű31.

[3] Inaba M, Hoshino Y, Nagasaka K, Ninomiya T, Kagami S, Inoue H (1996) A full-body tactile sensor using electrically conductive fabric and strings. Proc. of 9th Int. Conf. on Intelligent Robots and Systems. 450-457.

[4] Iwata H, Hoshino H, Morita T, Sugano S (2001)Force Detectable Surface Covers for Humanoid Robots. Proceedings of the 2001 IEEE/ASMA International Conference on Advanced Intelligent Mechantronics (AIM'01).

[5] Ohmura Y, Kuniyoshi Y, Nagakubo A. (2006) Conformable and scalable tactile sensor skin for curved surfaces. Proc. IEEE Int. Conf. on Robotics and Automation (ICRA).

[6] Ohmura Y, Kuniyoshi Y (2007) Humanoid robot which can lift a 30kg box by whole body contact and tactile feedback. Proc. IEEE/RSJ Int. Conf. on Intelligent Robots and Systems (IROS).

[7] Mukai T, Onishi M, Hirano S, Luo Z (2008) Development of the tactile sensor system of a human-interactive robot ŞRI-MANŤ. IEEE Transactions on Robotics. 24:505Ű512.

[8] Asfour T, Regenstein K, Schroder J, Dillmann R (2006) ARMAR-III: A humanoid platform for perception-action integration. Proc. Int. Workshop on Human-Centered Robotic Systems.

[9] Mizuuchi I, Yoshikai T, Nishino T, Inaba M (2006) Development of musculoskeletal humanoid Kotaro. Proc. IEEE Int. Conf. on Robotics and Automation (ICRA).

[10] Tajika T, Miyashita T, Ishiguro H (2006) Automatic categorization of haptic interactions - what are the typical haptic interactions between a human and a robot?. Proc. IEEE-RAS Int. Conf. on IEEE Conference on Humanoid Robots (Humanoids).

[11] Minato T, Yoshikawa Y, Ishiguro H, Asada M (2007) CB2: A child robot with biomimetic body for cognitive developmental robotics. Proc. IEEE-RAS 7th Int. Conf. on Humanoid Robots (Humanoids).

[12] Yoshikai T, Fukushima H, Hayashi M, Inaba M (2009)Development of Soft Stretchable Knit Sensor for Humanoids Whole-Body Tactile Sensitivity. Proceedings of the 2009 IEEE Conference on Humanoid Robotics (HUMANOIDS 2009).

[13] Shimojo M, Araki T, Ming A, Ishikawa A(2010) A High Speed Mesh of Tactile Sensors Fitting Arbitrary Surfaces. IEEE Sensors Journal. 10:822-831.

[14] Mittendorfer P, Cheng G (2011) Humanoid Multimodal Tactile-Sensing Modules. Robotics, IEEE Transactions on. 27:401-410.

[15] Metta G, Sandini G, Vernon D, Natale L, Nori F (2008) The iCub humanoid robot: an open platform for research in embodied cognition. In Proceedings of the 8th Workshop on Performance Metrics for Intelligent Systems (PerMIS '08).

[16] Dautenhahn K, Nehaniv C L, Walters M L, Robins B, Kose-Bagci H, Assif Mirza N, Blow M (2009) KASPAR Ű a minimally expressive humanoid robot for humanŰrobot interaction research. Applied Bionics and Biomechanics. 6:3-4.

[17] Gouaillier D, Hugel V, Blazevic P, Kilner C, Monceaux J, Lafourcade P, Marnier B, Serre J, Maisonnier B (2009)Mechatronic design of NAO humanoid. Robotics and Automation.IEEE International Conference on (ICRA). 769-774.

[18] Schmitz A, Maggiali M, Natale L, Bonino B, Metta G (2010) A tactile sensor for the fingertip of the humanoid robot iCub. Intelligent Robots and Systems (IROS), 2010 IEEE/RSJ International Conference on. 2212-2217.

Chapter 10

Ionic Polymer-Metal Composite Artificial Muscles in Bio-Inspired Engineering Research: Underwater Propulsion

Zheng Chen, T. Um and Hilary Bart-Smith

Additional information is available at the end of the chapter

1. Introduction

Electroactive polymers (EAPs), known as artificial muscles, can generate large deflections under an electrical stimulus (Bar-Cohen 2000, 2004). Due to the similarities with biological muscles, in terms of achievable stress and strain, EAPs have great potential to be used as actuators in bio-inspired robots, bio-medical devices, and micro/nano manipulation systems. Ionic polymer-metal composites (IPMCs) are an important category of ionic EAPs (Shahinpoor & Kim, 2001, 2005; Kim & Shahinpoor, 2003, 2003; Shahinpoor et al, 2003; Kim et al, 2007; Nemat-Nasser & Li, 2000; Nemat-Nasser, 2002), which can work well under a low actuation voltage (1 to 2 Volts) in a sodium salt-water environment.

As shown in Fig. 1, an IPMC consists of one ion exchange membrane, such as Nafion (DuPont), sandwiched in between two thin metal electrodes (Shahinpoor & Kim, 2001). When the IPMC is hydrated, the positive ions in the Nafion polymer, such as sodium and calcium ions, bring water molecules and migrate to the cathode side under an electric field, whereas the negative ions are permanently fixed to the carbon chain (Nemat-Nasser & Li, 2000). The ion transportation and water migration introduce swelling in the cathode side and shrinking in the anode side, which eventually causes bending motion of the IPMC (Shahinpoor & Kim, 2001).

IPMCs can also be used for electrolysis. Electrolysis of water occurs when a relatively high voltage (> 1.6 V) is applied between the cathode and anode (Shaaban, 1994; Shimizu et al, 2005). The water molecules are decomposed into hydrogen and oxygen gases. IPMC enhanced electrolysis of water is due to several reasons; (i) ion transport brings high dense of water close to cathode side (Shahinpoor & Kim, 2001); (ii) the strong electrical fields established on the boundary (Nemat-Nasser & Li, 2000; Nemat-Nasser, 2002) breaks down of the water molecules; and (iii) the porous platinum electrodes (Shahinpoor & Kim, 2001) on the surface acts as a stable electrolysis catalyst (Millet et al, 1989).

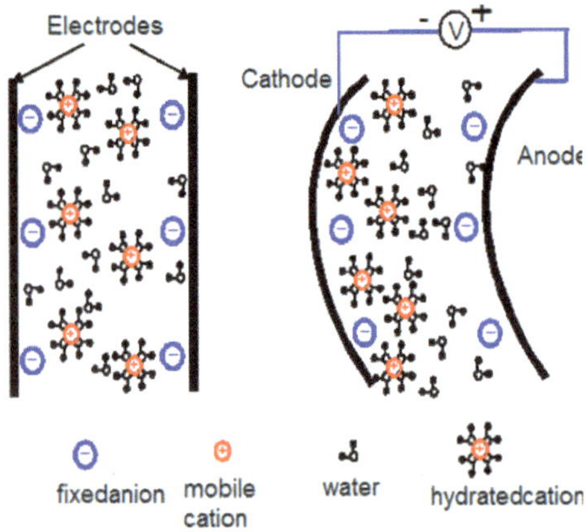

Figure 1. Actuation mechanism of IPMC (Chen & Tan, 2008).

In this chapter, the advantages and challenges of IPMCs as applied to bio-inspired engineering are discussed. A bio-inspired robotic manta ray integrated with IPMCs as artificial muscles and a novel buoyancy control device using IPMCs as an electrolysis generator are presented.

1.1. Introduction to bio-inspired robotic manta ray

Aquatic animals (e.g., fishes, cetaceans, etc.) are ultimate examples of superior swimmers as a result of millions of years of evolution, endowed with a variety of morphological and structural features for moving through water with speed, agility, and efficiency (Chen et al, 2010). The manta ray (*Manta birostris*, shown in Fig. 2) demonstrates excellent swimming capabilities; generating highly efficient thrust via flapping of dorsally flattened pectoral fins (Rosenberger, 2001). Many efforts have been directed towards building a bio-inspired pectoral fin structure to mimic the swimming behavior of the ray. Examples include rigid plates or tensegrity structures actuated by servo-motors (Gao et al, 2007; Moored et al, 2008; Moored & Bart-Smith, 2009; Zhou & Low, 2012) and flexible membrane actuated by shape memory alloy (SMA) (Wang et al, 2009). However, these methods are not suitable for small-scale robots (on the order of 5-10 cm) (Shahinpoor, 1992; Mojarrad & Shahinpoor, 1996; Tan et al, 2006; Guo et al, 2003; Punning et al, 2004) because of either the limitations in scaling or high power consumption. To construct a free swimming and small-scale robotic manta ray, there is a need for a bio-inspired actuating material that is lightweight, compliant, resilient, and capable of generating 3 dimensional (3D) deformations with portable power consumption.

Figure 2. Bio-inspiration: The Manta Ray (courtesy of www.elasmodivec.com).

In the past, IPMC actuators have been used as a caudal fin in bio-inspired robotic fishes (Shahinpoor, 1992; Mojarrad & Shahinpoor, 1996; Tan et al, 2006; Guo et al, 2003), where the propulsive fin mimics the bending motion observed in many biological fishes. In the propulsion mechanism of rays, undulatory and oscillatory flapping motions of the pectoral fin play an important role in generating highly efficient propulsion and maneuvering (Rosenberger, 2001). To fabricate an actuating membrane capable of generating complex deformations, lithography-based (Chen & Tan, 2010) and surface machining-based approaches (Kim et al, 2011) have been explored to pattern the electrodes of the IPMC. To create active and passive areas in a Nafion membrane, the electrodes on the membrane were separated. By independently controlling the bending of each active area, 3-dimensional deformations of the membrane have been achieved. However, the stiffness of the Nafion in the passive area limited its capability of generating large twisting motions. Punning et al, developed pectoral fins for ray-like underwater robot by assembling separated IPMC beams with a latex foil (Punning et al, 2004). However, this robot did not achieve free-swimming due to high power consumption and low propulsion efficiency. The challenge in this study is to fabricate a compliant actuating membrane cable of complex 3D kinematic motions capable of generating energy efficient locomotion. This will then be integrated in a small-scale robotic ray capable of free swimming.

1.2. Introduction to bio-inspired depth control device

A simple and efficient method for buoyancy control is critical in the design of an autonomous underwater device. The most common method used today to achieve depth control is to utilize a piston-cylinder assembly connected to a servomotor (Lin et al, 2010). A cylinder, usually containing air, is compressed and expanded to adjust the volume of the system. Another similar approach is to use a water-pump powered bladder to compress the air in a blaster tank (Zhou & Low, 2012). These two methods usually yield reliable results with relatively fast response times. However, in addition to the significant noise generated by the servomotors, there are limitations in scaling the servomotor. Consequently, this solution is not feasible for implementation in small devices. In order to build more efficient buoyancy control devices, researchers have turned to biology for inspiration for the next generation of autonomous underwater vehicles.

Biology has many novel and effective depth control mechanisms suitable for a variety of environments. For example, sperm whales (Fig. 3(a)) achieve buoyancy control by using their spermaceti oil. An adult sperm whale contains about 4 tons of spermaceti oils in their spermaceti organ, which represents approximately 8% of its total mass (Shibuya et al, 2006). The spermaceti oil has a low melting point and its density depends largely on the temperature of the oil. By manipulating the arterial blood flow through the spermaceti organ, the sperm whales can regulate the temperature of the oil and are thus able to control their buoyancy. There have been recent demonstrations of buoyancy control concepts manipulating temperature to change the density of oil (Shibuya et al, 2006) or wax (McFarland et al, 2003). However, the response times are slow (on the order of 10 minutes), and it is inefficient for small devices because a constant power must be supplied to maintain the temperature of the oil while cruising at a certain depth. Ray-finned fishes, such as one depicted in Fig. 3(b), change the buoyancy of their body using a swim bladder (Bond, 1996). Expansion of the bladder results in increased volume, thus making the body more positively buoyant and vice versa. Inspiration for the artificial bladder presented in this chapter comes from these ray-finned fishes. The challenges arise from how to generate and release the gas to control the volume of the bladder underwater.

(a) (b)

Figure 3. (a) Sperm whale (*Physeter macrocephalus*); (b) Golden fish (*Carassius auratus*).

The rest of this Chapter is organized as the follows. Section 2 is focused on the development of a bio-inspired robotic manta ray powered by IPMC artificial pectoral fin. Fabrication and characterization of the artificial pectoral fin and design of the robotic manta ray are presented. Section 3 is focused on bio-inspired depth control device enabled by IPMC enhanced water electrolysis, where the buoyancy control mechanism, device design, and open loop testing are demonstrated. Conclusions and future work on both studies are discussed in Section 4.

2. Bio-inspired robotic manta ray powered by IPMC artificial pectoral fin

In this section, we present a bio-inspired, free-swimming robotic manta ray propelled by artificial pectoral fins. In our previous work, we developed an assembly-based fabrication process to bond four IPMC actuators with Polydimethylsiloxane (PDMS) elastomer to create an actuating membrane capable of 3D kinematic motions (Chen et al, 2011). However, the control strategy for four IPMCs is too complicated to be implemented on-board. In this study, a pectoral fin, with the planform shape of a manta ray pectoral fin and consisting of one IPMC artificial muscle in the leading edge and a passive PDMS membrane in the trailing edge, has been designed. When the IPMC is actuated, the passive PDMS membrane follows the bending of IPMC with a phase delay, which causes an undulatory motion of the fin. The pectoral fin has been characterized by key factors as they relate to the function of the robot: tip deflection; twist angle; and power consumption. Since only one IPMC is needed for actuation in the pectoral fin, the control strategy is greatly simplified. To test the performance with a free-swimming robot, a light, compact on-board control unit with a lithium ion polymer battery has been developed. Experimental results have shown that the robot is capable of free swimming.

2.1. Fabrication of artificial pectoral fin

The proposed artificial pectoral fin must be able to generate oscillatory motion with a twist angle as observed in swimming of the manta ray, under hydrodynamic loads. An artificial pectoral fin was fabricated by combining one IPMC actuator with a PDMS elastomer in a mold to create a predefined planform shape. The design of pectoral fin is shown in Fig. 4(a). The outline shape of the fin mimics that of the manta ray. The fin is divided into two areas: IPMC beam on the leading edge and PDMS passive membrane on the trailing edge. Note that the size and shape of the IPMC is chosen to generate enough bending moment with limited power consumption. Optimal design of the fin will be focused in the future work.

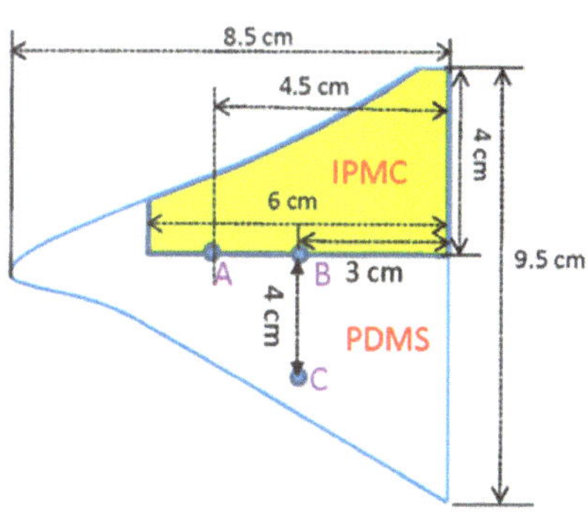

(a) Pectoral fin design

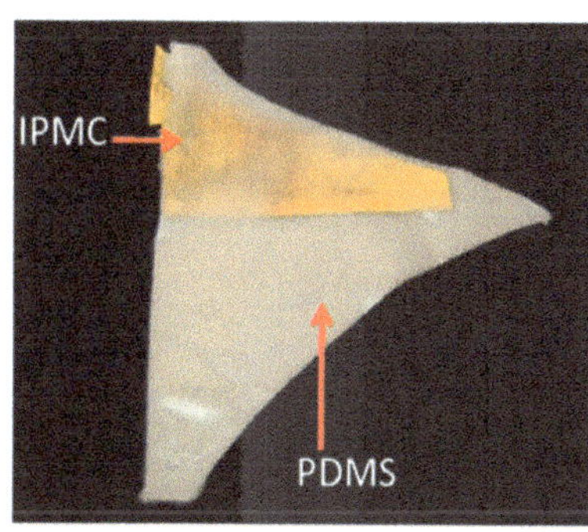

(b) Fabricated pectoral fin

Figure 4. IPMC powered artificial pectoral fin.

The first step in creating the artificial pectoral fin is to fabricate the IPMC actuator. Many groups have developed different IPMC fabrication processes for various purposes (Kim & Shahinpoor, 2003; Chung et al, 2006; Lee et al, 2006; Kim & Shahinpoor, 2002). In our fabrication method, we followed most of the procedure outlined by K. Kim and M. Shahinpoor (Kim & Shahinpoor, 2003) but added additional platinum/gold plating process to reduce the surface resistance of the electrodes (Lee et al, 2006). The following supplies are used to fabricate the IPMC beams: (1) Nafion ion exchange membrane Nafion 1110 (258 μm thick, DuPont); (2) tetraammineplatinum chloride 98% (Aldrich); (3) sodium borohydride ($NaBH_4$, Aldrich); (4) dilute ammonium hydroxide solution (NH_4OH 29% solution); 5) dilute hydrochloric acid (HCl aq, 1.0 N solution); and 6) de-ionized water. Fig. 5 shows the IPMC fabrication process. The major fabrication steps are as follows:

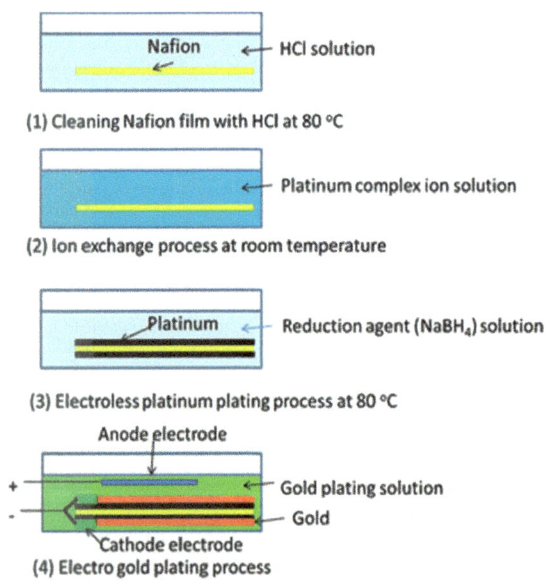

Figure 5. IPMC fabrication process.

1. Treatment of Nafion with HCl (cleaning). This step is to remove metal particles and other impurities from the film. The Nafion film (10 cm by 5 cm) is boiled in 1.0 N hydrochloride acid (HCl) at 80 °C for 30 min. The film is then rinsed with DI water to remove acid residue.

2. Ion exchange process. The purpose of the ion exchange process is to enable the Nafion film to absorb enough platinum complex ions $Pt(NH4)_2^{+2}$,which will be reduced to platinum particles to form electrodes in step 3. The membrane is then immersed in 50 ml tetraammineplatinum Chloride hydrate solution (3 mg/ml) mixed with 1 ml ammonium hydroxide (29%) to make a weak base environment. The ion exchange process takes about 3 hours to allow the Nafion film to fully absorb enough platinum complex ions.
3. Electroless platinum plating. In this step, the platinum complex ions are reduced to platinum particles to create the metal electrodes on the outer surfaces of the Nafion film. The ion-exchanged Nafion film is placed in a DI water bath. Then the bath is gradually heated to 40 °C. A sodium boron hydride solution ($NaBH_4$) (20 mg/ml) is prepared, which is used as a reducing agent. 5 ml $NaBH_4$ solution is added to the bath and the temperature of the bath is increased to 65 °C. 5 ml $NaBH_4$ solution is added to the bath every 30 min. After 3 h plating, approximately 6 μm of platinum is deposited on the surface with good polymer-metal adhesion.
4. Electro gold plating: To further improve the conductance of the electrode, about 1 μm thick gold was deposited on the surface via electroplating process. The sample was then submerged in a sodium solution (1 N) for one day to exchange H^+ with Na^+ to enhance the actuation performance of the IPMC.

After fabrication of the IPMC actuator, the next step is to bond the IPMC with a PDMS elastomer membrane. The PDMS bonding process (shown in Fig. 6) is described in the following 5 steps. (1) A Delrin polymer (McMaster) mold was made using a Computer Numerical Control (CNC) rapid milling machine (MDX-650, Roland). The mold has two concaved areas to house the PDMS passive membrane and IPMC actuator. The thicknesses of PDMS membrane and IPMC are 600 μm and 280 μm, respectively. (2) The IPMC was cut into the shape shown in Fig. 3(a). (3) About 3% glass bubbles (Glass bubble K37, 3M Inc) were added into PDMS gel (Ecoflex 0030, Smooth-on Inc.) to gain a neutrally buoyant pectoral fin. (4) The IPMC and the PDMS gel were then clamped with the mold and the PDMS was cured at room temperature for 3 hours. (5) The IPMC/PDMS artificial pectoral fin (shown in Fig. 4(b)) was removed from the mold. Note that there is about 150 μm thick PDMS covered on the top of IPMC to guarantee good bonding between passive and active areas. The additional layer of PDMS stiffens the IPMC active area. However, since the PDMS (Ecoflex) is an extremely compliant material (Young's modulus, 0.06 MPa), compared to the Nafion (Young's modulus 114 MPa), the stiffening effect can be neglected.

2.2. Characterization of pectoral fin

The pectoral fin was characterized in terms of tip deflection, twist angle, and power consumption. These characteristics are useful in providing comparison data in the design of the bio-inspired robot. To characterize the actuating response of the pectoral fin, three testing points (A, B, C) are defined on the membrane (shown in Fig. 4(a)).

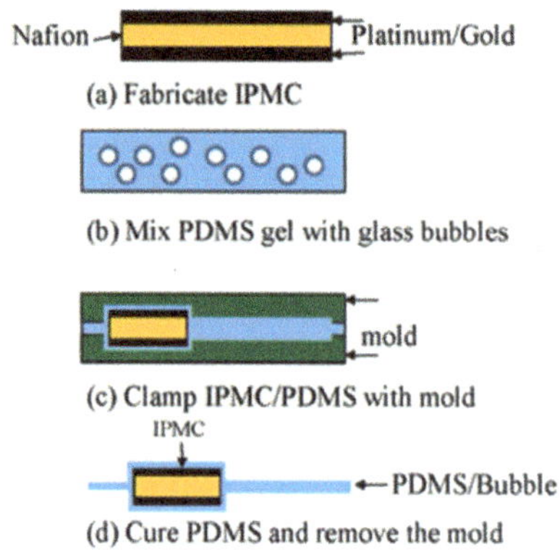

Figure 6. Fabrication process of pectoral fin.

2.2.1. *Tip displacement*

To measure the tip deflection, the fin was actuated in a transparent tank containing water. A laser sensor (OADM 20I6441/S14F, Baumer Electric) was fixed outside of the tank to measure the bending displacement at point A. Note that due to large deflection of the fin, the laser sensor was unable to capture the bending displacement of the tip. So we moved the measurement point from the tip to the point A. The tip deflection was normalized by dividing the bending displacement by the length at the point A. Fig. 7 shows the tip deflection when a square wave actuation voltage (0.09 Hz, 4 V) was applied to the IPMC. It shows a peak-to-peak deflection of 100% in the span-wise direction. One can achieve a larger deflection by applying a higher actuation voltage. But there is a limit to the size of voltage applied across the IPMC—anything greater than 6 V risks dielectric breakdown through the IPMC (Lee et al, 2006).

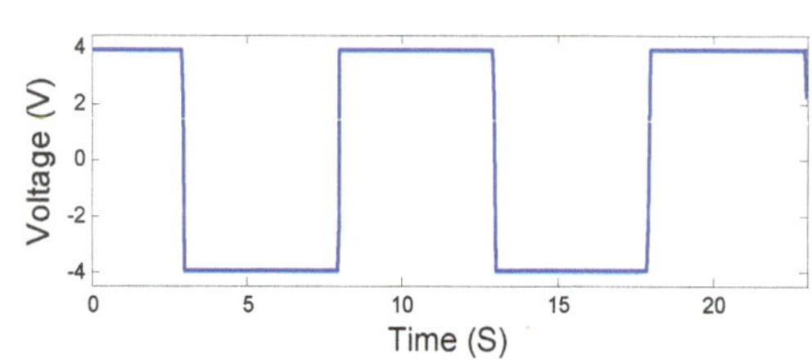

Figure 7. Tip deflection of the pectoral fin.

2.2.2. Bode plot

To capture the Bode plot of the pectoral fin, a series of sinusoidal actuation signals with amplitude of 4 V and frequencies ranging from 0.05 Hz to 0.9 Hz were applied to the IPMC. The tip deflection at the point B and the actuation voltage were measured. The magnitude and phase shift of the tip deflection over the actuation voltage were calculated. The Bode plot (Fig. 8) demonstrates that the actuation dynamics of the pectoral fin behaves as a low-pass filter with a 0.4 Hz cut-off frequency. This is to be expected as the ions in the IPMC cannot move very rapidly (Nemat-Nasser, 2002) and hydrodynamic force dumps high frequency vibration (Chen & Tan, 2010).

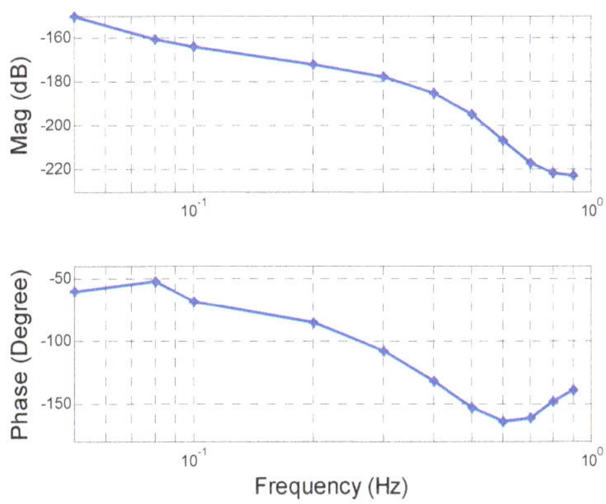

Figure 8. Bode plot of pectoral fin.

2.2.3. Characterization of twisting motion

In order to characterize the 3-dimensional (3D) kinematics of the fin, two laser sensors (OADM 20I6441/S14F, Baumer Electric) were used to measure the bending displacements at the points B and C, d_B and d_C respectively. The twist angle was calculated by

$$\alpha = \tan^{-1}\left(\frac{d_B - d_C}{BC}\right) \tag{1}$$

where BC=20 mm.

A series of square wave voltage signals were generated via LabVIEW (National Instruments), amplified using power amplifiers, and then applied to the IPMC actuator. All signals have the same amplitude of 4 V but varying frequencies from 0.06 Hz to 1 Hz. As the IPMC actuator is being used to generate thrust in underwater vehicles, the 3D kinematics was quantified in water. The fin was placed in a water tank and the laser sensors were fixed outside of the tank. Fig. 9 shows the twisting and flapping motion on the fin at f=0.09 Hz. The upper figure shows the bending displacements d_B and d_C, which indicates a phase delay and an amplified magnitude between the bending motions at the points B and C. The lower figure shows the calculated twist angle. The peak-to-peak twist angle is 40°. Fig. 10 shows the plot of twist angle versus operating frequency. The twist angle decreases as the frequency increases. The cut-off frequency of twist angle is around 0.1 Hz. Since the twist angle plays an important role in generating a thrust, the optimal operating frequency of the robot will be set around 0.1 Hz.

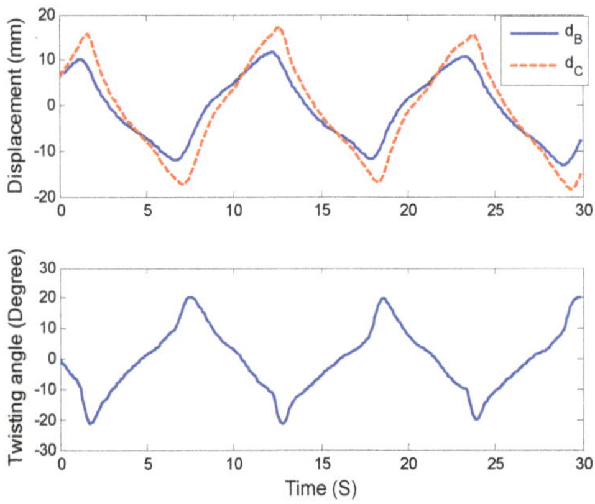

Figure 9. Twisted flapping motion on the fin.

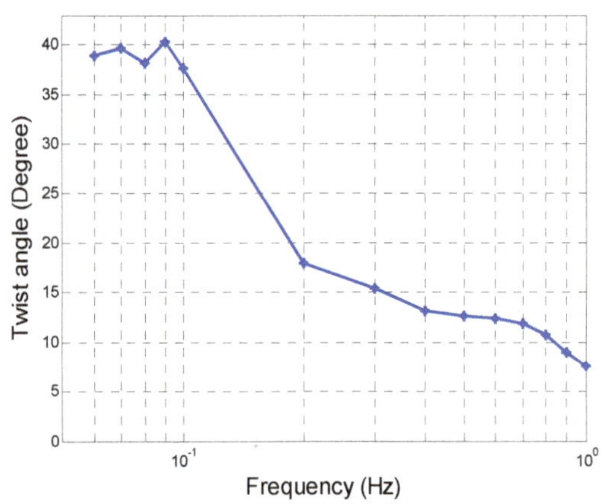

Figure 10. wist angle versus operating frequency.

2.2.4. *Power consumption*

For un-tethered bio-inspired robot applications, key questions regarding power consumption and optimal power management must be addressed. In this section, we study the power consumption of the pectoral fin under a square wave actuation voltage, which is easy to generate on board. The power consumed by the IPMC was calculated using the following equation:

$$P = \frac{1}{T}\int_0^T i(t)v(t)dt \tag{2}$$

where i(t), v(t), and T are measured actuation current, voltage, and period, respectively. A 4 V, 0.1 Hz square wave voltage was applied to the IPMC of the pectoral fin, which was fixed under water. Fig. 11 shows the actuation voltage and current. The power consumption at 0.1 Hz is 1.09 W.

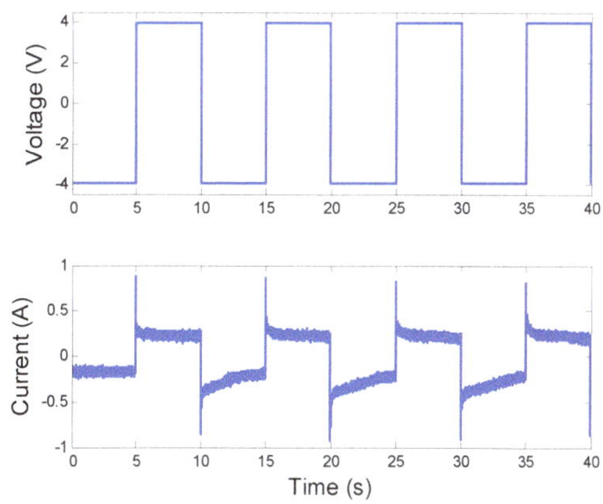

Figure 11. Actuation current and voltage under a 0.1 Hz square wave signal.

Under an electrical voltage signal, the IPMC behaves as a pseudo capacitor. When the applied voltage changes its polarization, the current reaches its peak, and then drops down to a steady-state current. The steady-state current is due to the DC resistance of the polymer and the water electrolysis on the electrodes. To reduce the power consumption of the pectoral fin, it would be better to eliminate the water electrolysis by coating a waterproof protecting layer on the IPMC, which will be our future focus. To study the relationship between power consumption and operating frequency, a series of square wave actuation signals with amplitude of 4 V and frequencies ranging from 0.1 Hz to 1 Hz were applied to the fin. Fig. 12 shows the power consumption versus operating frequency. It demonstrates that they are positively related and the power consumption is below 1.5 W.

2.3. Design of robotic manta ray

In our previous work, a free swimming and IPMC enabled robotic manta ray was developed for the first time (Chen et al, 2011). We used four IPMC beams bonded with thin PDMS membrane to create artificial pectoral fin. By independently controlling the bending of each IPMC beam, the fin can generate undulatory motions. However, in the free-swimming test of the robot, only a flapping motion was generated due to the complexity of generating the four control signals on-board. In this paper, we present an IPMC-enabled robotic manta ray capable of free swimming. Since there is only one IPMC beam in the leading edge, a single signal is generated on-board, greatly simplifying the control strategy.

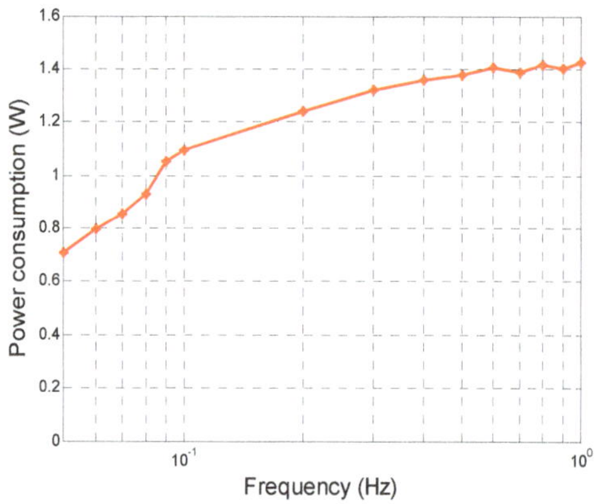

Figure 12. Power consumption versus operating frequency.

2.3.1. *Control circuit*

The on-board circuit, which was developed in our previous work (Chen et al, 2011), provided a square wave voltage signal to the IPMC actuator in the pectoral fin. Fig. 13(a) shows the schematic of the circuit. A 555-timer was used to generate a frequency adjustable square wave. The amplitude of the voltage signal, V_P, was controlled by an adjustable voltage regulator. An H-bridge driver was used to draw up to 2 A output peak current. A rechargeable 7.3 V Lithium Ion Polymer battery (400 mAh, AA Portable Power Corp) was selected as the power source for the robot. Fig. 13(b) shows the picture of PCB board and battery.

Figure 13. Control circuit and battery (Chen et al, 2011).

2.3.2. *Free swimming robot design*

Fig. 14 shows the overall shape of the robot. Two acrylic frames with gold electrodes were made to clamp the artificial wings to the body support. Gold electrodes were used to minimize corrosion. Polymer foam was put into the middle of the frame to make the robot slightly positively buoyant. A simple hinged plastic box was used to house the circuit board and battery. Once closed, only two wires were exposed outside the box. PDMS was used to seal the box waterproof. The plastic box was then glued on the top of the robot. With the robot was put under the water, the cover of the box was above the water level thus water could not leak into the box. The fully assembled robot was 11 cm long, 21 cm wide, and 2.5 cm thick with a mass of 55 grams. The free-swimming robot with the control unit is shown in Fig. 15. The total cost of the robot is about $200.

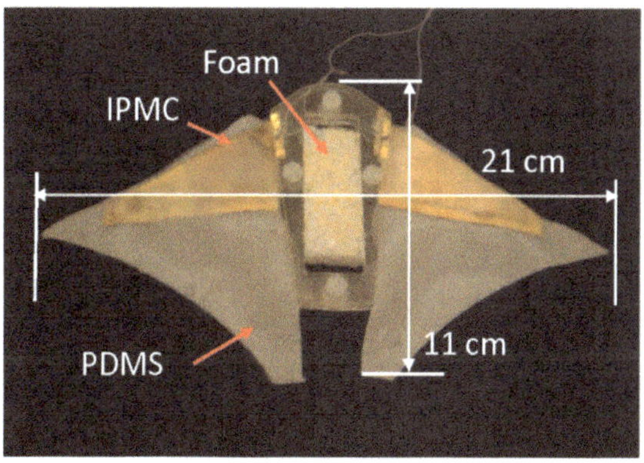

Figure 14. Robotic manta ray body.

Figure 15. Free swimming robotic manta ray body.

2.3.3. *Free-swimming test*

The robot was tested in a water tank (1.5 m wide, 4.7 m long, and 0.9 m deep). The robot was positively buoyant. The operating frequency of the square wave actuation voltage was tuned to 0.167 Hz and the amplitude was set to 4 V. Since each pectoral fin consumed less than 1.2 W (shown in Fig. 11), the overall power consumption of the robot was under 2.5 W. A digital video camera (VIXIA HG21, Canon) was used to capture the motion of the swimming robot. Fig. 16 presents six snap shots of the swimming robot from top view, where the trajectories of left and right fins are plotted in blue and red, respectively.

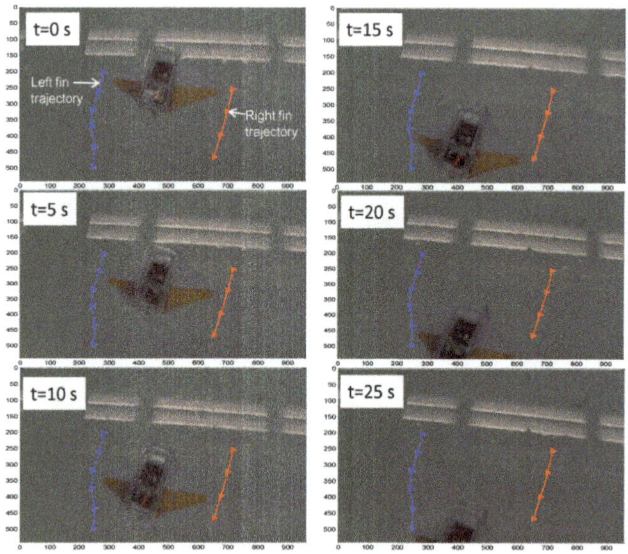

Figure 16. Snap shots of free-swimming robotic manta ray.

Each snap shot was taken every 5 seconds. A swim speed of 0.74 cm/s was calculated from the movie using the Edge Detection program in the LabVIEW. Since the body length was 11 cm, the speed in body length per second (BL/s) was 0.067.

2.4. Comparison of ray-like swimming robots

In our previous work, we developed an artificial pectoral fin that consists of four IPMCs bonded by PDMS membrane. The robot swam at 0.42 cm/s (0.053 BL/s) consuming 1 W in power. Many other research groups have been developed ray-like robots, using IPMCs, servomotors, and shape memory alloys as actuators. The comparison of ray-like swimming robots is shown in Table 1. It indicates that the IPMC powered ray-like robots are lighter and consume less power than the robots actuated by a servomotor or SMA. But they swim at lower speed for the reasons that IPMC is unable to flap at high frequency and generated force is also very low.

Reported Article	Speed (BL/s)	Power (W)	Weight (g)	Actuator
This chapter	0.067	2.5	55	IPMC
Chen et al. 2010	0.053	1	55	IPMC
Punning, et al. 2004	0.038	2.2	60	IPMC
Zhou et al, 2012	0.8	10.8	7300	Servo Motor
Wang et al, 2009	0.43	Not given	354	SMA

Table 1. Comparison of Ray-like Robots.

The advantages of using IPMCs as artificial muscle in ray-like swimming robots are: 1) low actuation voltage; 2) works well in wet conditions; 3) no gears and motors, 4) simple mechanical structure; 5) low noise; 6) able to be shaped in bio-inspired engineering design. However, IPMC can only generate low force with slow response time, which limits the swimming speed of the robot. To accommodate the disadvantages and utilize the advantages, an optimal design of the pectoral fin must be conducted, where the dimensions and location of the IPMC actuator are optimized. The challenge arises from modelling the IPMC actuating membrane to produce optimal 3D kinematic motions, which is a future focus. Another focus will be on modelling of the fluid dynamics introduced by the 3D kinematic motions of the fin, to understand how the thrust force is generated.

3. A novel buoyancy control device enabled by IPMC

In this study, a novel depth control device has been designed and built. The proof-of-concept device utilizes the principles of electrolysis of water, enhanced by the inclusion of an ionic polymer-metal composite (IPMC) membrane as a medium. The device design incorporates an artificial bladder where the volume of gas generated by electrolysis is controlled by a solenoid valve, thus changing the device's buoyancy. A set of gold electrodes, separated by an IPMC film, is used as a lightweight and compact electrolysis generator. IPMC acts as stable, low power, highly efficient and environmentally friendly gas generator. Experimental results using open-loop control show that the device is capable of controlling its buoyancy efficiently with no noise and low power consumption. Applications for this technology include integration into bio-inspired, unmanned underwater vehicles.

3.1. Mechanism of buoyancy control

3.1.1. *Gas generation by water electrolysis*

When an electric current is applied between positively and negatively charged electrodes in ionized water, a chemical reaction occurs, in which pure oxygen and hydrogen gases are produced. Hydrogen gas appears at the cathode electrode (3) and oxygen gas appears at the anode electrode (4). Equation (5) shows the net reaction (Shaaban, 1994).

$$2H_2O_2 + 2e^- \rightarrow H_2 + 2OH^- \quad (3)$$

$$4OH^- \rightarrow O_2 + 2H_2O + 4e^- \tag{4}$$

$$2H_2O(l) \xrightarrow{current} O_2(g) + 2H_2(g) \tag{5}$$

Two thin gold plates, each approximately 2 cm by 2 cm, are used as electrodes (Fig. 17). Typically, the two plates are placed 1-2 cm apart to allow the current to travel through the water. Most electrolysis experiments are performed in ionic solutions, which are usually prepared by adding salt, acid, or base. However, adding electrolytes is not feasible in this case because the device must be able to generate gas in regular tap water.

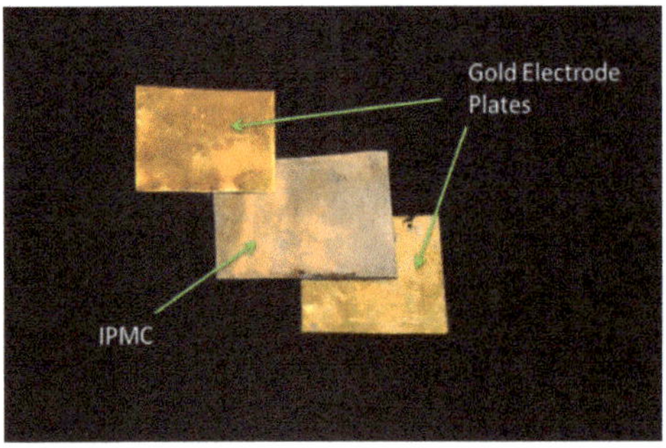

Figure 17. Gold electrodes and a sample of an ionic polymer-metal composite.

Electrolysis in tap water is much slower because of the limited number of ions present. In order to enhance the electrolysis process, an IPMC (150 μm thick) is placed directly in between the electrodes (Fig. 17). In this study, the IPMC is used only as a medium to contain electrolytes and provide channels for ion movement in the electrolysis process. By allowing the current to flow through the electrodes and IPMC, oxygen and hydrogen gases are generated. These gases are collected in a gas chamber, displacing the water in the artificial bladder. Because the gas mixture has a much lower density than that of water, the device becomes more positively buoyant. Care must be taken because the mixture contains a proportion of constituents such that it can burn instantly when ignited.

3.1.2. *Gas releasing mechanism by solenoid valve*

The mechanism to control the release of gas, and thereby depth, uses a two-way solenoid valve (Hargraves Tech. Corp, Part # 75M06U2.A005S). This valve is closed in its non-actuated state and is suitable for the size and power constraints of this proof-of-concept design. This particular valve can be opened by applying 6 VDC voltage and has a power consumption and mass of 0.5 W and 5 g, respectively. When the valve is actuated, the gas formed during electrolysis escapes from the device and water supersedes. As a result, the density of the device increases, causing it to become more negatively buoyant. When the valve is closed, water cannot enter through a bottom opening because the pressure inside the device is equal to that of outside. Thus the device is able to maintain the same depth by having a constant buoyant force. In order to travel up or downwards, gas is either produced or released in a controlled manner.

3.2. Device design

The device consists of three parts: bottom chamber, middle seal, and top chamber. The overall schematic of the device is shown in Fig. 18. The parts are drawn using Autodesk Inventor 2010 and printed using a Fused Deposition Modeling (FDM) machine (uPrint Plus by Dimension). The device is approximately 15 cm tall, 6.5 cm in diameter, and has a mass of 114 g.

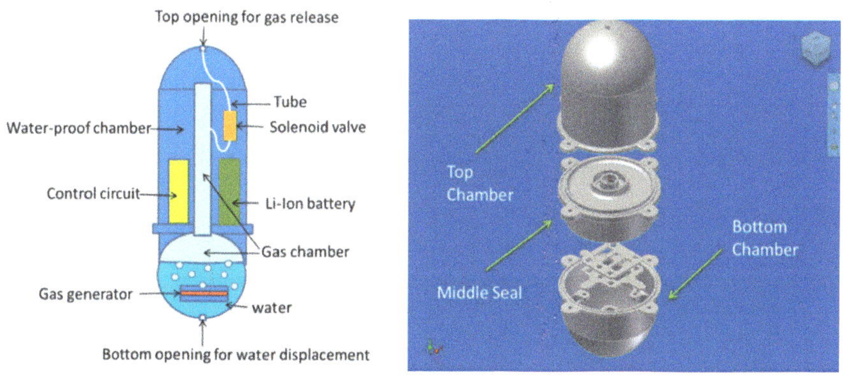

Figure 18. Schematic of the device (left); Computer drawing of the depth control device (right).

3.2.1. Bottom chamber

The bottom chamber encases the gas generator and holds water (Fig. 19). The water level varies depending on the depth location of the device. When the device is below the surface, the gas generator is submerged under water to ensure that the electrolysis can take place to allow the device to float up. The bottom chamber also has room to secure up to 8 metal washers as deadweights such that the device is initially about 80-90% submerged. On the bottom of the chamber, there is a small opening to allow water to enter or escape the device. This opening ensures that the inside and outside pressures are kept equal.

3.2.2. Middle seal

The purpose of the middle seal (Fig. 20) is to provide waterproofing of the electronics in the top chamber (Fig. 21). The bottom inner surface is concaved to direct the gas products to the gas chamber.

3.2.3. Top chamber

The top chamber is partitioned into a waterproof chamber and a gas chamber. The waterproof chamber contains all the electronics, Li-ion battery and Solenoid valve (Fig. 21).

The gas chamber contains a mixture of air and gases produced at the electrodes in the bottom chamber. A tube connects the gas chamber to the device surroundings via solenoid valve.

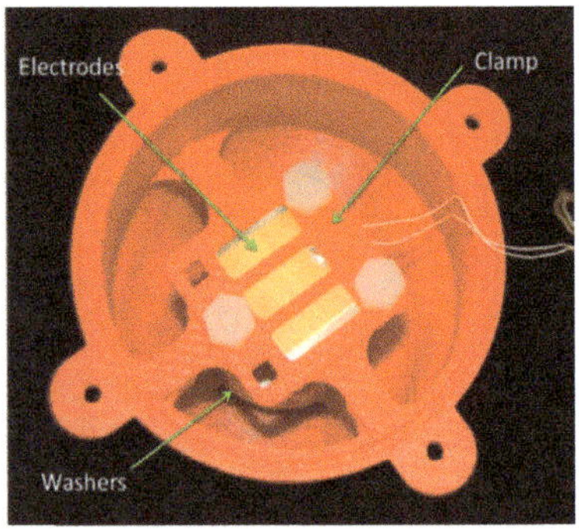

Figure 19. Bottom Chamber with electrodes assembly (Top View).

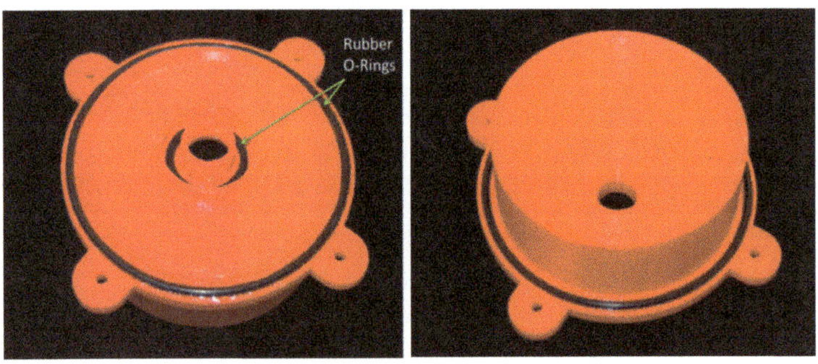

Figure 20. Middle Seal with rubber O-rings (Left: Top view; Right: Bottom view).

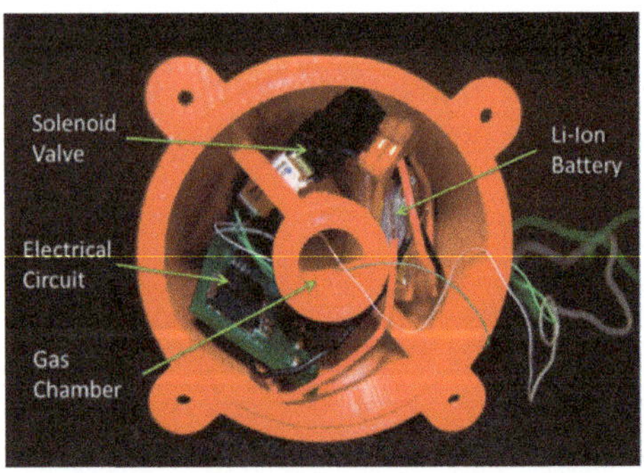

Figure 21. Top Chamber with electronic components and gas chamber.

3.2.4. *On-board electrical circuit design*

The on-board circuit (Fig. 22) provides actuation voltage signals to the electrodes and solenoid valve. A rechargeable 7.4 V, 400 mAh AA Portal Power Corp Lithium Ion Polymer battery is used as a power source and PIC12F508 microcontroller is used to generate two square wave control signals, S_1 and S_2. A square wave is chosen for simplicity. Since the microcontroller draws only 25 mA, two H-bridge drivers are used to provide up to 2 A peak current output to the electrodes and solenoid valve, which draw up to 500 mA and 80 mA, respectively. Amplitude of the voltage that controls the valve, V_{p1}, is 7.4 V. A voltage regulator sets the amplitude of the voltage applied to the electrodes and microcontroller, V_{p2}, to 5 V. Mass of the circuit and the battery are 11.5 g and 19.1 g, respectively.

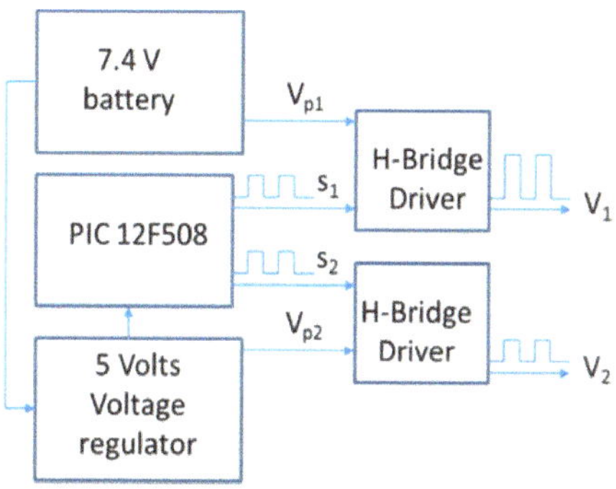

Figure 22. Circuit schematics.

3.3. Experimental results

3.3.1. *Gas generation rate*

An experiment is set up to measure the gas generation rate at different voltage levels. Voltages ranging from 2.0 V to 6.0 V at 0.5 V intervals are applied to the electrodes using an Agilent DC Power Supply (Model #E3646A). The hydrogen and oxygen gases generated are collected using water displacement technique with 50 mL graduated cylinder (Fig. 23(left)). A constant voltage is applied for 60 s then the average current and displaced volumes are recorded. The gas generation rate can be found by dividing the displaced volume by 60 s. Gas generation rate vs. power consumption is plotted (Fig. 23(right)) and a least squares regression line is fitted to find a correlation. The results indicate a fairly strong linear relationship between gases generated and power consumption. The proportionality constant is approximately 0.032 mL/J.

The on-board circuit is set up such that the output voltage is 5V. However the actual voltage measured across the electrodes is 4 V due to limited capacity of the battery. At this voltage, the average current and power consumption based on 5 trials are 0.3 A and 1.2 W, respectively. The average gas generation rate was 0.048 mL/s. The linear model predicts 0.040 mL/s at this power so the model and the result are in a fair agreement.

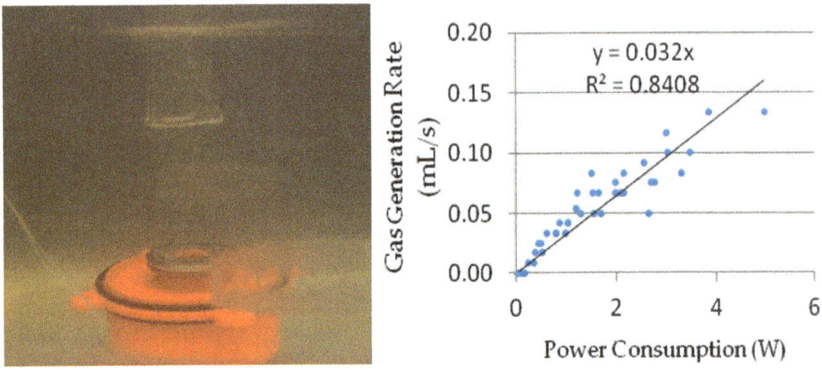

Figure 23. Gas generation rate experiment set-up; using water displacement technique to collect the gas generated (left); Gas generation rate vs. power consumption (right).

3.3.2. Diving test

The depth control device is tested in a water tank (1.5 m wide, 4.7 m long, and 0.9 m deep). The tank is filled with tap water at a room temperature. A critical mass is when the device is about 95% submerged under water, at which a slight decrease in buoyancy causes the device to sink. The critical mass of the device is experimentally found to be 283 g. Because the mass of the device is 114 g, 169 g of metal washers are added to the bottom chamber as deadweights. This mass is also equal to the payload for this particular device. The microcontroller in the circuit is programmed such that there is an initial 3-minute delay to allow time for assembly and fastening of the bolts. After the initial delay, the solenoid valve turns on for 12 s to allow gas to escape such that the device will sink to the bottom of the tank. Then the solenoid valve turns off and 5 V is applied to the electrode plates for up to 15 minutes. Gases generated by the electrolysis fill the gas chamber, which displaces the water inside the bottom chamber. Thus, the device becomes more positively buoyant. The time it takes to rises back up to the top is measured. Fig. 24 shows the timing of the control signals.

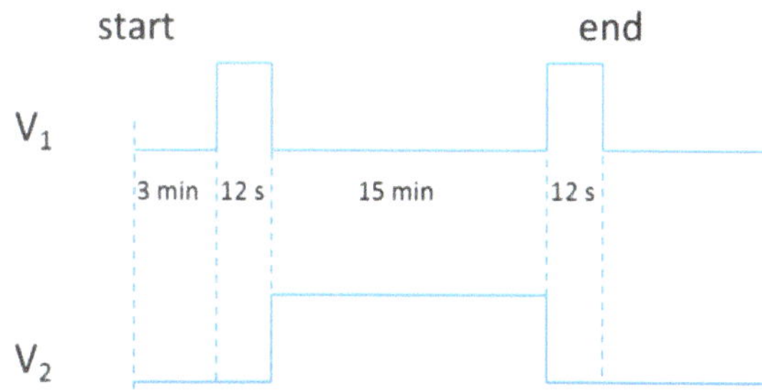

Figure 24. Timing of the control signals.

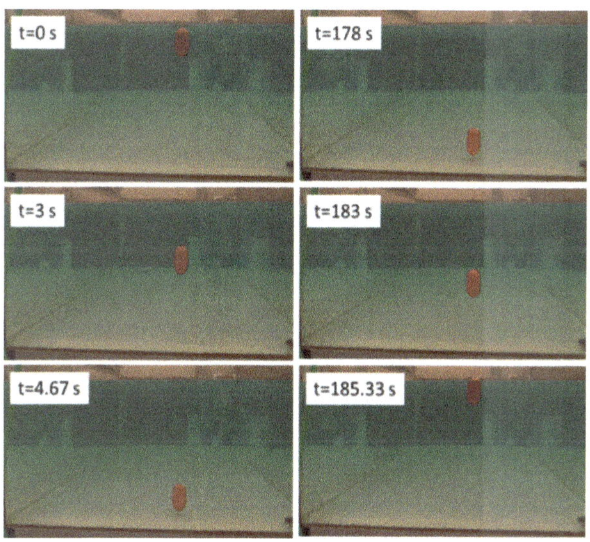

Figure 25. Snapshots of the device during open-loop testing.

It took approximately 4.7 s for the device to sink to the bottom of the tank and 180 s to rise back up to the top. The power consumption for sinking and rising are 0.5 W and 1.2 W, respectively. Fig. 25 shows snapshots of a successful demonstration of the open-loop proof-of-concept depth control device. In order to reduce the rising response time, multiple sets of electrodes can be implemented and actuated simultaneously at a higher voltage. To decrease the current rising time by a factor of 1/3 (60 s), a gas generation rate must triple to be at least 0.14 mL/s. This is achievable with two sets of electrodes with approximately 6 V actuation voltages.

3.3.3. Discussion

The advantages of using IPMC as electrolysis generator in depth control device are: 1) quiet gas generation; 2) low activation voltage; 3) compact design. The challenge arises from the low gas generation rate. In the close-loop depth control (Chen et al, 2012b), a fast gas generation rate is required to compensate the depth drop caused by disturbances and the volume decrease caused by water pressure, rapidly. A multi-IPMC electrodes design might be a good solution to increase the gas generation. However, it also increases the power consumption. In our future work, an additional volume-fixed gas storage chamber will be used to collect the generated gas with high pressure. Another solenoid valve will be used to release the high pressure gas into the gas chamber, which controls the volume of the device rapidly. One may be aware of hydrogen and oxygen explosive gas in the gas storage chamber. A two-gas-chamber design can solve this issue by separating oxygen and hydrogen gases. If a fuel cell is embedded, these gases can be recycled as fuel to recharge the battery, which makes the depth control more power efficient.

4. Chapter conclusion

In this Chapter, we presented two studies on IPMC artificial muscles in bio-inspired engineering research. In the first study, we developed a bio-inspired robotic manta ray powered by IPMC pectoral fin. We developed a novel synthesis process to fabricate IPMC pectoral fin that is capable of 3D kinematic motion and characterized the pectoral fin in terms of tip deflection (up to 100%), bode plot (0.4 Hz cut-off frequency), twist angle (up to 40o), and power consumption (below 1.5 W). A small size free-swimming robotic manta ray has been developed and experimental results have demonstrated its free-swimming capability with speed at 0.067 body length per second and 2.5 W power consumption. In the second study, we developed a novel buoyancy control device as an artificial swimming bladder. IPMC acted as an efficient, environmentally friendly water electrolysis generator to gain volume of the device while a solenoid valve was used as gas releasing controller. A compact and low power device has been assembled with an on-board open loop controller. Experimental results have shown that the device was able to control its 0.9 m depth within 3 minutes.

In both studies, advantages and challenges of IPMC in bio-inspired engineering research have been addressed. In this first study, the advantages of using IPMC as artificial muscle are: 1) low actuation voltage; 2) working well in wet condition; 3) low noise; 4) simple mechanical structure; 5) able to be shaped into bio-inspired engineering design. The disadvantages are: 1) low generated force; 2) slow response time. The challenge in this study comes from optimal design of the pectoral fin. In future research, we will focus on modeling an IPMC powered pectoral fin and modeling of the fluid dynamics introduced by the 3D kinematic motions of the fin. In the second study, the advantages of using IPMC as electrolysis generator are: 1) low noise; 2) compact design; 3) low activation voltage and low power consumption. The disadvantage arises from 1) slow gas generation rate that limits its capability of feedback depth control; 2) unstable nonlinear dynamics. The challenge in this study is feedback depth control that requires high gas generation rates and complex nonlinear control algorithms. In future work, we will focus on nonlinear feedback control and improved device design that can gain gas volume rapidly.

Author details

Zheng Chen, T. Um and Hilary Bart-Smith*
University of Virginia, USA

Acknowledgement

This research was supported in part by the Office of Naval Research (ONR) under the Multidisciplinary University Research Initiative (MURI) Grant N00014-08-1-0642 and the David and Lucille Packard Foundation.

5. References

Bar-Cohen, Y. (2004). Electric flex. *IEEE Spectrum,* Vol.41, No.6, pp. 29-33, ISSN 0018-9235

Bar-Cohen, Y. (2000). Electroactive Polymers as Artificial Muscles - Capabilities, Potentials and Challenges, In: *Handbook on Biomimetics,* Y. Osada, (Ed.), 936-950, NTS Inc.

Bond, C. (1996). Swim Bladder. In: *Biology of Fishes* (2nd edition). pp. 283-290, Saunders College Publishing.

Chen, Z. & Tan, X. (2008). A Control-oriented and Physics-based Model for Ionic Polymer-Metal Composite Actuators. *IEEE/ASME Transactions on Mechatronics,* Vol. 13, No. 5, pp. 519–529, ISSN 1083-4435.

Chen, Z. ; Shatara, S. & Tan, X. (2010). Modeling of Biomimetic Robotic Fish Propelled by an Ionic Polymer-Metal Composite Caudal Fin. *IEEE/ASME Transactions on Mechatronics,* Vol. 15, No. 3, pp. 448-459, ISSN 1083-4435.

Chen, Z & Tan, X. (2010). Monolithic Fabrication of Ionic Polymer-Metal Composite Actuators Capable of Complex Deformation, *Sensors and Actuators A: Physical,* Vol. 157, No. 2, pp. 246–257, ISSN 0924-4247.

Chen, Z.; Um, T. I. & Bart-Smith, H. (2011). A Novel Fabrication of Ionic Polymer-Metal Composite Membrane Actuator Capable of 3-dimensional Kinematic Motions. *Sensors and Actuators A: Physical,* Vol. 168, No. 1, pp. 131–139, ISSN 0924-4247.

Chen, Z.; Um, T. I. & Bart-Smith, H. (2012a). Bio-inspired Robotic Manta Ray Powered by Ionic Polymer-Metal Composite Artificial Muscles. *International Journal of Smart and Nano Materials,* to appear, ISSN 1947-5411.

---

* Corresponding Author

Chen, Z; Um, T. I. & Bart-Smith, H. (2012b). Modelling and Control of Artificial Bladder Enabled by Ionic Polymer-Metal Composite. *Proceedings of the 2012 American Control Conference*, to appear, June 27-29 2012, Montreal, Canada, 2012, pp. 3340-3345.

Chung, C.; Fung, P.; Hong, Y.; Ju, M.; Lin, C. & Wu, T. (2006). A Novel Fabrication of Ionic Polymer-Metal Composites (IPMC) Actuator with Silver Nano-powders. *Sensors and Actuators B: Chemical*, Vol. 117, No. 2, pp. 367–375, ISSN 0925-4005.

Gao, J.; Bi, S.; Xu, Y. & Liu, C. (2007). Development and Design of A Robotic Manta Ray Featuring Flexible Pectoral Fins. *Proceedings of the 2007 IEEE International Conference on Robotic and Biomimetics*, pp. 519-523, ISBN 978-1-4244-1761-2, Sanya, China, December 15-18, 2007.

Guo, S.; Fukuda, T. & Asaka, K. (2003). A New Type of Fish-like Underwater Microrobot. *IEEE/ASME Transactions on Mechatronics*, Vol. 8, No. 1, pp. 136–141, ISSN 1083-4435.

Kim, K. J. & Shahinpoor, M. (2002). A Novel Method of Manufacturing Three-dimensional Ionic Polymer-Metal Composites (IPMCs) Biomimetic Sensors, Actuators, and Artificial Muscles. *Polymer*, Vol. 43, No. 3, pp. 797–802, ISSN 0032-3861.

Kim, K. J. & Shahinpoor, M. (2003). Ionic polymer-metal composites: II. Manufacturing techniques. *Smart Materials and Structures*, Vol. 12, No. 1, pp. 65–79, ISSN 0964-1726.

Kim, K. J.; Pugal, D. & Leang, K. K. (2011). A Twistable Ionic Polymer-Metal Composite Artificial Muscle for Marine Applications. *Marine Technology Society Journal*, Vol. 45, No. 4, pp. 83-98, ISSN 0025-3324.

Kim, S. J.; Lee, I. T. & Kim. Y. H. (2007). Performance Enhancement of IPMC Actuator by Plasma Surface Treatment. *Smart Materials and Structures*, Vol. 16, No. 1, pp. N6–N11, ISSN 0964-1726.

Lee, S. J.; Han, M. J.; Kim, S. J.; Jho, J. Y.; Lee, H. Y. & Kim, Y. H. (2006). A New Fabrication Method for IPMC Actuators and Application to Artificial Fingers. *Smart Materials and Structures*, Vol. 15, No. 5, pp. 1217–1224, ISSN 0964-1726.

Lin, L.; Xu, H.; Xie, H. & Shen, L. (2010). Bionic Bladder Based Depth Control for Bionic Underwater Robots. *Proceedings of 2010 8th IEEE International Conference on Control and Automation*, pp. 932-936, ISBN 978-1-4244-5195-1, June 9-11, Xiamen, China, 2010.

McFarland, D.; Gihespy, I. & Honary, E. (2003). Divebot: A Diving Robot with A Whale-like Buoyancy Mechanism. *Robotica*, Vol. 21, No. 4, pp. 385-398, ISSN 0263-5747

Millet, P.; Pineri, M. & Durand, R. (1989). New Solid Polymer Electrolyte Composites for Water Electrolysis. *Journal of Applied Electrochemistry*, Vol. 19, No. 2, pp. 162-166, ISSN 0021-891X.

Mojarrad, M. & Shahinpoor, M. (1996). Noiseless Propulsion for Swimming Robotic Structures Using Polyelectrolyte Ion-exchange Membranes. *Proceedings of SPIE Conference on Smart Structures and Materials*, Vol. 2176, pp. 183, ISSN 0277-786X, February 26, San Diego, California, 1996.

Moored, K.; Smith, W.; Hester, J.; Chang, W. & Bart-Smith, H. (2008). Investigating the Thrust Production of A Myliobatoid-inspired Oscillating Wing. *Advances in Science and Technology*, Vol. 58, pp. 25–30, ISSN 1662-0356.

Moored, K. & Bart-Smith, H. (2009). Investigation of Clustered Actuation in Tensegrity Structures. *International Journal of Solids and Structures*, Vol. 46, No. 17, pp. 3272–3281, ISSN 0020-7683.

Nemat-Nasser, S. & Li, J. (2000). Electromechanical Response of Ionic Polymer–Metal Composites. *Journal of Applied Physics*, Vol. 87, No. 7, pp. 3321–3331, ISSN 0021-8979.

Nemat-Nasser, S. (2002). Micromechanics of Actuation of Ionic Polymer–Metal Composites. *Journal of Applied Physics*, Vol. 92, No. 5, pp. 2899–2915, ISSN 0021-8979.

Punning, A.; Anton, M.; Kruusmaa, M. & Aabloo, A. (2004). A Biologically Inspired Ray-like Underwater Robot with Electroactive Polymer Pectoral Fins. *Proceedings of the IEEE International Conference on Mechatronics and Robotics 2004*, pp. 241-245, ISBN 3-938153-50-X, September 13-15, Aachen, Germany, 2004.

Rosenberger, L. J. (2001). Pectoral Fin Locomotion in Batoid Fishes: Undulation versus Oscillation. *The Journal of Experimental Biology*, Vol. 204, pp. 379–394, ISSN 0022-0949.

Shahinpoor, M. (1992). Conceptual Design, Kinematics and Dynamics of Swimming Robotic Structures Using Ionic Polymeric Muscles. *Smart Materials and Structures*, Vol. 1, No. 1, pp. 91-94, ISSN 0964-1726.

Shahinpoor, M. & Kim, K. J. (2001). Ionic polymer-metal composites: I. Fundamentals. *Smart Materials and Structures*, Vol. 10, No. 4, pp. 819–833, ISSN 0964-1726.

Shahinpoor, M.; Kim, K. J. & Leo, D. (2003). Ionic Polymer-Metal Composites as Multifunctional Materials. *Polymer Composites*, Vol. 24, No. 1, pp. 24-33, ISSN 0272-8397

Shahinpoor, M & Kim K. J. (2005). Ionic polymer–metal composites: IV. Industrial and medical applications. *Smart Materials and Structures*, Vol. 14, No. 1, pp. 197–214 ISSN 0964-1726.

Shaaban, A. H. (1994). Pulsed DC and Anode Depolarization in Water Electrolysis for Hydrogen Generation. In: *StormingMedia*, 01.08.1994, Avalable from http://www.stormingmedia.com.

Shimizu, N.; Hotta, S.; Seekiya, T. & Oda, O. (2005). A Novel Method of Hydrogen Generation by Water Electrolysis Using an Altra-short-pulse Power Supply. *Journal of Applied Electrochemistry*, Vol. 36, No. 4, pp. 419-423, ISSN 0021-891X.

Tan, X.; Kim, D.; Usher, N.; Laboy, D.; Jackson, J.; Kapetanovic, A.; Rapai, J.; Sabadus, B, & Zhou, X. (2006). An Autonomous Robotic Fish for Mobile Sensing. *Proceedings of the 2006 IEEE/RSJ International Conference on Intelligent Robots and Systems*, pp. 5424-5429, ISBN 1-4244-0259-X, Oct. 9-15, Beijing, China, 2006.

Um, T. I.; Chen, Z. & Bart-Smith, H. (2011). A Novel Electroactive Polymer Depth Control Device for Bio-inspired Underwater Vehicles. *Proceedings of the IEEE International Conference on Robotics and Automation*, pp. 172-177, ISBN 978-1-61284-380-3, May 13-19, Shanghai, China, 2011.

Wang, Z.; Wang, Y.; Li, J. & Hang, G. (2009). A Micro Biomimetic Manta Ray Robot Fish Actuated by SMA. *Proceedings of the 2009 IEEE International Conference on Robotics and Biomimetics*, pp. 1809-1813, ISBN 978-1-4244-4774-9, Guilin, China, December 19-25, 2009.

Zhou, C. & Low, K. H. (2102). Design and Locomotion Control of A Biomimetic Underwater Vehicle with Fin Propulsion. *IEEE/ASME Transactions on Mechatronics*, Vol. 17, No. 1, pp. 25-35, ISSN 1083-4435.

Chapter 11

Simultaneous Smart Actuating-Sensing Devices Based on Conducting Polymers

José G. Martínez, Joaquín Arias-Pardilla and Toribio F. Otero

Additional information is available at the end of the chapter

1. Introduction

Towards the end of 1970s new artificial organic materials, conducting polymers (CP), were discovered (Chiang et al., 1977; Inzelt, 2011; Shirakawa et al., 1977). Since then, most of the scientists working on CP became interested by the fact that their conductivity can shift, in a reversible way, over several orders of magnitude by oxidation/reduction (also called doping/dedoping) processes. The availability of these new organic semiconductors has opened up possibilities to rebuilt electronics and microelectronics producing flexible devices (Guo et al., 2010; Klauk, 2006; Perepichka & Perepichka, 2009; So, 2010).

Besides conductivity other properties such as stored charge, stored chemicals, volume, porosity or colour also change during doping/dedoping, under electrochemical control, in parallel with the material composition (counterion content) along the redox reactions (Otero, 2008). Any intermediate oxidation state determines a chemical equilibrium characterized by an equilibrium potential. Any physical (temperature, pressure, applied current) or chemical (electrolyte concentration) magnitude affecting the chemical equilibrium modifies the electrical potential of the material that therefore can be used as a sensor of that magnitude (Otero, 2009).

These properties, whose value changes with the material composition, are allowing the development of different devices (Onoda et al., 1999a; Otero et al., 1992a; Otero et al., 1992c; Pei & Inganäs, 1992a). Volume variations driven by oxidation/reduction reactions are being used to generate new electrical motors having different configurations of the polymeric actuators (Alici et al., 2007; Smela et al., 1993; Wu et al., 2005).

We will present devices constructed with conducting polymers based on an electrochemical reaction and working, simultaneously, as an actuator and as several sensors of the surrounding conditions. During the movement they store and release charge working as a battery: they are multifunctional devices.

2. Electrochemical reactions in conducting polymers

Once generated (chemical or electrochemically), conducting polymers can be considered as both, classical unreactive materials having either a constant composition and a constant value of the magnitude of its physical and chemical properties; or new reactive materials varying their composition (polymeric chains, ions and solvent in different ratios) and properties along several orders of magnitude in a reversible way under electrochemical reaction (Otero, 1999), or promoted by redox agents present in solution (Kuttel et al., 2009).

2.1. p-doping

When the polymer chains are oxidized, consecutive electrons are removed from each chain generating an excess of positive charges (holes) along the chains. This excess of positive charges (lack of electrons) promotes the repulsion between the polymeric chains and the generation of free volume between them. This free space is occupied by anions arriving from the solution to compensate the emerging positive charges (keeping the electroneutrality) and solvent molecules to keep osmotic pressure balance (Huang et al., 1986; Otero, 1999; Tsai et al., 1988).

When the polymer is generated in the presence of small anions, they can be exchanged by other small anions present in solution by electrochemical reactions so a prevailing exchange of anion occurs during reaction:

$$\left(Pol^{0}\right)_{s}+n\left(A^{-}\right)_{sol}+m\left(Solv\right)\rightleftarrows\left[\left(Pol^{n+}\right)_{s}\left(A^{-}\right)_{n}\left(Solv\right)_{m}\right]_{gel}+n\left(e^{-}\right)_{metal} \quad (1)$$

where the different subscripts mean: *s*, solid; *sol*, in solution; *gel* indicates that the material is a gel formed by oxidized polymeric chains (Pol^{n+}) generated after the extraction of *n* electrons (e^-) through the metal (indicated by subscript *metal*) from neutral polymer chains (Pol^0), *n* anions (A^-) coming from the solution to keep the gel electroneutrality and m molecules of solvent (*Solv*) required to keep osmotic pressure balance.

When the polymer is generated in the presence of a macroanion, due to its volume and the interaction with polymer chains, this macroanion cannot be exchanged by the electrochemical reaction keeping trapped inside the polymer film. So, in order to keep the electroneutrality, smaller cations are exchanged with the solution during the reaction:

$$\left[\left(Pol^{0}\right)\left(MA^{-}\right)_{n}\left(C^{+}\right)_{n}\left(Solv\right)_{m}\right]_{gel}\rightleftarrows\left[\left(Pol^{n+}\right)\left(MA^{-}\right)_{n}\right]_{gel}+n\left(C^{+}\right)_{sol}+m\left(Solv\right)+n\left(e^{-}\right)_{metal} \quad (2)$$

where MA^- is the macroanion trapped inside the polymer film and C^+ are cations exchanged in order to keep the electroneutrality.

Usually, the real redox process is not as easy as expressed by reactions (1) and (2): anions and cations are exchanged simultaneously (Hillman et al., 1989; Inzelt, 2008). Usually one of the previous interchanges prevails supporting the greater percentage of charge balance (Kim et al., 2010; Lyutov et al., 2011; Orata & Buttry, 1987; Torresi & Maranhao, 1999).

2.2. n-doping

Some CP such as PEDOT (Ahonen et al., 2000), polythiophene (Arbizzani et al., 1995) or polyfluorenes (Ranger & Leclerc, 1998) have an electronic affinity high enough to allow transitions from the neutral state to a reduced state, storing negative charges (by electron injection) on the chains at high cathodic potentials. In this case, very stable solvents and salts are required, as electrolytes, to perform this redox reaction:

$$\left(Pol^{0}\right)_{s}+n\left(C^{+}\right)_{sol}+m\left(Solv\right)+n\left(e^{-}\right)_{metal}\rightleftarrows\left[\left(Pol^{n-}\right)_{s}\left(C^{+}\right)_{n}\left(Solv\right)_{m}\right]_{gel} \quad (3)$$

where Pol^{n-} represents the reduced polymer chains after insertion of n electrons. Here, in an analogous way compared to reaction (1), an excess of negatives charges promotes repulsion between polymeric chains generating the free volume that will be occupied by cations (exchanged with solution to keep electroneutrality) and solvent (exchanged with solution to keep pressure and osmotic balance).

2.3. Double doping

Some conducting polymers can be doped both, by p-doping and by n-doping. Thus, from their neutral state they can be reduced (suffering n-doping) or oxidized (suffering p-doping). In those polymers the energy difference between both processes is the electrochemical bandgap (Arias-Pardilla et al., 2010; Otero et al., 2011).

3. Reversible change of the electrochemical properties

Electrochemical reactions 1, 2 and 3 are reversible reactions moving through *n* consecutive oxidation states (Otero et al., 2012) having different content of the counterion. The value of the magnitude of those properties being a function of the material composition (electrochemical properties) can also be shifted in a reversible way under faradic control. As previously indicated, conducting polymers can be oxidized and/or reduced from their neutral state, with the entry/exit of ions and solvent. The most studied electrochemical properties of conducting polymers are: volume, colour, stored charge, stored chemicals, porosity or permselectivity, sensing responses or wetability, among others. The progressive and reversible variation of the value of these electrochemical properties allows the development of different devices and products. The change in volume will be reviewed in detail below: this is the most important property of conducting polymers for the development of actuators or artificial muscles.

3.1. Volume variations

A chain of conducting polymer in solution can be considered as an electrochemical molecular motor (Balzani et al., 2005; Davis, 1999; Otero, 2011): movements are produced by reversible conformational changes in chains originated by oxidation/reduction reactions.

The reversible conformational movement from a coil like structure to a rod like structure is produced by extraction (oxidation) or injection (reduction) of *n* electrons through *n* consecutive steps of one electron per step, together with movement of balancing counterions. This results in length variation of a free polymer chain in solution but, in polymer films three dimensional changes of volume are observed. The entanglement of the polymer chains in the film gives reversible swelling or shrinking changes of volume under reversible electrochemical stimulation (Fig. 1).

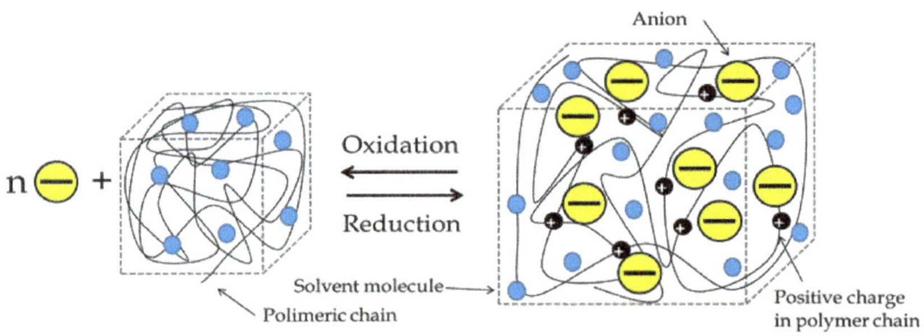

Figure 1. Schematic representation of the reversible volume change associated with the electrochemical reaction in conducting polymer chains during oxidation/reducction during p-doping exchanging anions.

Some mechanical test machines have been developed following length or thickness variations produced by submitting the film to different potential (Bay et al., 2003; Kiefer et al., 2007; Mazzoldi et al., 1998; Spinks et al., 2002) or current (Otero et al., 2006; Otero et al., 2007c) programs. In situ Atomic Force Microscopy (AFM) technique follows film thickness variation during reverse oxidation/reduction processes (Bieńkowski et al., 2011; Cho et al., 2011; Smela & Gadegaard, 2001). In this way, it has been possible to measure a volume difference between reduced and oxidized state up to 35% (Smela & Gadegaard, 1999). The volume change depends on multiple factors such as type of polymer, synthesis conditions (potential or current applied), electrolyte and solvents used.

4. Simultaneous sensing and actuating properties in conducting polymers

The driving reversible electrochemical reaction supports the development of simultaneous sensing and actuating properties by conducting polymer materials and by any electrochemical device based on those materials. The electrochemical working device (based on the electrochemical reaction) senses changes of any physical or chemical variable acting on the polymer reaction rate while working. This is the Le Chatelier principle applied outside equilibrium conditions (Smith, 2004). Therefore, for increasing values of the electrolyte concentrations or of the temperature working under flow of a constant current (constant reaction rate), lower values of the device potential are observed (the reaction is easier) during the transition between the same initial and final oxidation states of the materials. When a greater mechanical work is required (moving faster, applying a higher current, or a higher mechanical stress is needed to move the actuator) the reaction gives increasing potential when the device moves between the same initial and final oxidation states.

Those sensing abilities are intrinsic properties of the reaction. They are characteristics of the material reaction and of any device based on this electrochemical reaction. So, the dual and simultaneous sensing-actuation property is expected to be quantified from electrochemical equations.

The evolution of the conducting polymer film potential with time E(t) during the movement from the same initial oxidation/reduction state to the same final oxidation/reduction state driven by flow of a constant anodic current is given by a stair function (Otero et al., 2012).

$$E(t)=\sum E_n(t)p_n(t)=E_1(t)p_1(t)+E_2(t)p_2(t)+...+E_n(t)p_n(t) \tag{4}$$

where:

$$p_n(t)=u(t-t_n)-u(t-t_{n+1})=\begin{cases}1, t\in[t_n,t_{n+1}]\\0, t\notin[t_n,t_{n+1}]\end{cases} \tag{5}$$

Being t_n the time while the nth electron is removed from every polymeric chain and $E_n(t)$:

$$E_n(t)=E_0+(n-1)\Delta E+\frac{RT}{(1-\alpha)F}\left\{\ln\left(\frac{i_a}{FV}\right)-d\ln\left[A^-\right]-e\ln\left(\left[Pol^*\right]_{initial}-\frac{i_a t}{FV}\right)-\ln k_{a0}\right\} \tag{6}$$

where E_0 is the standard potential, i_a is the applied current; n is the number of consecutive electrons extracted from a chain; ΔE is the increment observed in the potential when a new electron is extracted from a polymeric chain, R is the universal gas constant (R = 8.314 J K^{-1} mol^{-1}); α is the symmetry factor; F is the Faraday constant (F=96485 C mol^{-1}); V, the volume of the film; $[A^-]$ the concentration of anions (counterion) in solution; t, the time of current flow; T is the experimental temperature; d and e are the reactions orders related with the concentration of anions in solution or to that of the active centres $[Pol^*]$ in the film (sites of the polymer where a positive charge will be stored after oxidation) and k_{a0} is the rate constant or rate coefficient for $E=E_0$.

Therefore, Eqs. 4 and 6 are the sensing equations: the evolution of the device potential during actuation is a function of either, driving (current) and working (temperature, electrolyte concentration and film volume) variables.

Being electrical machines, by integration of Eq. 6 the evolution of the electrical energy consumed by the electrochemical device (U_a) during the actuation time is attained:

$$U_a(t)=i_a\int E(t)dt=i_a t\{E_0+(n-1)\Delta E\}+\frac{RTi_a t}{(1-\alpha)F}\left\{\ln\left(\frac{i_a}{FV}\right)-d\ln\left[A^-\right]-\ln k_{a0}\right\}+$$
$$\frac{RTVe}{(1-\alpha)}\left\{\ln\left(\left[Pol^*\right]_{initial}-\frac{i_a t}{FV}\right)-1\right\}\left\{\left[Pol^*\right]-\frac{i_a t}{FV}\right\} \tag{7}$$

The consumed energy (U_a) after any constant time (t) of current flow is also a sensing function of the same variables. Fig. 2 shows the good agreement between experimental and theoretical results for the consumption of three different charges (from the same initial oxidation/reduction state, three different final oxidation/reduction states are obtained) at different experimental temperatures by flow of a constant anodic current for three different times of current flow.

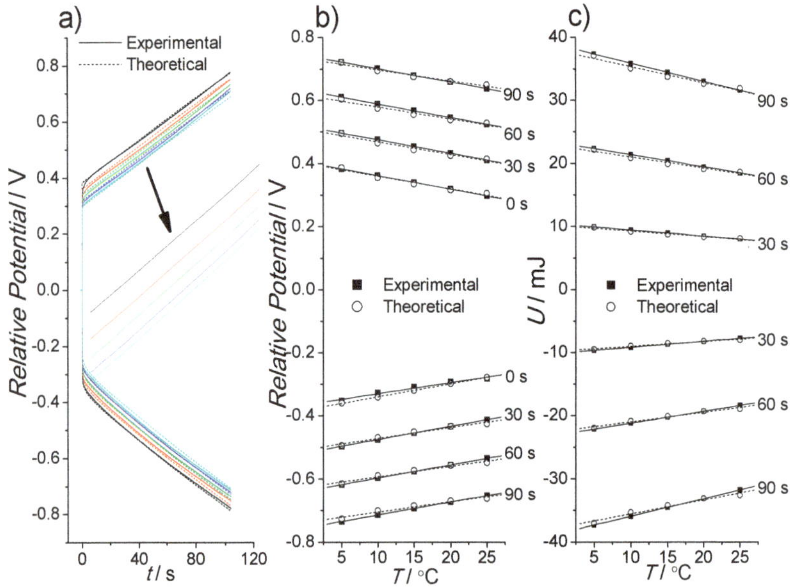

Figure 2. a) Anodic and cathodic experimental (full lines) and theoretical (dotted lines) chronopotentiograms obtained by flow of ±0.75 mA through a 1.6 mg pPy film (10.77 mm x 5.09 mm x 19 μm) at different temperatures (black line: 5ºC; red line: 10ºC; green line: 15ºC; blue line: 20ºC and cyan line: 25ºC) in 1 M $LiClO_4$ aqueous solution. b) Achieved potential after different times of anodic (positive) or cathodic (negative) current flow. c) Consumed electrical energy after the same times of current flow. Reproduced from (Otero et al., 2012), with permission of American Chemical Society).

5. Artificial muscles

In general, artificial muscles are devices attempting to reproduce composition, characteristics and capabilities of natural muscles. Different materials have been used as piezoelectrics (PZT), shape memory alloys (SMA) (Ouyang et al., 2008), carbon based materials as nanotubes (Baughman et al., 1999) or graphene (Huang et al., 2012) and polymeric gels. Also the pneumatic approach has been studied with contractile or extensional devices operated by pressurized air (Daerden & Lefeber, 2002).

Natural muscles transform chemical energy into mechanical energy and heat. Their actuation involves: a) aqueous media, b) an electric pulse arriving from the brain to the muscle through the nervous system, c) liberation of calcium ions inside the sarcomere, d) chemical reactions (ATP hydrolysis), e) conformational changes along natural polymeric chains (actin and myosin) with change of the sarcomere volume and f) water exchange. Among of the above mentioned materials for artificial muscles, conducting polymers are the most similar to natural devices including: electric pulses, polymeric chains, chemical reactions, aqueous solutions (ions and water), volume variations, and strain and stress changes. Two families of electrical polymeric actuators can be differentiated (Otero et al., 2007a; Otero, 2000; Otero, 2007):

- **Electromechanical actuators**: the polymer responds to electric fields (E) and the polymer chains do not participate in chemical reactions (the structure of its intramolecular chemical bonds doesn't change) during actuation. In this type of actuator the dimension variations are proportional to E^2 for electrostrictive actuators and to E for electrostatic and electrokinetic (movement of ions and/or solvent molecules) actuators.
- **Electrochemomechanical actuators**: the polymer chains respond to electric currents and participate in electrochemical or chemical reactions, changing the structure of its chemical bonds, varying their composition and originating volume changes. In this case dimensional variations are under control of the electrochemical reaction becoming proportional to consumed charge. Conducting polymers can be used for the production of this kind of actuators. Whether carbon nanotubes and graphenes based actuators are electromechanical or electrochemomechanical is still under discussion (Gimenez et al., 2012; Mukai et al., 2011). Any electrochemically reactive material that can be laminated as stable films can be used as part of electrochemomechanical actuators.

From now on we will focus on electrochemomechanical actuators. These actuators have been built using different configurations as discussed below, always containing one or several films of conducting polymers, where reaction 1, 2 or 3 takes place. Volume variations generated by reactions 1, 2 or 3 are almost isotropic, while natural muscles are anisotropic devices. So to produce anisotropy, only volume changes following length variation of the films are used, with the consequent efficiency reduction.

5.1. Bending artificial muscles

Historically the first way to transduce reversible length variations in films of conducting polymers into macroscopic movements was through a bilayer, or bimorph structure, i.e. CP/passive layer (Otero et al., 1992b; Otero et al., 1992a; Otero et al., 1992c; Pei & Inganäs, 1992a) (Fig. 3). The variation of the mechanical stress gradient generated across the bilayer interface by swelling/shrinking processes induced by the electrochemical reactions in the conducting polymer film develops a macroscopic movement of the bilayer free end by the progressive bending of the device. The direction (clockwise or anticlockwise) of the movement depends on the prevalent ionic exchange (anions or cations) of the conducting polymer film. Conducting polymers with a prevalent exchange of anions swell by oxidation, pushing the bilayer free end meanwhile conducting polymers with prevalent cation exchange shrink during oxidation, trailing the device. Different materials have been used as passive layer, for example a tape (Otero et al., 1992b; Otero et al., 1992a; Otero et al., 1992c; Pei & Inganäs, 1992a), a sputtered metal (Jager et al., 2000a; Jager et al., 2000b; Smela et al., 1993), a piece of paper (Deshpande et al., 2005b), a non conducting plastic (Higgins et al., 2003), a solid state electrolyte film (Alici et al., 2011; Baughman, 1996) or a thin film of any flexible material which is metal coated (i.e. by sputtering) (Deshpande et al., 2005a).

In a similar way it is possible to obtain bending movement from asymmetrical monolayers of the same conducting polymer, having an internal asymmetry capable of producing asymmetric swelling or shrinking across the film under the same electrochemical process (Okamoto et al., 2000; Onoda et al., 1999a; Onoda & Tada, 2004; Onoda et al., 1999b; Shakuda et al., 1993; Takashima et al., 2003; Takashima et al., 1997; Wang et al., 2002). Here half of the film has a prevalent anionic exchange, while the second half experiences a prevalent cationic exchange. These asymmetrical monolayers are obtained in two separate stages of electrogeneration using different salts and the same monomer. Other ways are being studied to produce asymmetric monolayers by physical means, for example, by growing the conducting polymer on adsorbed and porous materials (Li et al., 2004), or by electrochemical means generating a film of conducting polymer with a counterion concentration gradient (Okuzaki & Hattori, 2003; Sansiñena et al., 2003), conductivity (Nakano & Okamoto, 2001; Onoda et al., 2005) or morphology gradients (Han & Shi, 2006; Okamoto et al., 2001) by crosslinked networks.

All these bilayer devices require a counter-electrode in order to close the electrical circuit allowing the current flow. In this electrode (usually a metal) different electrochemical reactions as solvent oxidation, must occur during current flow consuming a major fraction of the electrical energy, resulting in pH variations and generating new chemicals, which can deteriorate progressively the bilayer device.

Trying to avoid the counter electrode and its associated problems, a three layer structure was proposed (Otero et al., 1992c). Initially, it was produced by using a central passive film (two sides tape) each side coated with a conducting polymer film (Fig. 3). The triple layer was immersed in an electrolyte allowing the current flow. One of the conducting polymer films acts as the anode while the second film acts as the cathode (Garcia-Cordova et al., 2011; John et al., 2008; Yao et al., 2008). But using this configuration it is also possible to obtain movement outside a liquid electrolyte media using an ionic conducting membrane to separate the two films of conducting polymers. This membrane can be obtained by solvent evaporation and UV irradiation (Blonsky & Meridian, 1997; Heuer et al., 2002; Sansinena et al., 1997; Song et al., 2002), or by formation of interpenetrated networks (Cho et al., 2007; Plesse et al., 2005; Vidal et al., 2009; Vidal et al., 2003). In this case, the two conducting polymer films are generated by chemical polymerization on the external part of the membrane. Using this approach, multilayer devices were constructed and characterized (Ikushima et al., 2009; Zainudeen et al., 2008). The three-layer configuration provides greater efficiencies of the consumed energy: the same current is used two times to produce opposite electrochemical reactions and volume variations in the conducting polymer films; the anode swells and pushes the device and the cathode shrinks and trails the device.

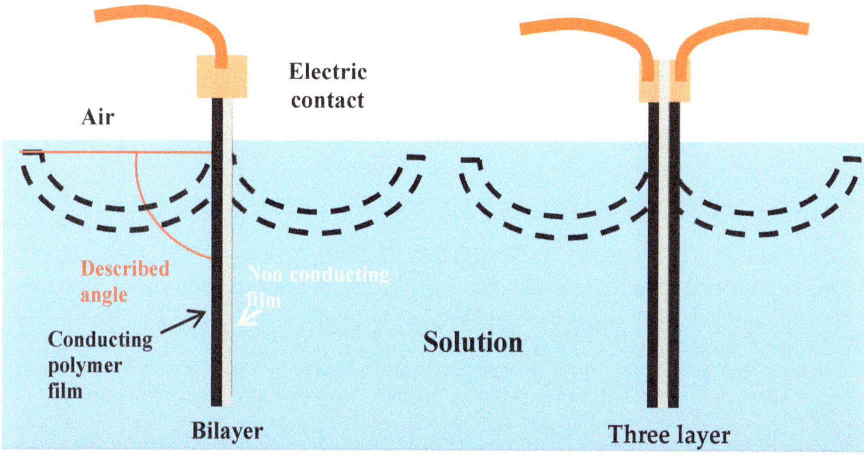

Figure 3. Bilayer and three layer devices in solution, formed by conducting polymer films and non-conductive films. Described angle is also shown.

5.2. Longitudinal or linear artificial muscles

Freestanding conducting polymer films are the simplest longitudinal actuators (DellaSanta et al., 1997). Its actuation principle is based on longitudinal expansion and contraction of the polymer during ionic exchange, although expansion and contraction occur in all three dimensions, as previously indicated. To improve the performance of these actuators overcoming problems of fragility, multilayered actuators were proposed (Hara et al., 2006; Kaneto et al., 2008; Kaneto et al., 2009), in which several thin conducting polymer films and an electrolyte (ionic liquid-soaked paper) are used to develop a compact and scalable longitudinal actuator with a high work output (Ikushima et al., 2009). Also folded films with Origami shapes provide good linear movements (Okuzaki, 2008).

Fibres of conducting polymer also can be considered as longitudinal actuators. Fibres can be obtained by extrusion (Mazzoldi et al., 1998) or by chemical polymerization over a fibre-shaped substrate (Ismail et al., 2011; Lu et al., 2002), or hollow fibre solid polymer electrolyte (Plesse et al., 2010) making it possible to obtain two concentric CP films separated by the electrolytic medium, allowing its movement in air (Dobbelin et al., 2010; Plesse et al., 2009; Vidal et al., 2010; Vidal et al., 2009). Microrods (Cho et al., 2011) or nanorods (Park et al., 2009; Vlad et al., 2012) of conducting polymers and bundles of films or fibres were investigated to produce vertical displacements of weights (Lu et al., 2002). Also, conducting polymer tubes were generated using springs and helical metallic wires (Ding et al., 2003; Hara et al., 2003; Hara et al., 2005; Spinks et al., 2003b) or zigzag metal wires (Hara et al.,

2004; Morita et al., 2010) as substrates, looking for uniform potential and current distribution. When individual fibres, bundles or tubes are used, a counter electrode is necessary, for the same reasons given above in the case of the bilayers.

Finally, it is possible to obtain linear displacement by combination of different bending structures as bilayers (Fuchiwaki et al., 2009; Naka et al., 2010; Otero & Broschart, 2006) or trilayers (Mutlu & Alici, 2010; Otero et al., 2007b) achieving longitudinal displacements over 60% of their original length.

6. Sensing and tactile muscles

As mentioned above, any reactive (electrochemical) device based on conducting polymers will sense every variable influencing the electrochemical reaction rate during actuation. Following this basic principle of the chemical kinetics sensing and tactile artificial muscles have been developed.

While a current is applied to the artificial muscle, producing a mechanical work, it is possible to follow the potential achieved in the muscle at every time. Under flow of a constant current (constant reaction rate) the achieved potential is lower when a variable which favours the electrochemical reaction increases. This is the case for temperature (Garcia-Cordova et al., 2011; Ismail et al., 2011; Otero & Cortes, 2003b; Valero Conzuelo et al., 2010; Valero et al., 2010) or electrolyte concentration (Arias-Pardilla et al., 2011; Garcia-Cordova et al., 2011; Otero & Cortes, 2003b; Otero et al., 2007b; Valero Conzuelo et al., 2010). On the other hand, the potential shifts to higher values when a variable makes the reaction harder: the muscle moves larger masses (Garcia-Cordova et al., 2011; Otero et al., 2007b; Valero Conzuelo et al., 2010) or moves the mass faster by applying now rising currents (Garcia-Cordova et al., 2011; Ismail et al., 2011; Valero et al., 2011).

Being the potential evolution a sensor of the working variables, the electrical energy (U) consumed by the device during actuation and obtained by integration is also sensor:

$$U(t) = \int E(t) I dt \tag{8}$$

Where t is the elapsed time, E(t) is the potential evolution during the actuation time and I is the constant applied current.

When a free muscle moves driven by a constant current finds an obstacle, the potential steps to higher values, trying to produce more energy and to shift the obstacle. The potential increment detects the object and its mechanical resistance to be shifted. Related with this property, artificial muscles with tactile sensitivity have been developed too (Otero & Cortes, 2003a).

7. Multi-devices: Actuators-sensors-battery-electrochromic

As mentioned above, different properties can be tuned simultaneously and in a reversible way during electrochemical reactions in conducting polymers. Electrochemical devices created with conducting polymers may change several of those properties, such as volume, stored charge or color, at the same time. For example, using the same configuration, changing the quantity of conducting polymer used in the device, it's possible to obtain different optical or mechanical devices (Vidal et al., 2010).

Also, as the configuration of a three layer is the same of a battery with two conducting polymer films as electrodes, electrical charge is trapped in the actuator during electrochemical reaction. A way to recover that charge may be developed in order to recover that energy, acting now as a battery and moving the actuator in the opposite direction until a uniform oxidation state is attained in both polymer films (vertical position in Fig. 3).

8. Theoretical models

Several models have been proposed to characterize the electrochemomechanical behaviour of artificial muscles. At the moment, there exist different approaches from different fields as mathematics, physics or chemical-physics.

8.1. Faradic control of the movement

As mentioned above, artificial muscles are electrochemomechanical machines controlled by the electrochemical reaction occurring while actuating. As any electrochemical reaction, actuation in conducting polymers is controlled by the charge consumed during actuation. Volume changes are not an exception: according to reactions 1, 2 or 3, volume variations will be given by the number of extracted (or inserted) charges from (or to) the polymeric chains, promoting the swelling/shrinking and the ionic exchange during reaction.

For bending bilayer or trilayer artificial muscles, it has been experimentally proven that the described angle (α in rad) follows a linear relationship with the consumed charge (Q in C):

$$\alpha(t) = kQ(t) \tag{9}$$

where k is a constant (rad C^{-1}) depending on every actuator system: the device (conducting polymer and isolating tape) and the electrolyte where is moving.

By definition of the angular rate of the movement (ω):

$$\omega(t) = \frac{d\alpha(t)}{dt} = k\frac{dQ(t)}{dt} = kI(t) \tag{10}$$

This expression confirms the faradic control of the movement: the angular rate of the movement is a linear function of the applied current. Any increment of the current produces (immediately, without any relaxing time) a faster movement of the actuator, by stopping the current flow the movement stops (the driving reaction and the film volume variation stops). Eq. 9 also indicates that the actuator is a positioning device: the same charge produces the same displacement and the charge consumed during description of a movement of one degree ($\alpha/Q = k$) is constant (independent of the applied current).

The above expressions can be normalized by mass unit of active conducting polymer reacting during actuation. This allows predicting the behaviour of every artificial muscle moving in a known electrolyte made of the same material whatever the geometry of the device is (shape, thickness, surface area, etc.). That means that the same change of the specific composition (according with reactions 1, 2 or 3) per unit time produce the same angular rate in devices having different geometry. This means that experiments from one muscle are only required in order to obtain this faradic characteristic of the CP.

The faradic control of the movement has been checked with different artificial muscles made of different polymers, exchanging both anions (Otero & Cortes, 2004) or cations (Valero et al., 2011).

8.2. Bending beam method

The bending beam method (Gere & Goodno, 2009; Timoshenko, 1925) is based on the analogy existing between a bending artificial muscle and a solid-state bending beam: the study of the forces generated at the interface between the non-conductive layer, keeping its volume constant during actuation, and the conducting polymer film, varying its volume locally.

This mechanical model assumes several characteristics related with the study of traditional mechanical bending beam: (I) the thickness of the beam is small compared to the minimum radius of curvature, (II) a linear relationship exists between stress and strain of the material and (III) the Young's modulus, Y, and the actuation expansion coefficient of the conducting polymer, α, keep constant: they do not depend on spatial location inside each layer.

The actuator curvature radius (R_∞ is the radius at the equilibrium and R_0 is the initial radius) is related to either, the Young's modulus (Y) and the thicknesses (h) of the conducting and non-conductive films (indicated by subscripts 1 and 2 respectively), and to the volume changes locally produced at the interface between both films α(t) (Pei & Inganäs, 1993b; Pei & Inganäs, 1993a; Pei & Inganäs, 1992b):

$$\frac{1}{R_\infty}-\frac{1}{R_0}=\frac{6\alpha_{(t)}}{\dfrac{\left(Y_1h_1^2-Y_2h_2^2\right)^2}{Y_1Y_2h_1h_2\left(h_1+h_2\right)}+4\left(h_1+h_2\right)} \tag{11}$$

Christophersen et al. (Christophersen et al., 2006) expanded the model by including strain and modulus variations along the direction of film thickness. Actuator's position, rate of the movement and force generated by the actuator (Alici & Huynh, 2006; Alici et al., 2006b) were simulated and applied to the design of biomimetic device (propulsion fins) (Alici et al., 2007). Du et al. (Du et al., 2010) have developed a general model for a multilayer system (N layers) to link the actuation strain of the actuator to the bending curvature.

8.3. Finite element method

The finite elements methodology is a well know mathematical treatment for engineering designs that can be applied to solve the movement of the artificial muscles too. Alici et al. (Alici et al., 2006a; Metz et al., 2006) developed a model based on a lumped-parameter mathematical model for trilayer actuators employing the analogy between thermal strain and the real strain (due to the insertion/extraction of ions inside the polymeric film) in polypyrrole actuators actuating in air. An optimization of the geometry was required, in order to obtain the greater output properties from a determined input voltage. Shapiro et al. (Shapiro & Smela, 2007) developed a two dimensional model (along a full area) to obtain curvature and angular moment from bilayer and trilayer actuators. Thus, they combined the results from the previous method (bending beam method) with finite element method to attain a solution. Another example of the employment of this method was carried out by Gutta et al. who applied it to the study the movement of a cylindrical ionic-polymer metal composite actuator (Gutta et al., 2011).

8.4. Equivalent transmission line model

Electrochemical systems, as many other systems, can be assimilated to electrical circuits and electrochemomechanical actuators have been treated by the equivalent transmission line method. This resource is a practical tool due to the great number of facilities available to the study of electrical circuits through different steps or modules. Such treatments are employed by engineers and physicists, or electrochemists, in order to explain the claimed capacitive behaviour of CP (Albery & Mount, 1993; Bisquert et al., 2000; Paasch, 2000). Ren et al. (Ren & Pickup, 1995) proposed equivalent electrical circuits to model the electron transport and electron transfer in composite pPy-PSS films based on Albery's works. Fang et al. (Fang et al., 2008; Yang et al., 2008) have developed a scalable method including dynamic actuation performance under a given voltage input, joining three different modules for different aspects of the actuator: electrochemical dynamics, stress-generation by charge and mechanical dynamics. Shoa et al. (Shoa et al., 2011) developed a dynamic electromechanical method for electrochemically driven conducting polymer actuators based on a 2-D impedance model using an RC transmission line equivalent circuit to predict the charge transfer during actuation. Besides, a mechanical model (based on the bending beam model) is considered after the equivalent circuit that simulates ion "diffusion" through the thickness and electronic resistance along the length (Shoa et al., 2010). If the angular movement is not linear along the full geometry of the actuator, the bending beam method has to be modified, for example for cantilever type conducting polymer actuators (Alici, 2009).

From all these kind of models, it is possible to employ only one or several of them at the same time in order to obtain the best required model (Woosoon et al., 2007).

9. Actuators applications

The investigation of these devices is mainly performed in academic laboratories. Nevertheless a rising number of applications and products are emerging with pioneering companies that are being incorporated by large multinationals. So Creganna Tactx Medical and Bayer MaterialScience recently acquired a pair of companies working in the field, indicating the potential of these technologies. Also EAMEX from Japan is developing actuators for biomedical and robotic applications. NASA and ESA space agencies consider polymeric actuators as preferential technologies, and the European Scientific Network for Artificial Muscles (ESNAM) has started funded by the European Union. Many different applications can be found in literature. The following is a summary of a few of them, both macroscopic and microscopic.

- **Smart electronic textiles**: Although different strategies have been used to overcome problems like short lifetimes and high response times (Carpi & De Rossi, 2005; De Rossi et al., 2009; Wallace et al., 2007), much work has to be done in order to endow textiles with efficient and reliable actuating functions.
- **Braille displays**: Conducting polymers is one of the materials explored to develop low-cost and efficient displays for Braille text. Actual displays are close to achieve the required specs, but they have to overcome short cycling life (Ding et al., 2003; Spinks et al., 2003a).
- **Micropump**: It is possible to find in literature different types of micropumps made with conducting polymer actuators able to transport fluids at a microflow rate with high precission with low energy consumption (Kiefer et al., 2008; Kim et al., 2008; Morita et al.,2010 ; Naka et al., 2010; Ramírez-García & Diamond, 2007; Wu et al., 2005).
- **Propulsion or locomotion devices**: Systems based on trilayer configuration were proposed in aquatic devices (Alici et al., 2007; James et al., 2007; McGovern & et al., 2009; McGovern et al., 2010) for creating enough thrust for propulsion. Also bending actuators like cilia can be used in mini-robotic devices (Alici & Gunderson, 2009).
- **Microcatheters**: By depositing two conducting polymer films around a passive catheter and using a fast conductor solid state electrolyte, it is possible to control the movement of the catheter tip (DellaSanta et al., 1996; Shoa et al., 2008). This kind of microcatheters can be used in Optical Coherence Tomography to enable high-resolution 3D imaging (Lee et al., 2009).
- **Microactuators**: The electrochemical synthesis of conducting films and their electrochemical actuation (Wilson et al., 2007) are suitable for the construction different microdevices (He et al., 2007; Jager et al., 2001; Pede et al., 1998; Roemer et al., 2002) and microtools constituted by bilayers (Jager et al., 2000a; Jager et al., 2001; Jager et al., 2000b; Jager et al., 1999; Smela, 1999; Smela & Gadegaard, 1999; Smela et al., 1995; Smela et al., 1993) or trilayers (Gursel et al., 2009; Kiefer et al., 2008) using microelectronic technologies. These microactuators can be integrated in CMOS chips containing all the elements required to transport, contain and nurture biological cells (Jager et al., 2002; Smela et al., 2007).

10. Published papers and patents evolution

An indication of the initiation and fast growth of the field is given by bibliographic and patents evolution. As can be observed, the literature related to conducting polymers actuators starts from 1990 (Fig. 4.a) reaching a maximum in 2006 with 45 papers, an excellent value for such a specific topic. After that, around 30 papers have been published every year. Fig. 4.b shows exponential evolution of citations with a value of almost 2000 in 2010. A similar evolution is observed in Fig. 4.c for patents, reaching a maximum in 2009 with 180 patents.

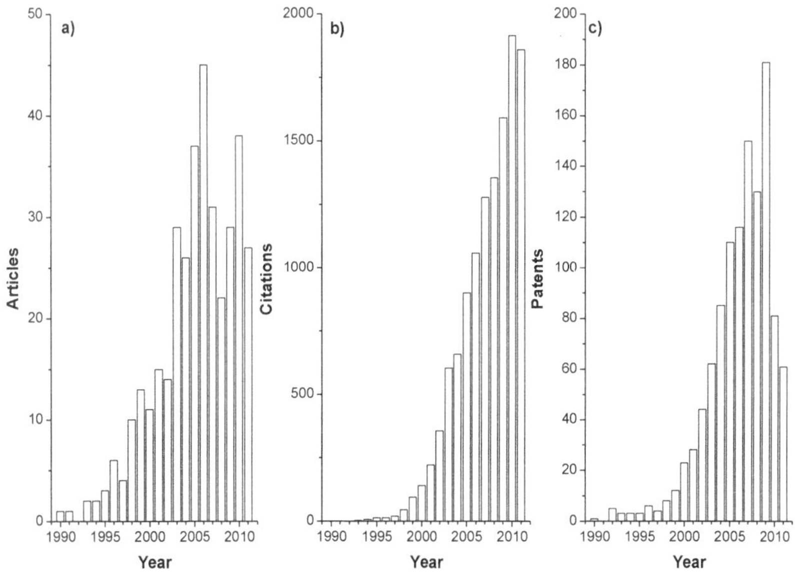

Figure 4. Evolution of a) published papers, b) citations and c) patents for "conducting polymer & actuator", from the ISI web of Knowledge® and Scopus®.

11. Future and challenges

Although a hard work has been performed, it does not exist yet an uncontroversial model for the description of the new reactive polymers and devices based on these reactive polymer gels whose composition and properties mimic those from artificial organs in mammals. New models should include concepts from very different fields of knowledge as electrochemistry, mechanics, polymer science and thermodynamics. In those systems, electrochemical reactions produce structural (conformational) changes in the polymeric chains and macroscopic volume changes related to the composition change of the material. Those aspects are not considered in the classical chemical kinetic models. Conducting polymer films should be used as models for the study and quantification of chemical kinetics under structural control with the aim to develop a new structural chemical model, able to quantify conformational changes and structural information in conducting polymers and in biological reactions originating life in living cells.

This model will allow the synthesis of new conducting polymers providing a more precise control of structural changes and intermolecular forces (polymer-ion interactions) with the reaction. Those materials should be used in a new generation of polymeric actuators able to overcome current limitations, opening possibilities for new applications.

Other factors, such as configuration, manufacture of the actuators, design of the electric contacts or electrolytic media (solid or liquid) are also very important for the improvement of the electrochemomechanical actuators. This is a multidisciplinary field, a lot of work performed by specialists from different disciplines is required in order to attain a good control and modelling of both, devices and soft robots.

Author details

José G. Martínez, Joaquín Arias-Pardilla and Toribio F. Otero
Universidad Politécnica de Cartagena, Spain

Acknowledgement

Authors acknowledge financial support from Spanish Government (MCINN) Projects MAT2008-06702, MAT2011-24973, Seneca Foundation Project 08684/PI/08 and to the European network for Artificial Muscles ESNAM. Jose G. Martinez acknowledges Spanish Education Ministry for a FPU grant (AP2010-3460).

12. References

Ahonen, H. J., Lukkari, J. & Kankare, J. (2000). n- and p-doped poly(3,4-ethylenedioxythiophene): Two electronically conducting states of the polymer. *Macromolecules*, Vol. 33, No. 18, pp. 6787-6793, ISSN 0024-9297.

Albery, W. J. & Mount, A. R. (1993). Application of A Transmission-Line Model to Impedance Studies on A Poly(Vinylferrocene)-Modified Electrode. *J.Chem.Soc.Faraday T.*, Vol. 89, No. 2, pp. 327-331, ISSN 0956-5000.

Alici, G. & Huynh, N. N. (2006). Predicting force output of trilayer polymer actuators. *Sens. Actuat. A-Phys*, Vol. 132, No. 2, pp. 616-625, ISSN 0924-4247.

Alici, G., Metz, P. & Spinks, G. M. (2006a). A methodology towards geometry optimization of high performance polypyrrole (PPy) actuators. *Smart Mater.Struct.*, Vol. 15, No. 2, pp. 243-252, ISSN 0964-1726.

Alici, G., Mui, B. & Cook, C. (2006b). Bending modeling and its experimental verification for conducting polymer actuators dedicated to manipulation applications. *Sens. Actuat. A-Phys*, Vol. 126, No. 2, pp. 396-404, ISSN 0924-4247.

Alici, G., Spinks, G., Huynh, N. N., Sarmadi, L. & Minato, R. (2007). Establishment of a biomimetic device based on tri-layer polymer actuators-propulsion fins. *Bioinspir. Biomim.*, Vol. 2, No. 2, pp. S18-S30, ISSN 1748-3182.

Alici, G. (2009). An effective modelling approach to estimate nonlinear bending behaviour of cantilever type conducting polymer actuators. *Sens. Actuat. B-Chem*, Vol. 141, No. 1, pp. 284-292, ISSN 0925-4005.

Alici, G. & Gunderson, D. (2009). A bio-inspired robotic locomotion system based on conducting polymer actuators. In: *2009 IEEE/ASME International Conference on Advanced Intelligent Mechatronics*. pp. 998-1004, IEEE, ISBN 978-1-4244-2852-6, New York.

Alici, G., Punning, A. & Shea, H. R. (2011). Enhancement of actuation ability of ionic-type conducting polymer actuators using metal ion implantation. *Actuat. B-Chem*, Vol. 157, No. 1, pp. 72-84, ISSN 0925-4005.

Arbizzani, C., Catellani, M., Mastragostino, M. & Mingazzini, C. (1995). N-doped and p-doped polydithieno[3,4-b-3',4'-d] thiophene - a narrow-band gap polymer for redox supercapacitors. *Electrochim. Acta*, Vol. 40, No. 12, pp. 1871-1876, ISSN 0013-4686.

Arias-Pardilla, J., Walker, W., Wudl, F. & Otero, T. F. (2010). Reduction and Oxidation Doping Kinetics of an Electropolymerized Donor-Acceptor Low-Bandgap Conjugated Copolymer. *J. Phys. Chem. B*, Vol. 114, No. 40, pp. 12777-12784, ISSN 1520-6106.

Arias-Pardilla, J., Plesse, C., Khaldi, A., Vidal, F., Chevrot, C. & Otero, T. F. (2011). Self-supported semi-interpenetrating polymer networks as reactive ambient sensors. *J. Electroanal. Chem.*, Vol. 652, No. 1-2, pp. 37-43, ISSN 1572-6657.

Balzani, V., Credi, A., Ferrer, B., Silvi, S. & Venturi, M. (2005). *Artificial molecular motors and machines: Design principles and prototype systems*, Springer-Verlag, ISBN 0340-1022, Berlin.

Baughman, R. H. (1996). Conducting polymer artificial muscles. *Synth. Met.*, Vol. 78, No. 3, pp. 339-353, ISSN 0379-6779.

Baughman, R. H., Cui, C. X., Zakhidov, A. A., Iqbal, Z., Barisci, J. N., Spinks, G. M., Wallace, G. G., Mazzoldi, A., De Rossi, D., Rinzler, A. G., Jaschinski, O., Roth, S. & Kertesz, M. (1999). Carbon nanotube actuators. *Science*, Vol. 284, No. 5418, pp. 1340-1344, ISSN 0036-8075.

Bay, L., West, K., Sommer-Larsen, P., Skaarup, S. & Benslimane, M. (2003). A conducting polymer artificial muscle with 12% linear strain. *Adv. Mater.*, Vol. 15, No. 4, pp. 310-313, ISSN 0935-9648.

Bieńkowski, K., Strawski, M. & Szklarczyk, M. (2011). The determination of the thickness of electrodeposited polymeric films by AFM and electrochemical techniques. *J. Electroanal. Chem.*, Vol. 662, No. 1, pp. 196-203, ISSN 1572-6657.

Bisquert, J., Belmonte, G. G., Santiago, F. F., Ferriols, N., Yamashita, M. & Pereira, E. C. (2000). Application of a distributed impedance model in the analysis of conducting polymer films. *Electrochem.Commun.*, Vol. 2, No. 8, pp. 601-605, ISSN 1388-2481.

Blonsky, P. M. & Meridian, I. (1997). Structurally stable gelled electrolytes, US Patent 5648011.

Carpi, F. & De Rossi, D. (2005). Electroactive polymer-based devices for e-textiles in biomedicine. *IEEE T. Inf. Technol. B*, Vol. 9, No. 3, pp. 295-318, ISSN 1089-7771.

Chiang, C. K., Fincher, C. R., Park, Y. W., Heeger, A. J., Shirakawa, H., Louis, E. J., Gau, S. C. & Macdiarmid, A. G. (1977). Electrical-Conductivity in Doped Polyacetylene. *Phys. Rev. Lett.*, Vol. 39, No. 17, pp. 1098-1101, ISSN 0031-9007.

Cho, M., Seo, H., Nam, J., Choi, H., Koo, J. & Lee, Y. (2007). High ionic conductivity and mechanical strength of solid polymer electrolytes based on NBr/ionic liquid and its application to an electrochemical actuator. *Sens. Actuat. B-Chem*, Vol. 128, No. 1, pp. 70-74, ISSN 0925-4005.

Cho, M. S., Choi, J. J., Kim, T. S. & Lee, Y. (2011). In situ three-dimensional analysis of the linear actuation of polypyrrole micro-rod actuators using optical microscopy. *Sens. Actuat. B-Chem*, Vol. 156, No. 1, pp. 218-221, ISSN 0925-4005.

Christophersen, M., Shapiro, B. & Smela, E. (2006). Characterization and modeling of PPy bilayer microactuators - Part 1. Curvature. *Sens. Actuat. B-Chem*, Vol. 115, No. 2, pp. 596-609, ISSN 0925-4005.

Daerden, F. & Lefeber, D. (2002). Pneumatic Artificial Muscles: actuators for robotics and automation. *Eur. J. Mech. Environ. Eng.*, Vol. 47, No. 1, pp. 10-21, ISSN 1371-6980.

Davis, A. P. (1999). Nanotechnology - Synthetic molecular motors. *Nature*, Vol. 401, No. 6749, pp. 120-121, ISSN 0028-0836.

De Rossi, D., Carpi, F. & Galantini, F. (2009). Functional Materials for Wearable Sensing, Actuating and Energy Harvesting. In: *Biomedical Applications of Smart Materials, Nanotechnology and Micro/Nano Engineering*. pp. 247-256, Trans Tech Publications Ltd, ISBN 978-3-908158-23-3, Stafa-Zurich.

DellaSanta, A., Mazzoldi, A. & de Rossi, D. (1996). Steerable Microcatheters Actuated by Embedded Conducting Polymer Structures. *J. Intel. Mat. Syst. Str.*, Vol. 7, No. 3, pp. 292-300, ISSN 1045-389X.

DellaSanta, A., DeRossi, D. & Mazzoldi, A. (1997). Performance and work capacity of a polypyrrole conducting polymer linear actuator. *Synth. Met.*, Vol. 90, No. 2, pp. 93-100, ISSN 0379-6779.

Deshpande, S. D., Kim, J. & Yun, S. R. (2005a). New electro-active paper actuator using conducting polypyrrole: actuation behaviour in LiClO4 acetonitrile solution. *Synth. Met.*, Vol. 149, No. 1, pp. 53-58, ISSN 0379-6779.

Deshpande, S. D., Kim, J. & Yun, S. R. (2005b). Studies on conducting polymer electroactive paper actuators: Effect of humidity and electrode thickness. *Smart Mater. Struct.*, Vol. 14, No. 4, pp. 876-880, ISSN 0964-1726.

Ding, J., Liu, L., Spinks, G. M., Zhou, D., Wallace, G. G. & Gillespie, J. (2003). High performance conducting polymer actuators utilising a tubular geometry and helical wire interconnects. *Synth. Met.*, Vol. 138, No. 3, pp. 391-398, ISSN 0379-6779.

Dobbelin, M., Marcilla, R., Pozo-Gonzalo, C. & Mecerreyes, D. (2010). Innovative materials and applications based on poly(3,4-ethylenedioxythiophene) and ionic liquids. *J. Mater. Chem.*, Vol. 20, No. 36, pp. 7613-7622, ISSN 0959-9428.

Du, P., Lin, X. & Zhang, X. (2010). A multilayer bending model for conducting polymer actuators. *Sens. Actuat. A-Phys*, Vol. 163, No. 1, pp. 240-246, ISSN 0924-4247.

Fang, Y., Tan, X. O., Shen, Y. T., Xi, N. & Alici, G. (2008). A scalable model for trilayer conjugated polymer actuators and its experimental validation. *Mater. Sci. Eng., C*, Vol. 28, No. 3, pp. 421-428, ISSN 0928-4931.

Fuchiwaki, M., Tanaka, K. & Kaneto, K. (2009). Planate conducting polymer actuator based on polypyrrole and its application. *Sens. Actuat. A-Phys*, Vol. 150, No. 2, pp. 272-276, ISSN 0924-4247.

Garcia-Cordova, F., Valero, L., Ismail, Y. A. & Otero, T. F. (2011). Biomimetic polypyrrole based all three-in-one triple layer sensing actuators exchanging cations. *J. Mater. Chem.*, Vol. 21, No. 43, pp. 17265-17272, ISSN 0959-9428.

Gere, J. M. & Goodno, B. J. (2009). *Mechanics of Materials 4th Ed.*, Cengage Learning, ISBN 0534553974, Toronto.

Gimenez, P., Mukai, K., Asaka, K., Hata, K., Oike, H. & Otero, T. F. (2012). Capacitive and faradic charge components in high-speed carbon nanotube actuator. *Electrochim. Acta*, Vol. 60, No. 177-183, ISSN 0013-4686.

Guo, Y., Yu, G. & Liu, Y. (2010). Functional Organic Field-Effect Transistors. *Adv. Mater.*, Vol. 22, No. 40, pp. 4427-4447, ISSN 1521-4095.

Gursel, A., Valerie, D., Philippe, R. & Geoff, S. (2009). Conducting polymer microactuators operating in air. *J. Micromech. Microeng.*, Vol. 19, No. 2, pp. 025017, ISSN 0960-1317.

Gutta, S., Realmuto, J., Woosoon, Y. & Kim, K. J. (2011). *Dynamic model of a cylindrical ionic polymer-metal composite actuator*. Proceedings of 8th International Conference on Ubiquitous Robots and Ambient Intelligence (URAI), Incheon.

Han, G. & Shi, G. (2006). Electrochemical actuator based on single-layer polypyrrole film. *Sens. Actuat. B-Chem*, Vol. 113, No. 1, pp. 259-264, ISSN 0925-4005.

Hara, S., Zama, T., Sewa, S., Takashima, W. & Kaneto, K. (2003). Polypyrrole-metal coil composites as fibrous artificial muscles. *Chem. Lett.*, Vol. 32, No. 9, pp. 800-801, ISSN 0366-7022.

Hara, S., Zama, T., Ametani, A., Takashima, W. & Kaneto, K. (2004). Enhancement in electrochemical strain of a polypyrrole-metal composite film actuator. *J. Mater. Chem.*, Vol. 14, No. 18, pp. 2724-2725, ISSN 0959-9428.

Hara, S., Zama, T., Tanaka, N., Takashima, W. & Kaneto, K. (2005). Artificial fibular muscles with 20% strain based on polypyrrole-metal coil composites. *Chem.Lett.*, Vol. 34, No. 6, pp. 784-785, ISSN 0366-7022.

Hara, S., Zama, T., Takashima, W. & Kaneto, K. (2006). Tris (trifluoromethylsulfonyl)methide-doped polypyrrole as a conducting polymer actuator with large electrochemical strain. *Synth.Met.*, Vol. 156, No. 2-4, pp. 351-355, ISSN 0379-6779.

He, X. M., Li, C., Chen, F. G. & Shi, G. Q. (2007). Polypyrrole microtubule actuators for seizing and transferring microparticles. *Adv. Funct. Mater.*, Vol. 17, No. 15, pp. 2911-2917, ISSN 1616-301X.

Heuer, H. W., Wehrmann, R. & Kirchmeyer, S. (2002). Electrochromic window based on conducting poly (3,4-ethylenedioxythiophene)poly(styrene sulfonate). *Adv. Funct. Mater.*, Vol. 12, No. 2, pp. 89-94, ISSN 1616-301X.

Higgins, S. J., Lovell, K. V., Rajapakse, R. M. G. & Walsby, N. M. (2003). Grafting and electrochemical characterisation of poly-(3,4-ethylenedioxythiophene) films, on Nafion and on radiation-grafted polystyrenesulfonate-polyvinylidene fluoride composite surfaces. *J. Mater. Chem.*, Vol. 13, No. 10, pp. 2485-2489, ISSN 0959-9428.

Hillman, A. R., Loveday, D. C., Swann, M. J., Eales, R. M., Hamnett, A., Higgins, S. J., Bruckenstein, S. & Wilde, C. P. (1989). Charge Transport in Electroactive Polymer-Films. *Faraday Discuss. Chem. Soc.*, Vol. 88, No. 151-163, ISSN 0301-7249.

Huang, W. S., Humphrey, B. D. & MacDiarmid, A. G. (1986). Polyaniline, A Novel Conducting Polymer - Morphology and Chemistry of Its Oxidation and Reduction in Aqueous-Electrolytes. *J. Chem. Soc., Faraday Trans. I*, Vol. 82, No. 2385-2400, ISSN 0300-9599.

Huang, Y., Liang, J. & Chen, Y. (2012). The application of graphene based materials for actuators. *J. Mater. Chem.*, Vol. 22, No. 9, pp. 3671-3679, ISSN 0959-9428.

Ikushima, K., John, S., Yokoyama, K. & Nagamitsu, S. (2009). A practical multilayered conducting polymer actuator with scalable work output. *Smart. Mater. Struct.*, Vol. 18, No. 9, pp. 9, ISSN 0964-1726.

Inzelt, G. (2008). Redox Transformations and Transport Processes. In: *Conducting Polymers*, Scholz, F. pp. 169-224, Springer-Verlag Berlin, ISBN 9783540759294, Heidelberg (Germany).

Inzelt, G. (2011). Rise and rise of conducting polymers. *J. Solid State Electrochem.*, Vol. 15, No. 7-8, pp. 1711-1718, ISSN 1432-8488.

Ismail, Y. A., Martínez, J. G., Harrasi, A. S. A., Kim, S. J. & Otero, T. F. (2011). Sensing characteristics of a conducting polymer/hydrogel hybrid microfiber artificial muscle. *Actuat. B-Chem*, Vol. No. 0, pp. ISSN 0925-4005.

Jager, E. W. H., Smela, E., Inganäs, O. & Lundstrom, I. (1999). Polypyrrole microactuators. *Synth. Met.*, Vol. 102, No. 1-3, pp. 1309-1310, ISSN 0379-6779.

Jager, E. W. H., Inganäs, O. & Lundstrom, I. (2000a). Microrobots for micrometer-size objects in aqueous media: Potential tools for single-cell manipulation. *Science*, Vol. 288, No. 5475, pp. 2335-2338, ISSN 0036-8075.

Jager, E. W. H., Smela, E. & Inganäs, O. (2000b). Microfabricating conjugated polymer actuators. *Science*, Vol. 290, No. 5496, pp. 1540-1545, ISSN 0036-8075.

Jager, E. W. H., Inganäs, O. & Lundstrom, I. (2001). Perpendicular actuation with individually controlled polymer microactuators. *Adv. Mater.*, Vol. 13, No. 1, pp. 76-79, ISSN 0935-9648.

Jager, E. W. H., Immerstrand, C., Peterson, K. H., Magnusson, K.-E., Lundström, I. & Inganäs, O. (2002). The Cell Clinic: Closable Microvials for Single Cell Studies. *Biomed. Microdev*, Vol. 4, No. 3, pp. 177-187, ISSN 1387-2176.

James, T., Patrick, A., Timothy, F., Angela, C., Mike Del, Z. & Ian, H. (2007). The application of conducting polymers to a biorobotic fin propulsor. *Bioinspir. Biomim.*, Vol. 2, No. 2, pp. S6, ISSN 1748-3190.

John, S., Alici, G. & Cook, C. (2008). *Frequency response of polypyrrole trilayer actuator displacement*. Proceedings of Electroactive Polymer Actuators and Devices (Eapad) 2008, San Diego (USA).

Kaneto, K., Suematsu, H. & Yamato, K. (2008). Training effect and fatigue in polypyrrole-based artificial muscles. *Bioinspir. Biomim.*, Vol. 3, No. 3, pp. 035005, ISSN 1748-3182.

Kaneto, K., Suematsu, H. & Yamato, K. (2009). Conducting Polymer Soft Actuators based on Polypyrrole-Training Effect and Fatigue. In: *Artificial Muscle Actuators Using Electroactive Polymers*, Vincenzini, P., Bar-Cohen, Y. & Carpi, F. pp. 122-130, Trans Tech Publications Inc., ISBN 1662-0356, Stafa-Zuerich, Switzerland.

Kiefer, R., Chu, S. Y., Kilmartin, P. A., Bowmaker, G. A., Cooney, R. P. & Travas-Sejdic, J. (2007). Mixed-ion linear actuation behaviour of polypyrrole. *Electrochim. Acta*, Vol. 52, No. 7, pp. 2386-2391, ISSN 0013-4686.

Kiefer, R., Mandviwalla, X., Archer, R., Tjahyono, S. S., Wang, H., MacDonald, B., Bowmaker, G. A., Kilmartin, P. A. & Travas-Sejdic, J. (2008). *The application of polypyrrole trilayer actuators in microfluidics and robotics*. Proceedings of Electroactive Polymer Actuators and Devices (Eapad).

Kim, J. H., Lau, K. T., Shepherd, R., Wu, Y., Wallace, G. & Diamond, D. (2008). Performance characteristics of a polypyrrole modified polydimethylsiloxane (PDMS) membrane based microfluidic pump. *Sensors and Actuators A: Physical*, Vol. 148, No. 1, pp. 239-244, ISSN 0924-4247.

Kim, L. T. T., Sel, O., Biemme-Chouvy, C., Gabrielli, C., Laberty-Robert, C., Perrot, H. & Sanchez, C. (2010). Proton transport properties in hybrid membranes investigated by ac-electrogravimetry. *Electrochem.Commun.*, Vol. 12, No. 8, pp. 1136-1139, ISSN 1388-2481.

Klauk, H., Ed. (2006). *Organic Electronics: Materials, Manufacturing, and Applications* Wiley-VCH Verlag, ISBN 978-3-527-31264-1, Weinheim.

Kuttel, C., Stemmer, A. & Wei, X. (2009). Strain response of polypyrrole actuators induced by redox agents in solution. *Sens. Actuat. B-Chem*, Vol. 141, No. 2, pp. 478-484, ISSN 0925-4005.

Lee, K. K. C., Munce, N. R., Shoa, T., Charron, L. G., Wright, G. A., Madden, J. D. & Yang, V. X. D. (2009). Fabrication and characterization of laser-micromachined polypyrrole-based artificial muscle actuated catheters. *Sens. Actuat. A-Phys*, Vol. 153, No. 2, pp. 230-236, ISSN 0924-4247.

Li, W. G., Johnson, C. L. & Wang, H. L. (2004). Preparation and characterization of monolithic polyaniline-graphite composite actuators. *Polymer*, Vol. 45, No. 14, pp. 4769-4775, ISSN 0032-3861.

Lu, W., Fadeev, A. G., Qi, B. H., Smela, E., Mattes, B. R., Ding, J., Spinks, G. M., Mazurkiewicz, J., Zhou, D. Z., Wallace, G. G., MacFarlane, D. R., Forsyth, S. A. & Forsyth, M. (2002). Use of ionic liquids for pi-conjugated polymer electrochemical devices. *Science*, Vol. 297, No. 5583, pp. 983-987, ISSN 0036-8075.

Lyutov, V., Tsakova, V. & Bund, A. (2011). Microgravimetric study on the formation and redox behavior of poly(2-acrylamido-2-methyl-1-propanesulfonate)-doped thin polyaniline layers. *Electrochim. Acta*, Vol. 56, No. 13, pp. 4803-4811, ISSN 0013-4686.

Mazzoldi, A., Degl'Innocenti, C., Michelucci, M. & De Rossi, D. (1998). Actuative properties of polyaniline fibers under electrochemical stimulation. *Mater. Sci. Eng., C*, Vol. 6, No. 1, pp. 65-72, ISSN 0928-4931.

McGovern, S. & et al. (2009). Finding NEMO (novel electromaterial muscle oscillator): a polypyrrole powered robotic fish with real-time wireless speed and directional control. *Smart Mater. Struct.*, Vol. 18, No. 9, pp. 095009, ISSN 0964-1726.

McGovern, S. T., Abbot, M., Emery, R., Alici, G., Truong, V. T., Spinks, G. M. & Wallace, G. G. (2010). Evaluation of thrust force generated for a robotic fish propelled with polypyrrole actuators. *Polym. Int.*, Vol. 59, No. 3, pp. 357-364, ISSN 0959-8103.

Metz, P., Alici, G. & Spinks, G. M. (2006). A finite element model for bending behaviour of conducting polymer electromechanical actuators. *Sens. Actuat. A-Phys*, Vol. 130, No. 1-11, ISSN 0924-4247.

Morita, T., Chida, Y., Hoshino, D., Fujiya, T. & Nishioka, Y. Fabrication and characterization of a polypyrrole soft actuator having corrugated structures. *Mol. Cryst. Liq. Cryst.*, Vol. 519, No. 121-127, ISSN 1542-1406.

Morita, T., Chida, Y., Hoshino, D., Fujiya, T. & Nishioka, Y. (2010). Fabrication and Characterization of a Polypyrrole Soft Actuator Having Corrugated Structures. *Mol. Cryst. Liq. Cryst.*, Vol. 519, No. 121-127, ISSN 1542-1406.

Mukai, K., Asaka, K., Hata, K., Otero, T. F. & Oike, H. (2011). High-Speed Carbon Nanotube Actuators Based on an Oxidation/Reduction Reaction. *Chem.-Eur. J.*, Vol. 17, No. 39, pp. 10965-10971, ISSN 1521-3765.

Mutlu, R. & Alici, G. (2010). A Multistable Linear Actuation Mechanism Based on Artificial Muscles. *J. Mech. Des.*, Vol. 132, No. 11, pp. 111001, ISSN 1050-0472.

Naka, Y., Fuchiwaki, M. & Tanaka, K. (2010). A micropump driven by a polypyrrole-based conducting polymer soft actuator. *Polym. Int.*, Vol. 59, No. 3, pp. 352-356, ISSN 1097-0126.

Nakano, T. & Okamoto, Y. (2001). Synthetic Helical Polymers: Conformation and Function. *Chem. Rev.*, Vol. 101, No. 12, pp. 4013-4038, ISSN

Okamoto, T., Tada, K. & Onoda, M. (2000). Bending machine using anisotropic polypyrrole films. *Jpn. J. Appl. Phys. 1*, Vol. 39, No. 5A, pp. 2854-2858, ISSN 0021-4922.

Okamoto, T., Kato, Y., Tada, K. & Onoda, M. (2001). Actuator based on doping/undoping-induced volume change in anisotropic polypyrrole film. *Thin Solid Films*, Vol. 393, No. 1-2, pp. 383-387, ISSN 0040-6090.

Okuzaki, H. & Hattori, T. (2003). Electrically induced anisotropic contraction of polypyrrole films. *Synth. Met.*, Vol. 135-136, No. 45-46, ISSN 0379-6779.

Okuzaki, H. (2008). A biomorphic origami actuator fabricated by folding a conducting paper. *J. Phys. Conf. Ser.*, Vol. 127, No. 1, pp. 012001, ISSN 1742-6596.

Onoda, M., Okamoto, T., Tada, K. & Nakayama, H. (1999a). Polypyrrole films with anisotropy for artificial muscles and examination of bending behavior. *Jpn. J. Appl. Phys. 2*, Vol. 38, No. 9AB, pp. L1070-L1072, ISSN 0021-4922.

Onoda, M., Tada, K. & Nakayama, H. (1999b). Polypyrrole films with anisotropy. *Synth. Met.*, Vol. 102, No. 1-3, pp. 1321-1322, ISSN 0379-6779.

Onoda, M. & Tada, K. (2004). Anisotropic bending machine using conducting polypyrrole. *IEICE Trans. Electron.*, Vol. E87C, No. 2, pp. 128-135, ISSN 0916-8524.

Onoda, M., Shonaka, H. & Tada, K. (2005). A self-organized bending-beam electrochemical actuator. *Curr. Appl. Phys.*, Vol. 5, No. 2, pp. 194-201, ISSN 1567-1739.

Orata, D. & Buttry, D. A. (1987). Determination of ion populations and solvent content as functions of redox state and pH in polyaniline. *J. Am. Chem. Soc.*, Vol. 109, No. 12, pp. 3574-3581, ISSN 0002-786.

Otero, T., Cascales, J. & Fernandez-Romero, A. (2007a). *Attempting a classification for electrical polymeric actuators*. Proceedings of Electroactive Polymer Actuators and Devices (EAPAD), San Diego.

Otero, T. F., Angulo, E., Rodriguez, J. & Santamaria, C. (1992a). Electrochemomechanical Properties from A Bilayer - Polypyrrole Nonconducting and Flexible Material Artificial Muscle. *J. Electroanal. Chem.*, Vol. 341, No. 1-2, pp. 369-375, ISSN 0022-0728.

Otero, T. F., Angulo, E., Rodriguez, J. & Santamaria, C. (1992b). Dispositivos laminares que emplean polímeros conductores capaces de provocar movimientos mecánicos, ES Patent 2 048 086.

Otero, T. F., Rodriguez, J. & Santamaria, C. (1992c). Músculos artificiales formados por multicapas: polímeros conductores-no conductores, ES Patent 2 062 930.

Otero, T. F. (1999). Conducting polymers, electrochemistry, and biomimicking processes. In: *Modern Aspects of Electrochemistry*, Bockris, J.O.M., White, R.E. & Conway, B.E. pp. 307-434, Kluwer Academic/Plenum Publ, ISBN 0076-9924, New York.

Otero, T. F. (2000). Biomimicking materials with smart polymers. In: *Structural biological materials. Design and structure-properties relationships*, Elices, M. & Cahn, R.W. pp. 187-220, Pergamon Materials Series, ISBN 0080434169, Amsterdam (The Netherlands).

Otero, T. F. & Cortes, M. T. (2003a). Artificial muscles with tactile sensitivity. *Adv. Mater.*, Vol. 15, No. 4, pp. 279-282, ISSN 0935-9648.

Otero, T. F. & Cortes, M. T. (2003b). A sensing muscle. *Sens. Actuators, B*, Vol. 96, No. 1-2, pp. 152-156, ISSN 0925-4005.

Otero, T. F. & Cortes, M. T. (2004). Artificial muscle: movement and position control. *Chem. Commun.*, Vol. No. 3, pp. 284-285, ISSN 1359-7345.

Otero, T. F., Arenas, G. V. & Cascales, J. J. L. (2006). Effect of the doping ion on the electrical response of a free-standing polypyrrole strip subjected to different preloads: Perspectives and limitations associated with the use of these devices as actuators. *Macromolecules*, Vol. 39, No. 26, pp. 9551-9556, ISSN 0024-9297.

Otero, T. F. & Broschart, M. (2006). Polypyrrole artificial muscles: a new rhombic element. Construction and electrochemomechanical characterization. *J. Appl. Electrochem.*, Vol. 36, No. 2, pp. 205-214, ISSN 0021-891X.

Otero, T. F. (2007). Artificial muscles. In: *Handbook of Conducting Polymers*, Skotheim, T.A., Elsenbaumer, R.L. & Reynolds, J.R. pp. 591-623, CRC Press, ISBN Boca Raton.

Otero, T. F., Cortes, M. T. & Arenas, G. V. (2007b). Linear movements from two bending triple-layers. *Electrochim. Acta*, Vol. 53, No. 3, pp. 1252-1258, ISSN 0013-4686.

Otero, T. F., Lopez Cascales, J. J. & Vazquez Arenas, G. (2007c). Mechanical characterization of free-standing polypyrrole film. *Mater. Sci. Eng., C*, Vol. 27, No. 1, pp. 18-22, ISSN 0928-4931.

Otero, T. F. (2008). Artificial Muscles, Sensing and Multifunctionality. In: *Intelligent Materials*, Shahinpoor, M. & Schenider, H.-J. pp. 142-190, Royal Society of Chemistry, ISBN 9780854043354, Cambridge (U.K.).

Otero, T. F. (2009). Soft, wet, and reactive polymers. Sensing artificial muscles and conformational energy. *J. Mater. Chem.*, Vol. 19, No. 681-689, ISSN 0959-9428.

Otero, T. F. (2011). From Electrochemically-Driven Conformational Polymeric States to Macroscopic Sensing and Tactile Muscles. In: *From Non-Covalent Assemblies to Molecular Machines*, Sauvage, J.-P. & Gaspard, P. pp. Wiley-VCH Verlag & Co., ISBN Weinheim, Germany.

Otero, T. F., Arias-Pardilla, J., Herrera, H., Segura, J. L. & Seoane, C. (2011). Electropolymerization of naphthaleneamidinemonoimide-modified poly(thiophene). *Phys. Chem. Chem. Phys.*, Vol. 13, No. 37, pp. 16513-16516, ISSN 1463-9076.

Otero, T. F., Sánchez, J. J. & Martinez, J. G. (2012) Biomimetic Dual Sensing-Actuators based on Conducting Polymers. Galvanostatic Theoretical Model for Actuators Sensing Temperature. J. Phys. Chem. B, Vol 116, No. 17, pp. 5279-5290, ISSN 1089-5647.

Ouyang, P. R., Tjiptoprodjo, R. C., Zhang, W. J. & Yang, G. S. (2008). Micro-motion devices technology: The state of arts review. *Int. J. Adv. Manuf. Technol.*, Vol. 38, No. 5-6, pp. 463-478, ISSN 0268-3768.

Paasch, G. (2000). The transmission line equivalent circuit model in solid-state electrochemistry. *Electrochem. Commun.*, Vol. 2, No. 5, pp. 371-375, ISSN 1388-2481.

Park, S. J., Cho, M. S., Nam, J. D., Kim, I. H., Choi, H. R., Koo, J. C. & Lee, Y. (2009). The linear stretching actuation behavior of polypyrrole nanorod in AAO template. *Sens. Actuat. B-Chem*, Vol. 135, No. 2, pp. 592-596, ISSN 0925-4005.

Pede, D., Smela, E., Johansson, T., Johansson, M. & Inganäs, O. (1998). A general-purpose conjugated-polymer device array for imaging. *Adv. Mater.*, Vol. 10, No. 3, pp. 233-237, ISSN 0935-9648.

Pei, Q. & Inganäs, O. (1993a). Electrochemical applications of the bending beam method; a novel way to study ion transport in electroactive polymers. *Solid State Ionics*, Vol. 60, No. 1-3, pp. 161-166, ISSN 0167-2738.

Pei, Q. & Inganäs, O. (1993b). Electrochemical applications of the bending beam method. 2. Electroshrinking and slow relaxation in polypyrrole. *J. Phys. Chem.*, Vol. 97, No. 22, pp. 6034-6041, ISSN 0022-3654.

Pei, Q. B. & Inganäs, O. (1992a). Conjugated Polymers and the Bending Cantilever Method - Electrical Muscles and Smart Devices. *Adv. Mater.*, Vol. 4, No. 4, pp. 277-278, ISSN 0935-9648.

Pei, Q. B. & Inganäs, O. (1992b). Electrochemical Applications of the Bending Beam Method .1. Mass-Transport and Volume Changes in Polypyrrole During Redox. *J. Phys. Chem.*, Vol. 96, No. 25, pp. 10507-10514, ISSN 0022-3654.

Perepichka, I. F. & Perepichka, D. F., Eds. (2009). *Handbook of Thiophene-Based Materials: Applications in Organic Electronics and Photonics.* John Wiley & Sons Ltd., ISBN 978-0-470-05732-2, Chichester (UK).

Plesse, C., Vidal, F., Randriamahazaka, H., Teyssie, D. & Chevrot, C. (2005). Synthesis and characterization of conducting interpenetrating polymer networks for new actuators. *Polymer*, Vol. 46, No. 18, pp. 7771-7778, ISSN 0032-3861.

Plesse, C., Vidal, F., Teyssie, D. & Chevrot, C. (2009). Conducting IPN Fibers: a new design for linear actuation in open air. In: *Artificial Muscle Actuators using Electroactive Polymers* Vincenzini, P., Bar-Cohen, Y. & Carpi, F. pp. 53-58, Trans Tech Publicactions Ltd., ISBN 1662-0356, Stafa-Zurich (Switzerland).

Plesse, C., Vidal, F., Teyssie, D. & Chevrot, C. (2010). Conducting polymer artificial muscle fibres: toward an open air linear actuation. *Chem. Commun.*, Vol. 46, No. 17, pp. 2910-2912, ISSN 1359-7345.

Ramírez-García, S. & Diamond, D. (2007). Biomimetic, low power pumps based on soft actuators. *Sens. Actuat. A-Phys*, Vol. 135, No. 1, pp. 229-235, ISSN 0924-4247.

Ranger, M. & Leclerc, M. (1998). Optical and electrical properties of fluorene-based pi-conjugated polymers. *Can. J. Chem.*, Vol. 76, No. 11, pp. 1571-1577, ISSN 0008-4042.

Ren, X. & Pickup, P. G. (1995). Impedance measurements of ionic conductivity as a probe of structure in electrochemically deposited polypyrrole films. *J. Electroanal. Chem.*, Vol. 396, No. 1–2, pp. 359-364, ISSN 1572-6657.

Roemer, M., Kurzenknabe, T., Oesterschulze, E. & Nicoloso, N. (2002). Microactuators based on conducting polymers. *Anal. Bioanal. Chem.*, Vol. 373, No. 8, pp. 754-757, ISSN 1618-2642.

Sansinena, J. M., Olazabal, V., Otero, T. F., daFonseca, C. N. P. & DePaoli, M. A. (1997). A solid state artificial muscle based on polypyrrole and a solid polymeric electrolyte working in air. *Chem. Commun.*, Vol. No. 22, pp. 2217-2218, ISSN 1359-7345.

Sansiñena, J. M., Gao, J. B. & Wang, H. L. (2003). High-performance, monolithic polyaniline electrochemical actuators. *Adv. Funct. Mater.*, Vol. 13, No. 9, pp. 703-709, ISSN 1616-301X.

Shakuda, S., Morita, S., Kawai, T. & Yoshino, K. (1993). Dynamic Characteristics of Bimorph with Conducting Polymer Gel. *Jpn. J. Appl. Phys. 1*, Vol. 32, No. 11A, pp. 5143-5146, ISSN 0021-4922.

Shapiro, B. & Smela, E. (2007). Bending Actuators with Maximum Curvature and Force and Zero Interfacial Stress. *J. Intel. Mat. Syst. Str.*, Vol. 18, No. 2, pp. 181-186, ISSN 1045-389X.

Shirakawa, H., Louis, E. J., Macdiarmid, A. G., Chiang, C. K. & Heeger, A. J. (1977). Synthesis of Electrically Conducting Organic Polymers - Halogen Derivatives of Polyacetylene, (Ch)X. *J. Chem. Soc., Chem. Commun.*, Vol. No. 16, pp. 578-580, ISSN 0022-4936.

Shoa, T., Madden, J. D., Fekri, N., Munce, N. R. & Yang, V. X. D. (2008). *Conducting polymer based active catheter for minimally invasive interventions inside arteries*. Proceedings of 30th Annual International Conference of the IEEE Engineering in Medicine and Biology Society, 2008.

Shoa, T., Madden, J. D. W., Munce, N. R. & Yang, V. (2010). Analytical modeling of a conducting polymer-driven catheter. *Polym.Int.*, Vol. 59, No. 3, pp. 343-351, ISSN 1097-0126.

Shoa, T., Yoo, D. S., Walus, K. & Madden, J. D. W. (2011). A Dynamic Electromechanical Model for Electrochemically Driven Conducting Polymer Actuators. *IEEE ASME Trans. Mechatronics*, Vol. 16, No. 1, pp. 42-49, ISSN 1083-4435.

Smela, E., Inganäs, O., Pei, Q. B. & Lundstrom, I. (1993). Electrochemical Muscles - Micromachining Fingers and Corkscrews. *Adv. Mater.*, Vol. 5, No. 9, pp. 630-632, ISSN 0935-9648.

Smela, E., Inganäs, O. & Lundstrom, I. (1995). Controlled Folding of Micrometer-Size Structures. *Science*, Vol. 268, No. 5218, pp. 1735-1738, ISSN 0036-8075.

Smela, E. (1999). A microfabricated movable electrochromic "pixel" based on polypyrrole. *Adv. Mater.*, Vol. 11, No. 16, pp. 1343-1345, ISSN 0935-9648.

Smela, E. & Gadegaard, N. (1999). Surprising volume change in PPy(DBS): An atomic force microscopy study. *Adv. Mater.*, Vol. 11, No. 11, pp. 953-957, ISSN 0935-9648.

Smela, E. & Gadegaard, N. (2001). Volume change in polypyrrole studied by atomic force microscopy. *J. Phys. Chem. B*, Vol. 105, No. 39, pp. 9395-9405, ISSN 1089-5647.

Smela, E., Christophersen, M., Prakash, S. B., Urdaneta, M., Dandin, M. & Abshire, P. (2007). *Integrated cell-based sensors and "cell clinics" utilizing conjugated polymer actuators.* Proceedings of SPIE - The International Society for Optical Engineering.

Smith, E. B. (2004). *Basic Chemical Thermodynamics*, Imperial College Press., ISBN 1-86094-445-0, London.

So, F., Ed. (2010). *Organic electronics : materials, processing, devices and applications.* CRC Press, ISBN 978-1-4200-7290-7, Boca Raton (USA).

Song, M. K., Cho, J. Y., Cho, B. W. & Rhee, H. W. (2002). Characterization of UV-cured gel polymer electrolytes for rechargeable lithium batteries. *J. Power Sources*, Vol. 110, No. 1, pp. 209-215, ISSN 0378-7753.

Spinks, G. M., Liu, L., Wallace, G. G. & Zhou, D. (2002). Strain Response from Polypyrrole Actuators under Load. *Adv. Funct. Mater.*, Vol. 12, No. 6-7, pp. 437-440, ISSN 1616-3028.

Spinks, G. M., Wallace, G. G., Ding, J., Zhou, D., Xi, B. & Gillespie, J. (2003a). *Ionic liquids and polypyrrole helix tubes: bringing the electronic Braille screen closer to reality.* Proceedings of Smart Structures and Materials 2003: Electroactive Polymer Actuators and Devices (EAPAD), San Diego, USA.

Spinks, G. M., Zhou, D. Z., Liu, L. & Wallace, G. G. (2003b). The amounts per cycle of polypyrrole electromechanical actuators. *Smart. Mater. Struct.*, Vol. 12, No. 3, pp. 468-472, ISSN 0964-1726.

Takashima, W., Uesugi, T., Fukui, M., Kaneko, M. & Kaneto, K. (1997). Mechanochemoelectrical effect of polyaniline film. *Synth. Met.*, Vol. 85, No. 1-3, pp. 1395-1396, ISSN 0379-6779.

Takashima, W., Pandey, S. S. & Kaneto, K. (2003). Bi-ionic actuator by polypyrrole films. *Synth. Met.*, Vol. 135, No. 1-3, pp. 61-62, ISSN 0379-6779.

Timoshenko, S. (1925). Analysis of bi-metal thermostats. *J. Opt. Soc. Am.*, Vol. 11, No. 3, pp. 233-255, ISSN 0093-4119.

Torresi, R. M. & Maranhao, S. L. D. (1999). Anion and solvent exchange as a function of the redox states in polyaniline films. *J. Electrochem. Soc.*, Vol. 146, No. 11, pp. 4179-4182, ISSN 0013-4651.

Tsai, E. W., Pajkossy, T., Rajeshwar, K. & Reynolds, J. R. (1988). Anion-Exchange Behavior of Polypyrrole Membranes. *J. Phys. Chem.*, Vol. 92, No. 12, pp. 3560-3565, ISSN 0022-3654.

Valero Conzuelo, L., Arias-Pardilla, J., Cauigh-Rodríguez, J. V., Smit, M. A. & Otero, T. F. (2010). Sensing and Tactile Artificial Muscles from Reactive Materials. *Sensors*, Vol. 10, No. 2638-2674, ISSN 1424-8220.

Valero, L., Arias-Pardilla, J., Smit, M., Cauich-Rodríguez, J. & Otero, T. F. (2010). Polypyrrole Free-Standing electrodes sense temperature or current during reaction. *Polym. Int.*, Vol. 59, No. 337-342, ISSN 1097-0126.

Valero, L., Arias-Pardilla, J., Cauich-Rodríguez, J., Smit, M. A. & Otero, T. F. (2011). Characterization of the movement of polypyrrole-dodecylbenzenesulfonateperchlorate/tape artificial muscles. Faradaic control of reactive artificial molecular motors and muscles. *Electrochim. Acta*, Vol. 56, No. 10, pp. 3721-3726, ISSN 0013-4686.

Vidal, F., Popp, J. F., Plesse, C., Chevrot, C. & Teyssie, D. (2003). Feasibility of conducting semi-interpenetrating networks based on a poly(ethylene oxide) network and poly(3,4-ethylenedioxythiophene) in actuator design. *J. Appl. Polym. Sci.*, Vol. 90, No. 13, pp. 3569-3577, ISSN 0021-8995.

Vidal, F., Plesse, C., Palaprat, G., Juger, J., Citerin, J., Kheddar, A., Chevrot, C. & Teyssie, D. (2009). Synthesis and Characterization of IPNs for Electrochemical Actuators. In: *Artificial Muscle Actuators Using Electroactive Polymers*, Vincenzini, P., BarCohen, Y. & Carpi, F. pp. 8-17, Trans Tech Publications Ltd, ISBN 978-3-908158-27-1, Stafa-Zurich.

Vidal, F., Plesse, C., Aubert, P. H., Beouch, L., Tran-Van, F. o., Palaprat, G., Verge, P., Yammine, P., Citerin, J., Kheddar, A., Sauques, L., Chevrot, C. & Teyssie, D. (2010). Poly(3,4-ethylenedioxythiophene)-containing semi-interpenetrating polymer networks: a versatile concept for the design of optical or mechanical electroactive devices. *Polym. Int.*, Vol. 59, No. 3, pp. 313-320, ISSN 1097-0126.

Vlad, A., Dutu, C. A., Jedrasik, P., Sodervall, U., Gohy, J. F. & Melinte, S. (2012). Vertical single nanowire devices based on conducting polymers. *Nanotechnology*, Vol. 23, No. 2, pp. 5, ISSN 0957-4484.

Wallace, G., Campbell, T. & Innis, P. (2007). Putting function into fashion: Organic conducting polymer fibres and textiles. *Fiber. Polym.*, Vol. 8, No. 2, pp. 135-142, ISSN 1229-9197.

Wang, H. L., Gao, J. B., Sansiñena, J. M. & McCarthy, P. (2002). Fabrication and characterization of polyaniline monolithic actuators based on a novel configuration: Integrally skinned asymmetric membrane. *Chem. Mater.*, Vol. 14, No. 6, pp. 2546-2552, ISSN 0897-4756.

Wilson, S. A., Jourdain, R. P. J., Zhang, Q., Dorey, R. A., Bowen, C. R., Willander, M., Wahab, Q. U., Al-hilli, S. M., Nur, O., Quandt, E., Johansson, C., Pagounis, E., Kohl, M., Matovic, J., Samel, B., van der Wijngaart, W., Jager, E. W. H., Carlsson, D., Djinovic, Z., Wegener, M., Moldovan, C., Iosub, R., Abad, E., Wendlandt, M., Rusu, C. & Persson, K. (2007). New materials for micro-scale sensors and actuators: An engineering review. *Mat. Sci. Eng. R*, Vol. 56, No. 1-6, pp. 1-129, ISSN 0927-796X.

Woosoon, Y., Joonsoo, L. & Kwang, J. K. (2007). An artificial muscle actuator for biomimetic underwater propulsors. *Bioinspir. Biomim.*, Vol. 2, No. 2, pp. S31, ISSN 1748-3190.

Wu, Y., Zhou, D., Spinks, G. M., Innis, P. C., Megill, W. M. & Wallace, G. G. (2005). TITAN: a conducting polymer based microfluidic pump. *Smart. Mater. Struct.*, Vol. 14, No. 6, pp. 1511-1516, ISSN 0964-1726.

Yang, F., Xiaobo, T. & Alici, G. (2008). Robust Adaptive Control of Conjugated Polymer Actuators. *IEEE Trans. Control Syst. Technol.*, Vol. 16, No. 4, pp. 600-612, ISSN 1063-6536.

Yao, Q., Alici, G. & Spinks, G. A. (2008). Feedback control of tri-layer polymer actuators to improve their positioning ability and speed of response. *Sens. Actuat. A-Phys*, Vol. 144, No. 1, pp. 176-184, ISSN 0924-4247.

Zainudeen, U. L., Careem, M. A. & Skaarup, S. (2008). PEDOT and PPy conducting polymer bilayer and trilayer actuators. *Sens. Actuat. B-Chem*, Vol. 134, No. 2, pp. 467-470, ISSN 0925-4005.

Chapter 12

Sensing and Rheological Capabilities of MR Elastomers

Weihua Li, Tongfei Tian and Haiping Du

Additional information is available at the end of the chapter

1. Introduction

Magnetorheological elastomers (MREs) are smart materials where polarized particles are suspended in a non-magnetic solid or gel-like matrix. Two kinds of MREs, namely anisotropic and isotropic, are fabricated either under a magnetic field or without a field [1,2]. In anisotropic MREs, polarised particles are arranged in chains within a polymer media such as silicon rubber or natural rubber. The shear modulus of MRE can be controlled by the external magnetic field, which has led to many applications, such as tuned vibration absorbers, dampers and sensors [3].

Additives are used to adjust the mechanical properties or electrical performance of MREs. Silicone oil is an additive which is used to increase the gaps between the matrix molecules and to decrease the gaps between the conglutination of molecules. Apart from increasing the plasticity and fluidity of the matrix, the additives can average the distribution of internal stress in the materials, which makes them ideal for fabricating MRE materials. Graphite powder is a kind of additive which can affect the magnetorheology and electrical conductivity of MREs [4,5]. By introducing graphite microparticles into the elastic matrix, MREs exhibit a lower electrical conductive and a different magnetorheological response.

When the material in the matrix is magnetic, the polarization of the particles is less effective and the magnetorheological response is therefore smaller. The addition of magnetically active additives (other than MR particles) also decreases the magnetorheology [6, 7]. The overall properties of the elastomer composite are also influenced by the additives, as the filler material causes the volume to increase, so the previous effect also increases. Lokander et al. [7, 8] have shown that the absolute effect of MR (the difference between the zero-field modulus and modulus measured under an external magnetic field) is independent of the matrix material. However, the zero-field modulus can be much higher for hard matrix material (for instance, where materials with high volume fraction of iron specifically possess a high zero-field modulus), which means that the relative magnetorheological effect is quite low.

MREs feature viscoelastic properties and magnetorheology [9]. The magnetorheology of MREs is described as a reversible change in modulus in an applied magnetic field. Aligned MREs have mostly been characterized at relatively low frequencies (1 to 20 Hz) to measure the changes in the dynamic shear modulus induced by the external magnetic field [9, 10]. Ginder et al. found a substantial magnetorheology over the entire frequency range studied. The increase in the shear modulus varied initially with the strength of the magnetic field but saturated at higher field strength. When the magnetic field was increased from 0 to 0.56 Tesla the consequent increase in shear modulus was nearly 2 MPa and the frequency of the resonance was shifted upward by over 20% [10]. Zhou et al. [9] stated that the changes of dynamic shear storage modulus can be over 50%, while Gong et al. [11] said that it can be over 100%. Lokander et al. studied the dynamic shear modulus for isotropic MR elastomers with different filler particles and matrix materials. They measured the magnetorheology as a function of the amplitude of strain and found that the magnetorheology decreases rapidly with increasing strain within the measured range, and is not dependent on the frequency of testing. The fact that the absolute magnetorheology is independent of the matrix material means that softer matrix materials will show a greater relative magnetorheology [7, 8].

The effect of additives on the sensing capabilities were studied by a few groups. Kchit and Bossis [7] found that the initial resistivity of metal powder at zero pressure is about 108 Ωcm for pure nickel powder and 106 Ωcm for silver coated nickel particles. The change in resistance with pressure was found to be an order of magnitude larger for a MRE composite than for the same volume fraction of fillers dispersed randomly in the polymer. Wang et al. [8] proposed a phenomenological model to understand the impedance response of MREs under mechanical loads and magnetic fields. Their results showed that MRE samples exhibit significant changes in measured values of impedance and resistance in response to compressive deformation, as well as applied magnetic field. Bica [9] found that MRE with graphite micro particles (~14%) is electroconductive. These MREs possess an electric resistance whose value diminishes with both the increase of the intensity of the magnetic field and with the compression force. Such a variation of resistance with magnetic field intensity is due to the compression of MRE with graphite microparticles. In the approximation of the perfect elastic body, the sums of the main deformations and the compressibility module of MRE with graphite microparticles, depend on the magnetic field intensity. Li et al. [10] introduced graphite into conventional MREs and found that a MRE sample with 55% carbonyl iron, 20% silicon rubber and 25% graphite powder exhibits the best performance. The test result showed that at a normal force of 5 N, the resistance decreases from 4.62 kΩ without a magnetic field to 1.78 kΩ at a magnetic field of 600 mT. The decreasing rate is more than 60%. This result also demonstrated the possibility of using MREs to develop a sensor for measuring magnetic fields. This result indicates that the detection is very sensitive to the normal force. When the normal force is 15N, the field-induced resistance only has less than 28% change from 0.65 kΩ at 0 mT to 0.47 kΩ at 600 mT.

Depending on an elasto-plastic asperity microcontact model for contact between two nominally flat surfaces, Kchit and Bossis developed a model to analyse the contact of two rough surfaces. They used two kinds of magnetic particles: nickel and nickel coated with silver which are dispersed in a silicone polymer as the polarised particles [12]. To understand the complex conductivity of particle embedded composites, quantitative or semi-quantitative models can be used [16]. The Maxwell–Wagner and the Bruggeman symmetric and asymmetric media equations were introduced by McLachlan [17] to model the electrical behaviour of conductor-insulator composites. The microstructures for which these effective media equations apply are considered in simulating the measured impedance and modular spectra of these composites. Woo et al. [18] developed a universal equivalent circuit model in modelling the impedance response of composites with insulating or conductive particles or fibres. Based on the microstructure of MREs, Wang et al. proposed an equivalent circuit model to interpret the impedance measurement results [13].

This chapter is organized as follows. Section 2 describes the fabrication of graphite based MRE samples with various weight fractions. The morphology of the MRE samples and rheological properties of graphite based MREs were presented in Section 3 and 4, respectively. Section 5 presents the theoretical development based on a representative volume for investigating MRE electric properties. The main findings are summarized in Section 6.

2. Fabrication of graphite based MREs

The materials used for the Graphite based Magnetorheological Elastomers (Gr-MREs) are: silicone rubber (Selleys Pty. LTD), silicone oil, type 378364 (Sigma-Aldrich Pty. LTD), carbonyl iron particles, type C3518 (Sigma-Aldrich Pty. LTD) and graphite powder, type 282863 (Sigma-Aldrich Pty. LTD). The particle sizes of graphite powder are less than 20 μm, the iron particles' diameter is between 3 μm and 5 μm.

The MREs with various graphite weight fractions (Gr %) were fabricated to compare the effect of graphite on to MREs. Table 1 shows the compositions of all Gr MRE samples. all samples contain 10g carbonyl iron particles, 3g silicone rubber and 3g silicone oil. The only difference is the graphite weight fraction, which is from 0% to 23.81%. For each composition, the mixture is fabricated under a constant magnetic field of 165 mT to form anistropic MRE samples.

3. Microstructure observation

To see microstructure of Gr-MREs, LV-SEM (JSM 6490LV SEM) is used. Figures 1- 3 show the surface imaging for MRE microstructures. Specifically, Fig. 1 shows the microstructure of isotropic and anisotropic Gr-MREs without graphite. The carbonyl iron particles are arranged in chains in the anisotropic sample, where they are dispersed randomly in the isotropic sample. According to Fig. 2 (a) and Fig. 3 (a), in the anisotropic Gr-MREs, the carbonyl iron particles are arranged in chains, whereas the graphite powders dispersed randomly in the matrix. The reason of this phenomenon is that the magnetic field only affects the carbonyl iron particles, but not the graphite. So upon magnetisation, the carbonyl iron particle chains tend to orient along the same direction as the magnetic field in the matrix.

By comparing Fig. 1 (a), Fig. 2 (a) and Fig. 3 (ba), we can see that the carbonyl iron chains in the sample without graphite have the best alignment performance. Further, the carbonyl iron chains in Fig. 2 (a) are aligned better than those in Fig. 3 (a). The reason is that when the mixture of carbonyl iron, silicone rubber, silicone oil and graphite is curing under the magnetic field, the graphite powders in Gr-MREs affect the carbonyl iron particles' movement. The more graphite in the mixture, the more effects are applied on to the carbonyl iron chains, which influence the magnetorheology of MREs.

	Graphite based MREs							
Sample No.	1	2	3	4	5	6	7	8
Carbonyl iron	10g	10g	10g	10g	10g	10g	10g	10g
Silicone oil	3g	3g	3g	3g	3g	3g	3g	3g
Silicone rubber	3g	3g	3g	3g	3g	3g	3g	3g
Graphite	0g	1g	2g	3g	3.5g	4g	4.5g	5g
Graphite weight fraction (Gr %)	0%	5.88%	11.11%	15.79%	17.95%	20%	21.95%	23.81%

Table 1. Components of Gr MRE samples

4. MRE rheology

A rotational Rheometer (MCR 301, Anton Paar Companies, Germany) and a Magneto Rheolgical Device (MRD 180, Anton Paar Companies, Germany) were used to measure the MREs' mechanical properties. The Magneto Rheolgical Device is equipped with an electromagnetic kit which can generate a magnetic field perpendicular to the direction of the shear flow. Specifically, a 20mm diameter parallel-plate measuring system with 1 mm gap was used. The samples were sandwiched between a rotary disk and a base placed in parallel. In this study both a steady-state rotary shear and an oscillatory shear were applied for the experiments.

The schematic of the experimental setup is shown in Fig. 4. The stress and stain signals were output from two ports which were detected through a Data Acquisition (DAQ) board (Type: LABVIEW PCI-6221, National Instruments Corporation. U.S.A) and transferred to a computer by which the data was processed.

In this experiment the magnetic flux density of the sample of MRE (B_{MRE}) in the measuring gap depends not only on the current (I) applied to the samples, but also on the magnetic properties of MRE materials. The relationship between B_{MRE} versus I is found to be: $B_{MRE} = 220\,I$, where the units of B_{MRE} and I are in mT and Ampere (A), respectively.

In the following test, the test current varies from 0A to 2A with the increment 0.5A, for which the intensity of magnetic field is 0 mT to 440 mT with the increment 110 mT.

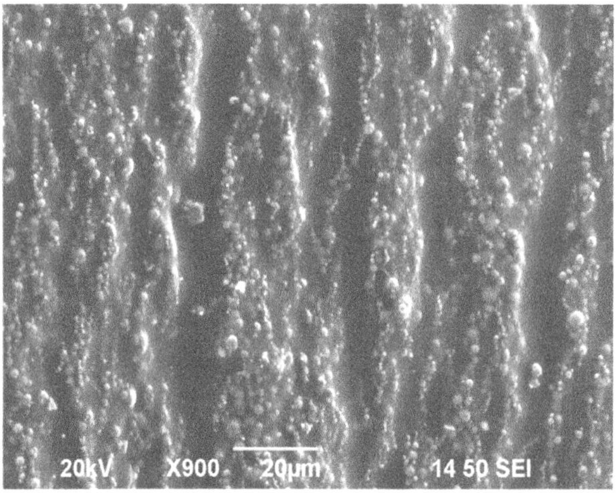

(a)

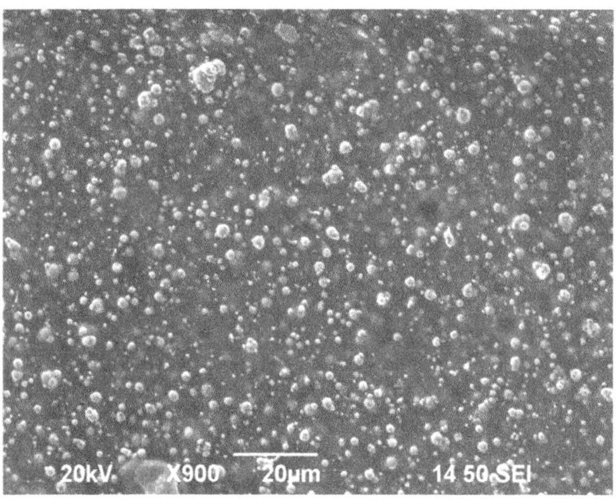

(b)

Figure 1. Microstructure of Gr-MREs (Gr 0%) (a) anisotropic (b) isotropic

4.1. Steady state results

Under rotary shear the shear stress and shear strain of MREs under fields varying from 0~440mT were measured. The MR effect was evaluated by measuring the shear strain-stress curve of the sample with and without a magnetic field applied.

Figures 5 and 6 show the strain-stress curve of different samples at 5 different magnetic field intensities ranging from 0 to 440mT.

The slope of the strain-stress curve is the shear modulus of the material. As can be seen in the figures, all the samples' shear modulus show an increasing trend with magnetic field before they reach magnetic saturation at high field strength, which proves that all the MRE samples exhibit obvious MR effects.

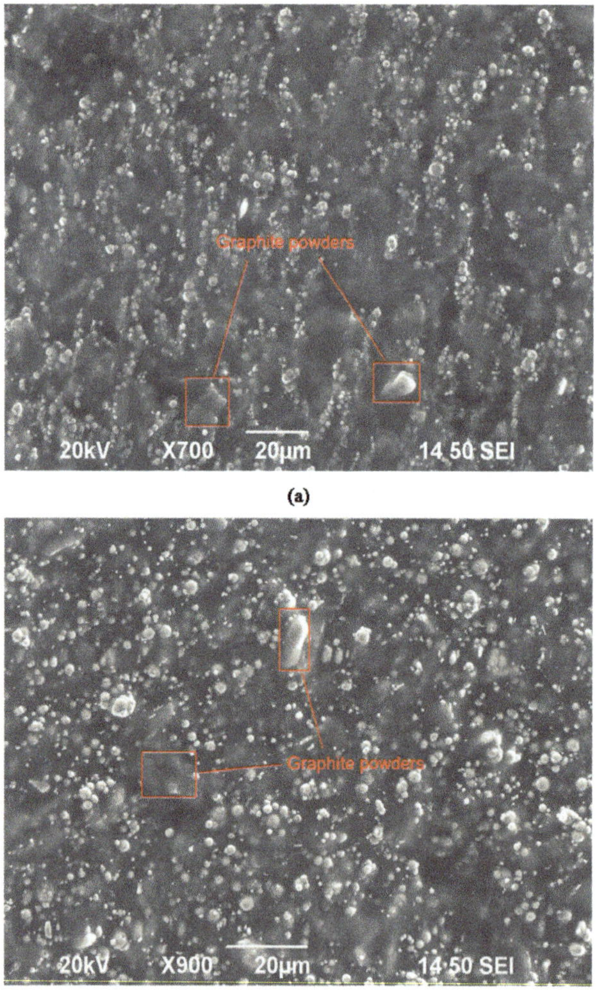

(b)

Figure 2. Microstructure of Gr-MREs (Gr 11.11%) (a) anisotropic (b) isotropic

From Figures 5 and 6, the shear stress shows a linear relationship with the shear strain when the strain is within a range. This means the MRE acts with linear viscoelastic properties when the strain is below a certain limit. For conventional MREs, the limitation is around

50% shear strain. When the graphite weight fraction increases from 0 to 15.79%, the range of linearity decreases from 50% to around 10%. For the samples with higher graphite weight fraction alike 23.81%, the linearity ranges are only 6% and 4% for isotropic and anisotropic samples, respectively. When the strain is above the limitation, the shear modulus reaches a saturation (maximum value) and decrease steadily. This could be due to sliding effect. The higher magnetic field intensity leads to higher steady shear stress. Fig. 7 shows the linear ranges of different samples at various magnetic field intensities.

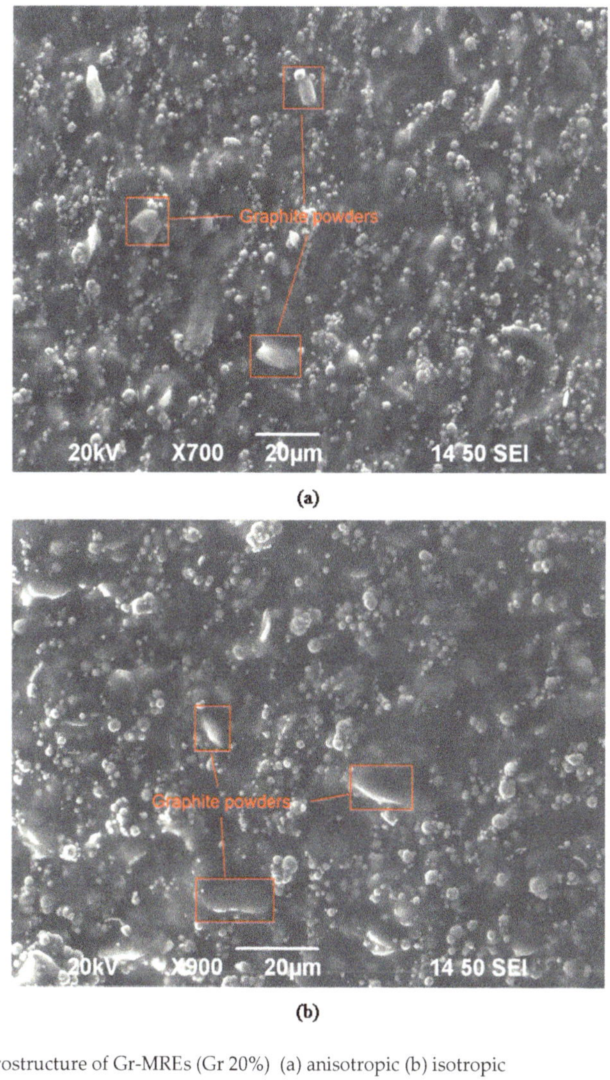

Figure 3. Microstructure of Gr-MREs (Gr 20%) (a) anisotropic (b) isotropic

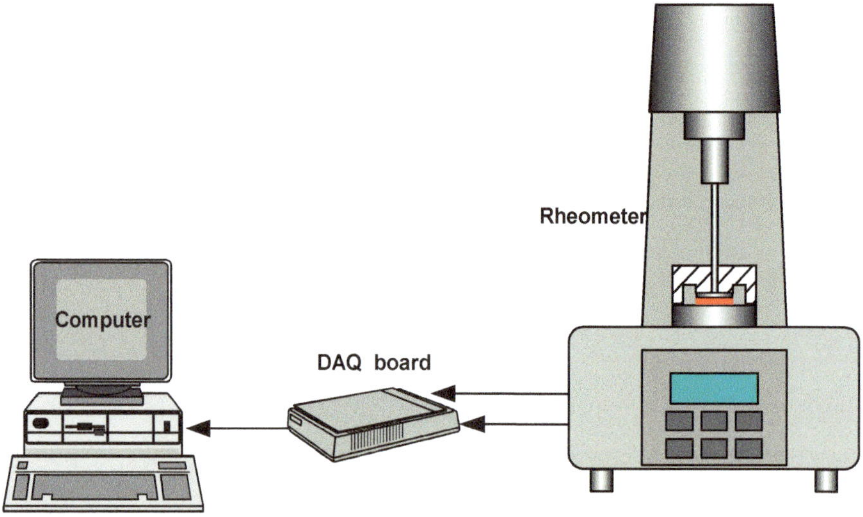

Figure 4. A diagram of the experimental setup

For the isotropic and anisotropic samples with same compositions, the isotropic samples always have the bigger linearity ranges and steady shear stress than chose of anisotropic samples.

For each curve, the slope equals the ratio between peak shear stress and relevant shear strain. By analyzing the slope of these curves, it is easy to see that the more graphite in the material, the smaller increment in slopes occurs when the magnetic field increase from 0 to 440mT. This is due to the contributions of graphite powders to the stiffness of the samples. The graphite increases the initial stiffness of Gr-MREs, thus the stiffness change induced by the MR effect can not be comparable to that conventional MREs. Fig. 8 shows the peak stresses of different samples for a magnetic field intensity equalling 220mT, 330mT and 440mT, respectively.

The relative MR effect ($\Delta G_{max}/G_0$) of several samples is shown in Fig. 9. G_0 denotes the MRE samples' zero-field modulus, ΔG_{max} denotes the saturated field-induced modulus, and $\Delta G_{max}/G_0$ denotes the relative MR effect.

It can be seen from Fig. 9 that G_0 is enhanced with the increase in graphite powders content. This result indicates that graphite powders can modify particle properties and, consequently, influenced the MR effect. From the abovementioned results, the exhibited MR effects correspond well with the microstructures of Gr-MREs.

4.2. Dynamic tests result

The dynamic mechanical behavior of MREs was studied by using the strain amplitude sweep tests. Five sets of data were collected for different amplitudes of oscillation, according to various magnetic fields applied to the samples of MREs. Similar to the steady state tests, five different magnetic field intensities, 0, 110, 220, 330 and 440mT, were used in this experiment. A constant frequency of 5Hz was selected for the strain amplitude sweep tests.

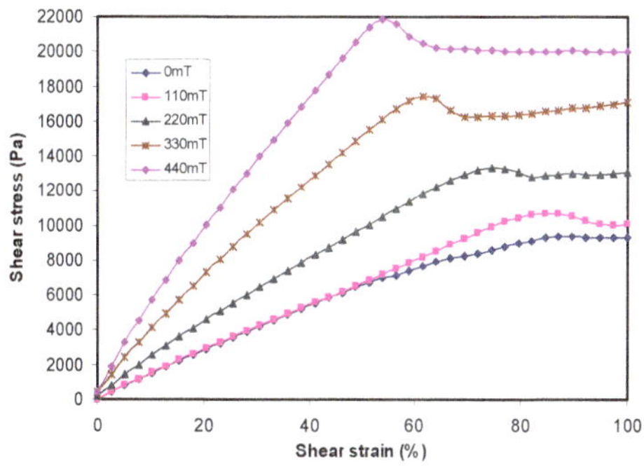

(a)

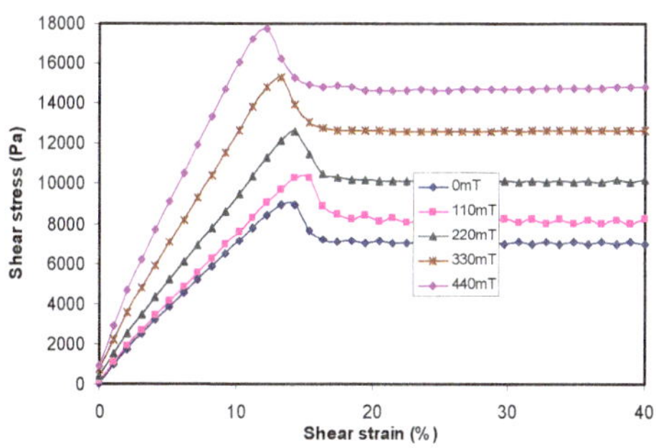

(b)

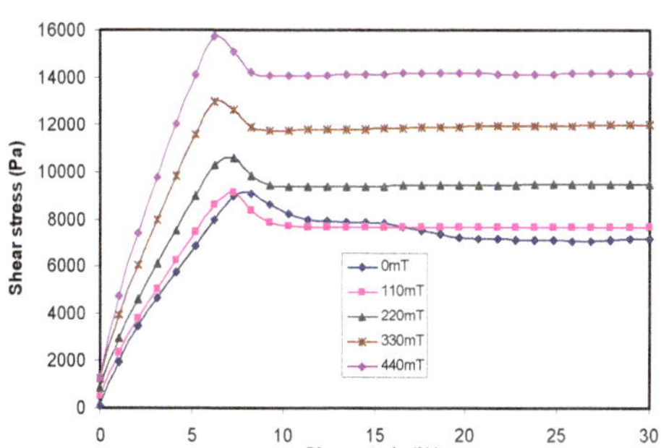

(c)

Figure 5. Strain-stress curve versus magnetic field (isotropic MRE) (a) Gr 0% (b) 15.79% (c) Gr 23.81%)

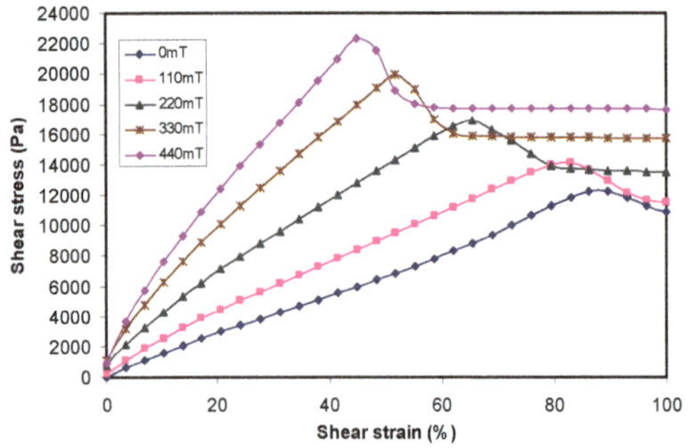

(a)

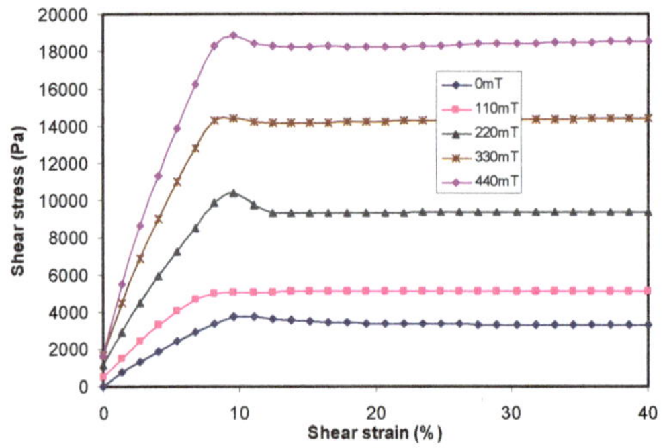

(b)

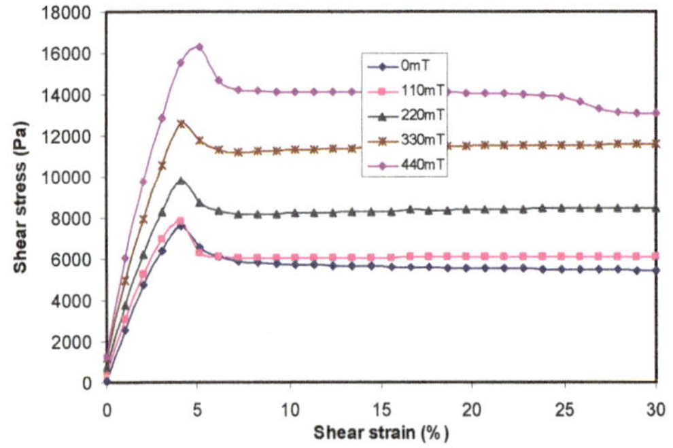

(c)

Figure 6. Strain-stress curve versus magnetic field (anisotropic MRE) (a) Gr 0% (b) 15.79% (c) Gr 23.81%)

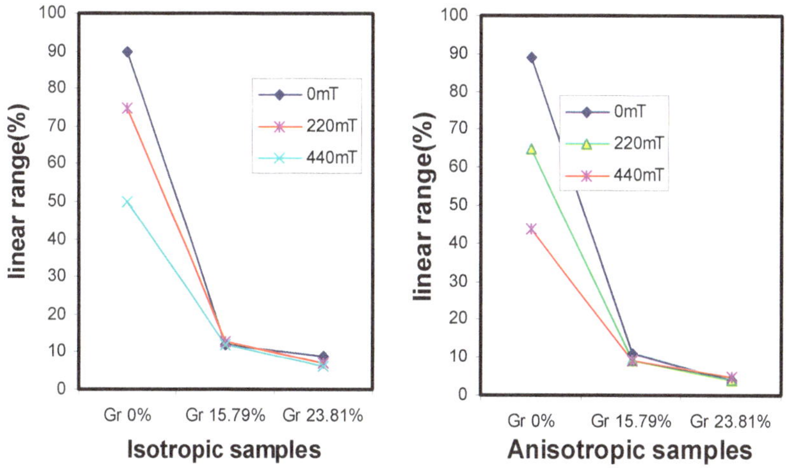

(a) (b)

Figure 7. Linear ranges versus different samples (a) isotropic samples (b) anisotropic samples

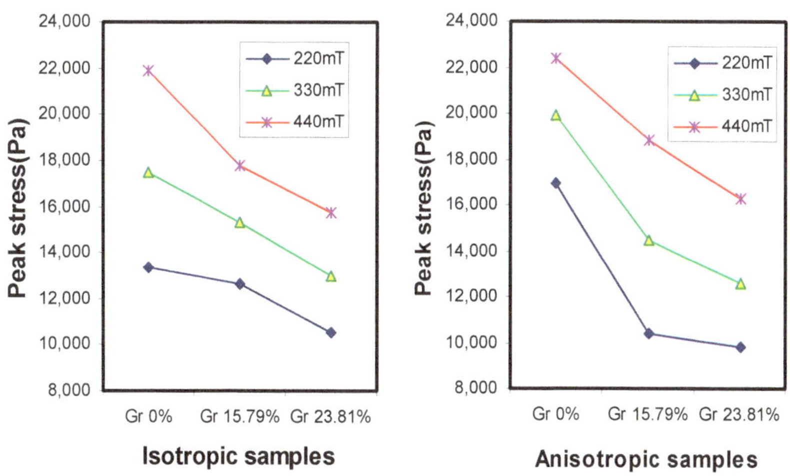

(a) (b)

Figure 8. Peaks stresses versus different samples (a) isotropic samples (b) anisotropic samples

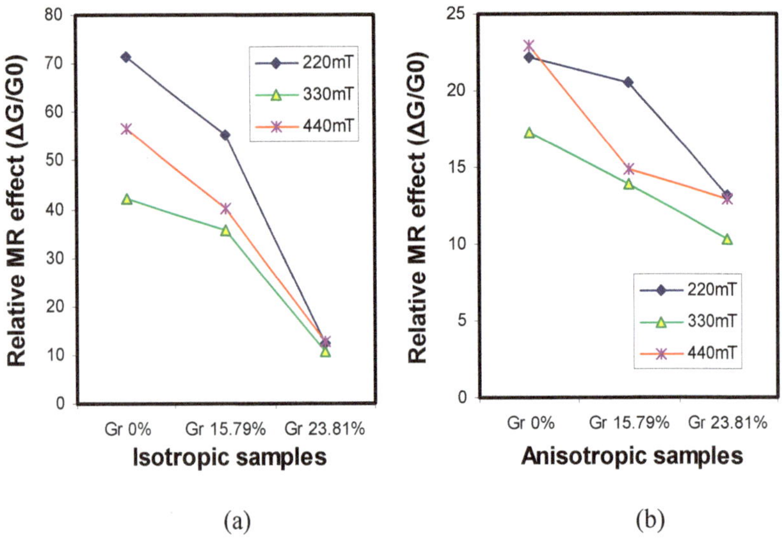

Figure 9. Relative MR effects versus different samples (a) isotropic samples (b) anisotropic samples

In the strain sweep test, the storage and loss moduli were measured by varying strain from 0.01% to 100% at different magnetic fields. Figs. 10-13 show the variation of storage and loss moduli with the strain amplitude sweep.

As shown in Figures from 10 to 13, the overall trend of storage modulus is decreasing with strain amplitude. The storage modulus goes down smoothly until 10% shear strain and begins to drop significantly beyond 10% shear strain. Except for isotropic MREs without graphite, the Loss modulus has almost the same trend as the storage modulus. This means that at the higher shear strains, the storage and loss moduli are much smaller than those at lower shear strains.

Figures 14 and 15 show the storage modulus of different samples at 0mT, 220mT and 440mT magnetic field. The data are collected at 10% and 87.5% shear strain, respectively.

As can be seen from Fig. 14, the storage modulus of all samples shows an increasing trend with graphite weight fraction at 10% shear strain which is in the linear range for most of the samples. In Fig. 15, it turns to a diminishing trend with graphite weight fraction at 87.5% shear strain which is out of the linear range. This means that in the linear range of shear strain, the samples with higher graphite weight faction have the bigger storage modulus.

Figures 16 and 17 show the storage modulus vs. magnetic field at 0.1% and 10% shear strain, respectively. The two shear strains are the beginning and the end of the linear range. From these figures we can see that the storage modulus shows an increasing trend with the intensity of magnetic field. The ratio of storage modulus at 440mT to that at 0mT is indicative of the MR effect. The MR effect of isotropic MREs with 0% graphite is around 4.5, when the graphite weight fraction increases to 15.79% and 23.81%, the MR effect decreases down to around 2.8 and 2.8, respectively. For anisotropic samples, the MR effects of 0%, 20% and 23.81% Gr-MREs are 4.8, 3.2 and 2.2, respectively. This proves again that with the increase of graphite weight fraction, the MR effect decreases.

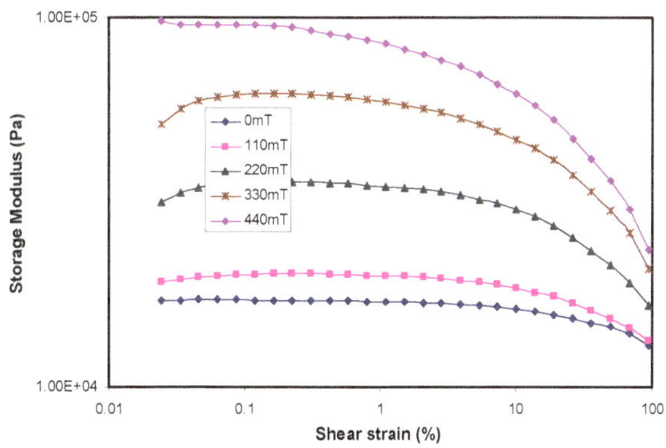

(a)

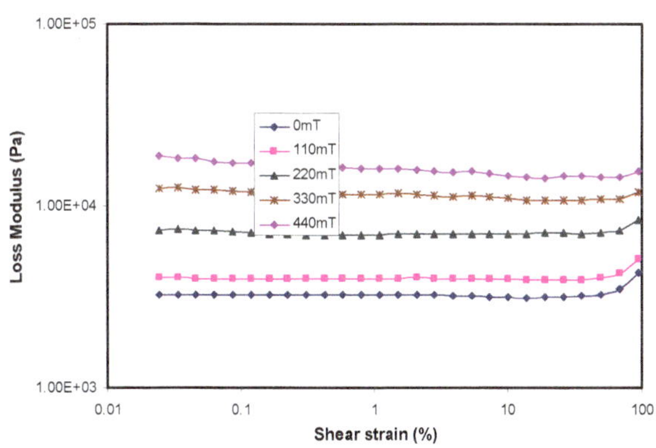

(b)

Figure 10. Storage and Loss Modulus versus strain amplitude sweep (isotropic MRE Gr 0%) (a) Storage modulus vs. shear strain (b) Loss modulus vs. shear strain

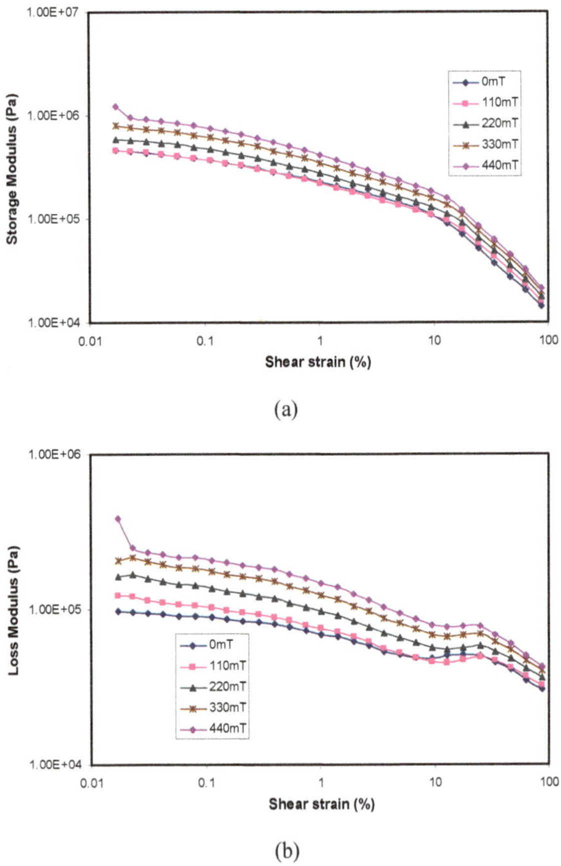

Figure 11. Storage and Loss Modulus versus strain amplitude sweep (isotropic MRE Gr 20%) (a) Storage modulus vs. shear strain (b) Loss modulus vs. shear strain

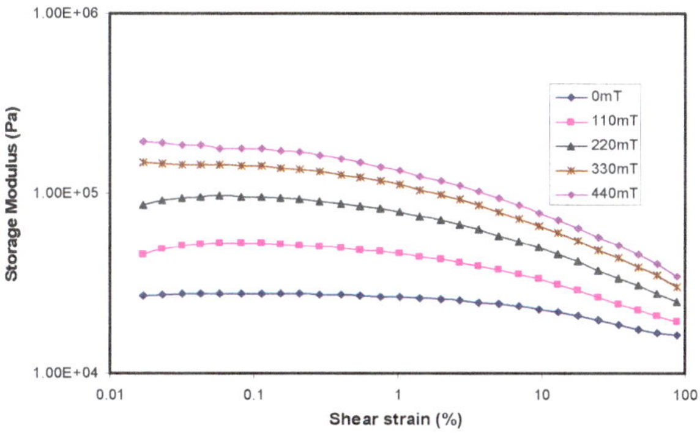

(a)

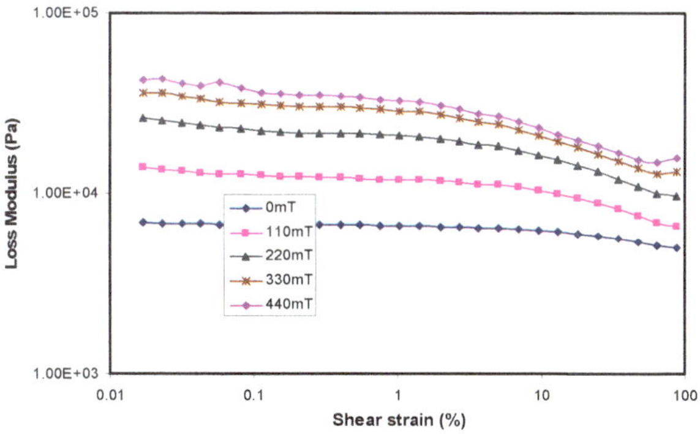

(b)

Figure 12. Storage and Loss Modulus versus strain amplitude sweep (anisotropic MRE Gr 0%) (a) Storage modulus vs. shear strain (b) Loss modulus vs. shear strain

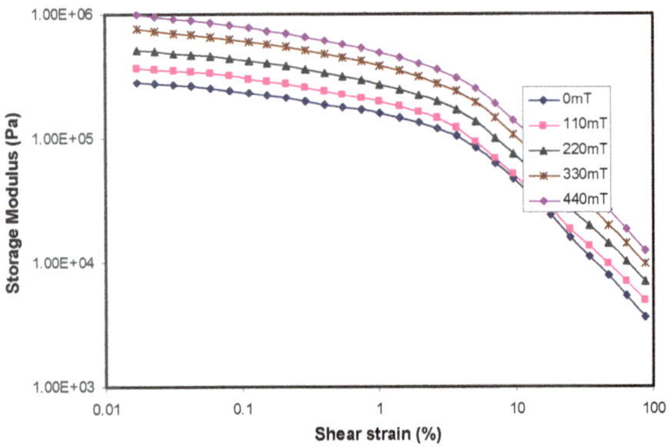

(a)

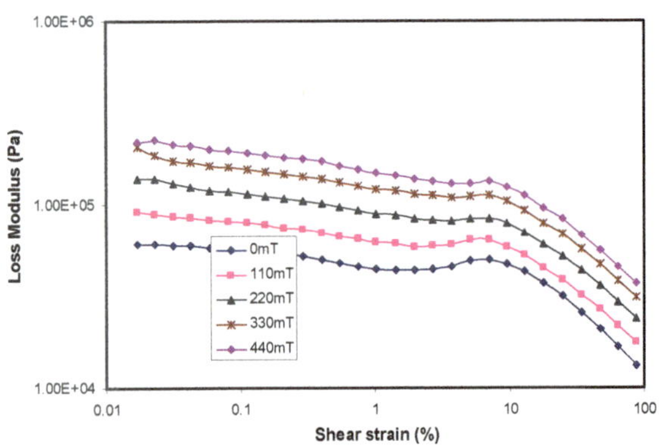

(b)

Figure 13. Storage and Loss Modulus versus strain amplitude sweep (anisotropic MRE Gr 20%) (a) Storage modulus vs. shear strain (b) Loss modulus vs. shear strain

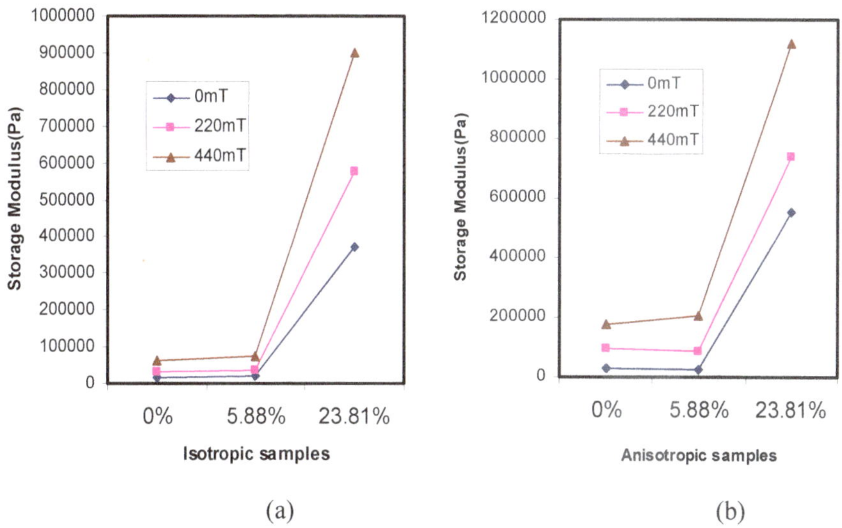

Figure 14. Storage Modulus of different samples at 10% shear strain (a) isotropic samples (b) anisotropic samples

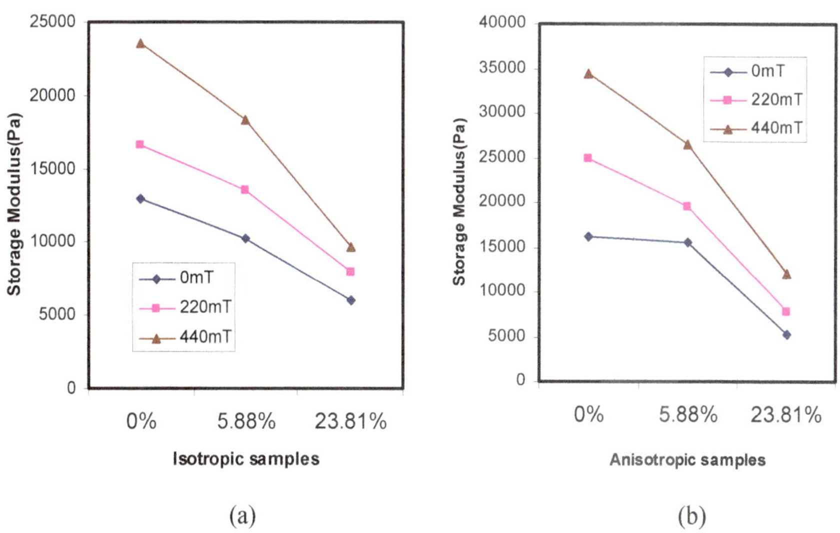

Figure 15. Storage Modulus of different samples at 87.5% shear strain (a) isotropic samples (b) anisotropic samples

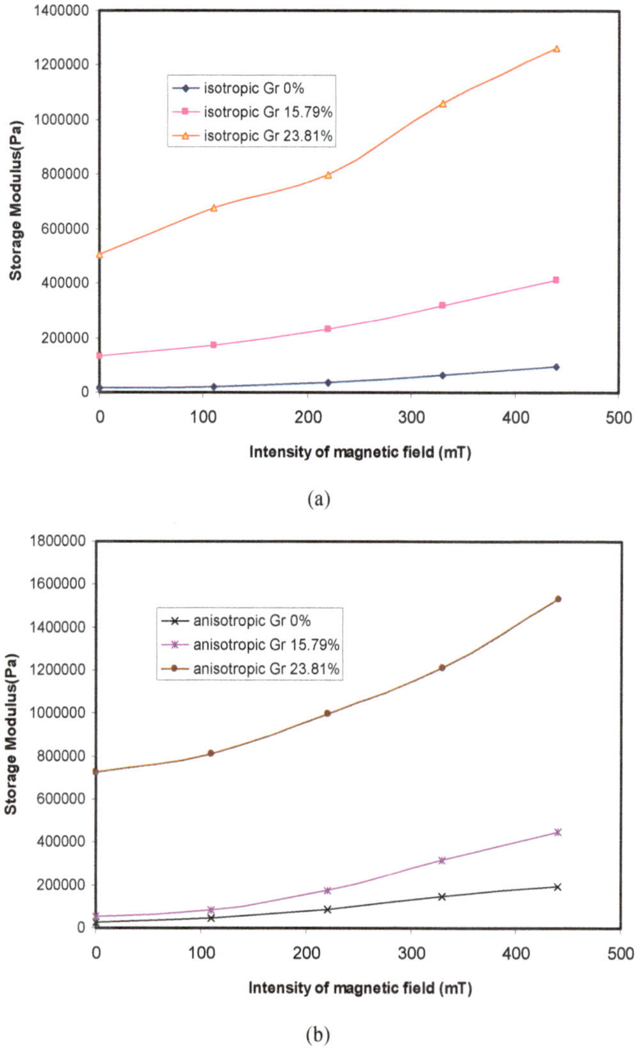

Figure 16. Storage Modulus versus magnetic field at 0.1% shear strain (a) isotropic samples (b) anisotropic samples

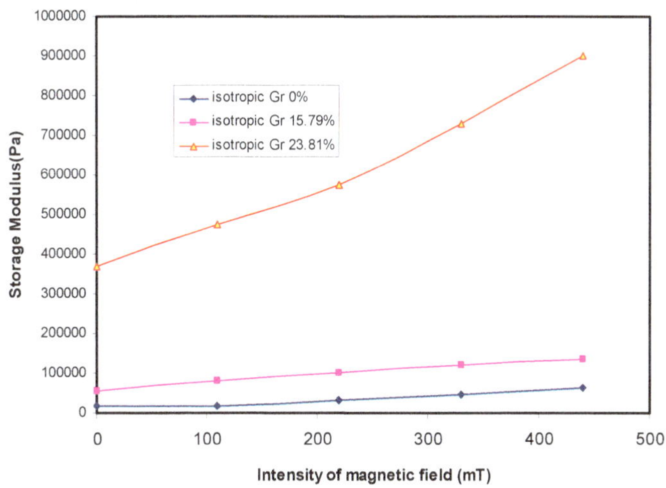

(a)

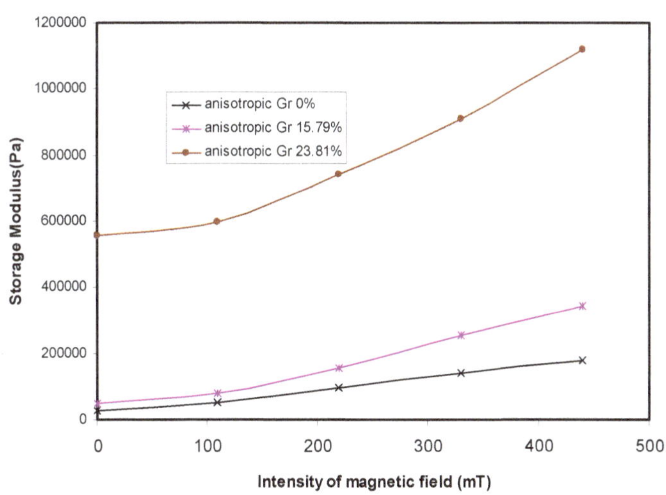

(b)

Figure 17. Storage Modulus versus magnetic field at 10% shear strain (a) isotropic samples (b) anisotropic samples

5. MRE sensing capabilities

5.1. Measurement of the relationship between the current and magnetic field

Figure 18 shows a schematic of the experimental device used. In this setup a long plastic plate is used to hold the weight which is applied load to the Gr-MRE samples. A Gaussmeter (HT201, Hengtong magnetoelectricity CO., LTD) is employed to test the intensity of the magnetic field. A multimeter (Finest 183, Fine Instruments Corporation) measures the resistance in the Gr-MRE samples.

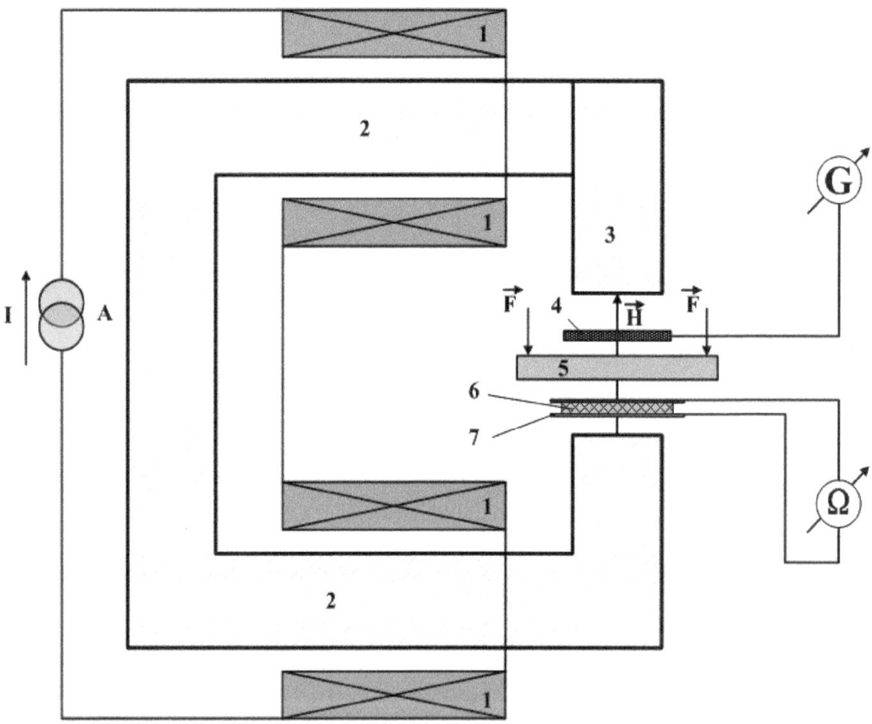

Figure 18. Sketch of the experimental device 1- coils; 2- electrical magnetic; 3- moving magnetic pole; 4- Gaussmeter probe; 5- plastic plate; 6- Gr MRE sample; 7- measuring copper plates; G- Gaussmeter; Ω- Multimeter; F- external force applied on Gr MRE samples

During the testing of the resistance of all the anisotropic MRE samples only the samples whose graphite weight fraction is higher than 15% are detectable by Finest 183 multimeter. Thus, among the tested anisotropic MRE samples only three of them are considered in Table 1, namely those whose graphite weight fractions are 20%, 21.95% and 23.81%. The results of these tests are shown in Figures 19, 20, and 21, respectively for the following data of anisotropic MRE with graphite weight fraction 20%, 21.95% and 23.81%.

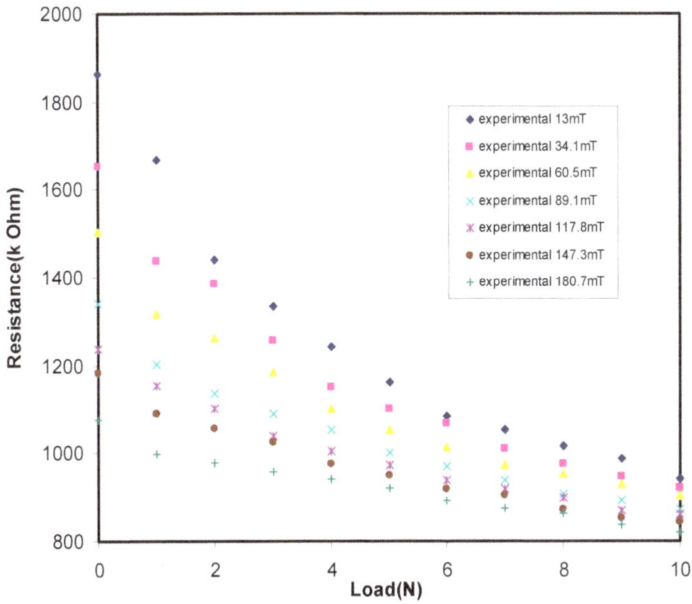

Figure 19. Resistance versus load (anisotropic MRE Gr 20%)

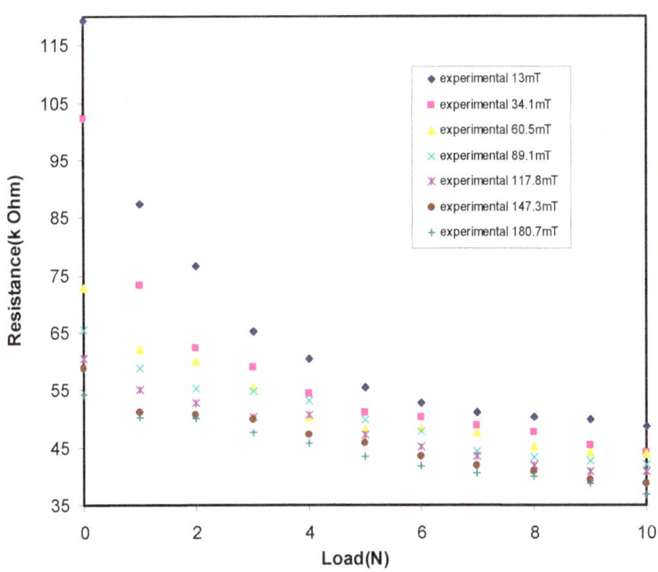

Figure 20. Resistance versus load (anisotropic MRE Gr 21.95%)

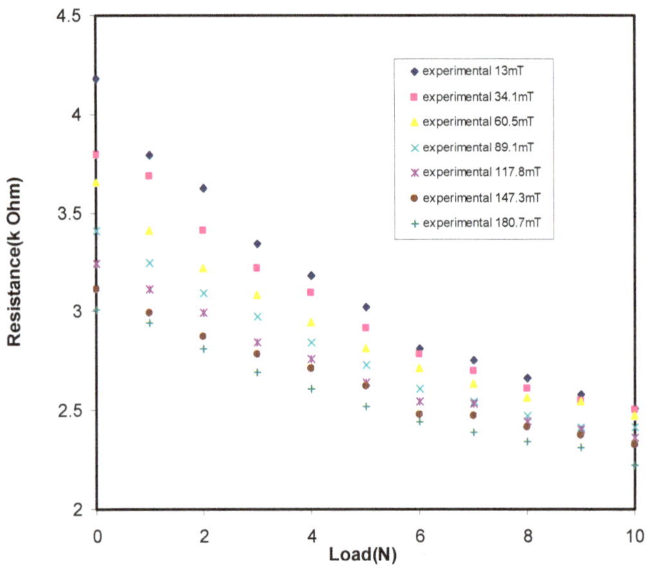

Figure 21. Resistance versus load (anisotropic MRE Gr 23.81%)

As shown in the plots similar trends are observed for the three samples. Specifically, in a fixed magnetic field, when the external load increases from 0 to 10 N, the resistance reduces in all three samples. With small loads, the resistance changes significantly but it decreases slowly when the load is more than 5 N. According to the absolute values, the sample with higher graphite weight fraction shows the higher electrical conductivity and the smaller decline in resistance. For instance, the resistance of the sample with graphite weight fraction 20% drops from 1862 kΩ at 0 N to 942 kΩ at 10 N, whereas a decline from 4.18 kΩ at 0 N to 2.51 kΩ at 10 N for the sample with graphite weight fraction 23.81%.

Besides, the resistance of each sample at a fixed external load decreases with increase in magnetic field intensity. Considering the sample with 21.95% graphite weight fraction as an example, at 5 N external force, the resistance 55.4 kΩ without the magnetic field decreases to 43.5 kΩ at a 440 mT magnetic field. This trend is shown in detail in Fig. 22.

5.2. A representative volume unit based mathematical model for investigating the magnetic field dependent sensing capabilities

5.2.1. Theoretical approach

From the Dipole Model a Representative Volume Unit (RVU) is derived. A RVU consists of two neighbouring hemispheres and the surrounding polymer matrix, which can be regarded as the minimum volume element in the conventional MRE. Fig. 23 shows the position of RVU in carbonyl iron particle chains and a longitudinal section of the unit is shown in Fig. 24.

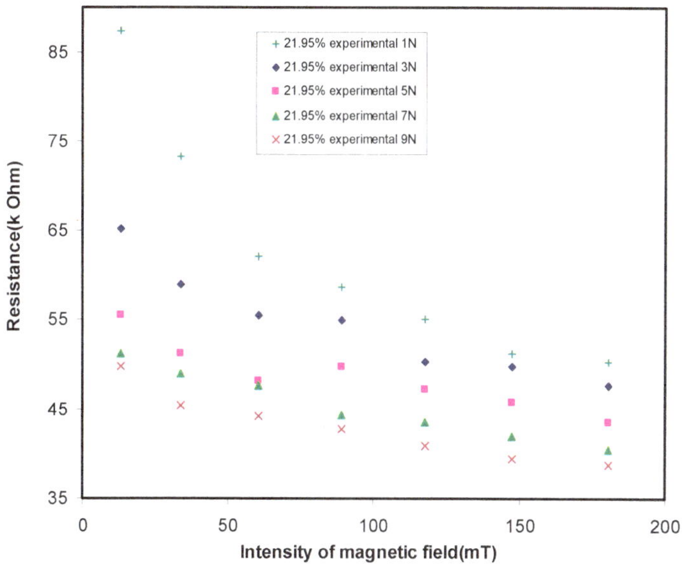

Figure 22. Resistance versus magnetic field (anisotropic MRE Gr 21.95%)

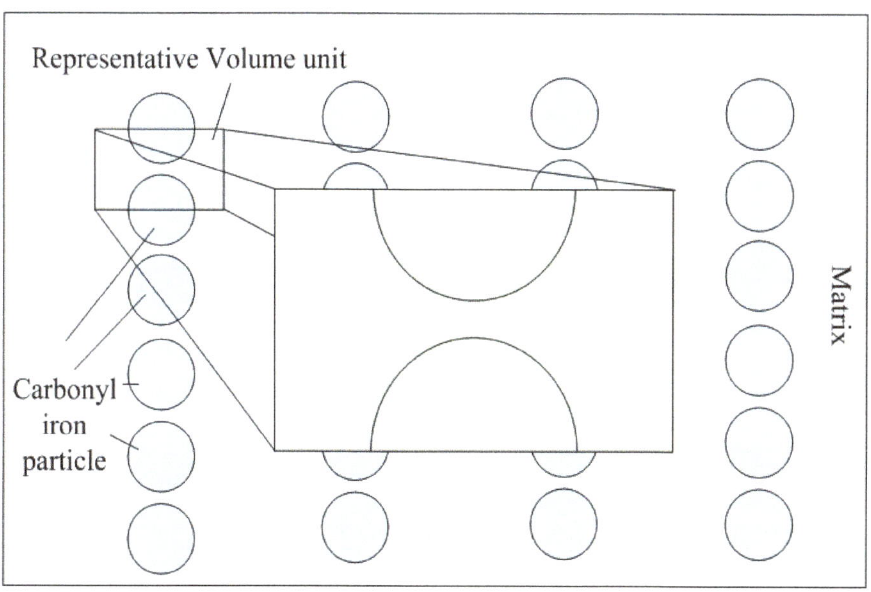

Figure 23. RVU in carbonyl iron chains

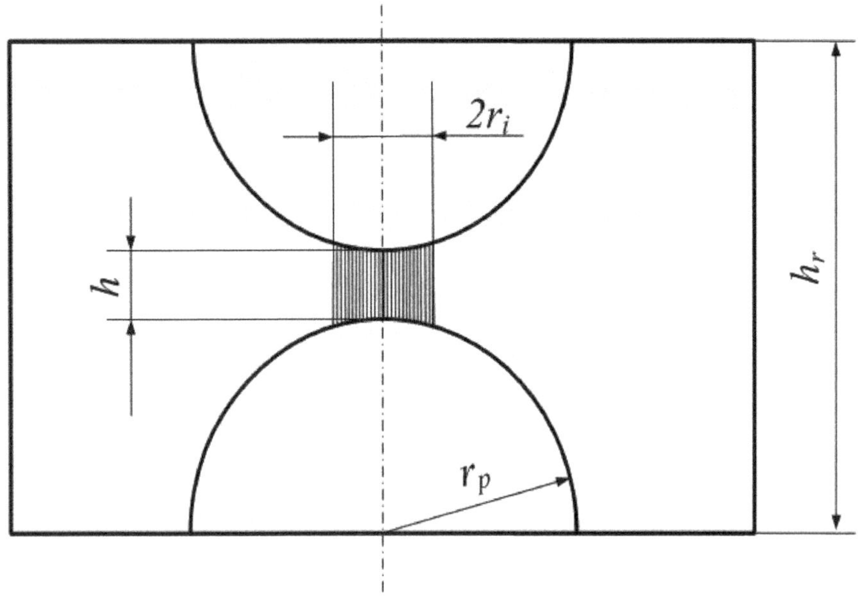

Figure 24. The longitudinal section of RVU

The RVU is a model for ideal anisotropic MREs which is supposed to have a chain structure.

Given the iron particle volume fraction ϕ in the RVU, the area of cross section can be expressed as

$$S_r = \frac{V_r}{h_r} = \frac{V_s}{\phi h_r} = \frac{4\pi r_p{}^3}{3\phi\left(2r_p + h\right)} \approx \frac{2\pi r_p{}^2}{3\phi} \tag{1}$$

where ϕ is the iron particle volume fraction, r_p is the particle radius, h is the particle distance, h_r is the distance between two half spheres, V_s is the particle volume, V_r is volume between two particles, and S_r is the cross section area.

In the RVU, the conductivity of the iron particles is much higher than that of the polymer, therefore the electric potential drops within the particles can be neglected. Due to system geometry and electrical properties, most of the current flowing through the RVU concentrates on the small area between the two adjacent hemispheres. Specifically, given the intensity of current flowing through the polymer is E, the intensity of the local electric field is

$$E_{loc} = \frac{h_r E}{h} = \frac{(2r_p + h)E}{h} \approx 2r_p \frac{E}{h} \tag{2}$$

Owing to the magnetic attraction during MRE preparation, the insulating polymer film between two neighbouring iron particles (here described by h) is very thin, and it is across this film that the electrical field induced tunnel current can occur. The Fowler-Nordheim equation [19-21] can be used to express the tunnel current. In addition, the iron particles dispersed in the polymer matrix contribute to the conductivity of the polymer and then the total current density is the sum of the tunnel density j_t and the conduction density of the polymer j_c

According to the Fowler-Nordheim equation, the relation between the tunnel density j_t and electric field intensity E is

$$j_t = \alpha E_{loc}^2 \exp\left(-\frac{\beta}{E_{loc}}\right) \tag{3}$$

in which α and β are constants determining the tunnel current. So the total current density j is

$$j = j_t + j_c = \alpha E_{loc}^2 \exp\left(-\frac{\beta}{E_{loc}}\right) + \sigma_f E_{loc} \tag{4}$$

in which σ_f is the conductivity of the polymer film.

The total current density j is for the current flowing through the small area between the tips of two adjacent iron particles, however, the overall density of RVU j_r should be derived from its entire cross section. So from the total density of RVU j_r and the electric field intensity E, the global conductivity of typical MRE σ_r can be represented as

$$\sigma_r = \frac{j_r}{E} = \frac{\pi \cdot r_i^2 \cdot j}{S_r \cdot E} = 3\phi \cdot r_i^2 \left[\frac{2\alpha}{h^2} E \exp\left(-\frac{h\beta}{2r_p E}\right) + \frac{\sigma_f}{r_p h}\right] \tag{5}$$

in which r_i is the radius of the circular section across which the tunnelling current flow, as shown in Fig. 24.

MRE also exhibits piezoresistivity. When a MRE sample is compressed, its conductivity increases. This phenomenon is explained by two factors, one of which is the increments of the conductive area induced by the deformation of MREs and the other is the reduction of the thickness of the polymer membrane between the two adjacent iron particles. Because of the large ratio of r_i/h, it is difficult to compress the membrane further and thus the increment of the conductive area is the significant reason for the conductivity increasing.

From the Hertz Theory [22-24], when the initial pressure applied on MRE is σ_0 through a constant loading, corresponding to which there is an initial contact area radius r_{i0}

$$r_{i0} = \left[\frac{3\pi\sigma_0\left(1-\nu^2\right)}{2E_p}\right]^{1/3} \cdot r_p \tag{6}$$

Where ν is the Poisson's ratio, E_p is the Young's Modulus.

So the radius r_i increases along with the increment of pressure

$$r_i = r_{i0} + r_p\left(\left(\sigma_0 + \sigma\right)^{1/3} - \sigma_0{}^{1/3}\right)\cdot\left(\frac{3\pi\left(1-\nu^2\right)}{2E_p}\right)^{1/3} \tag{7}$$

The magnetic field also contributes to the electrical resistance of MREs. When the external magnetic field is applied to MREs, the carbonyl iron particles are attracted by the poles of magnetic field, with closer magnetic pole providing more powerful magnetic force than the other pole. So the attraction from the farther magnetic pole can be neglected.

For the two iron particles in each RVU, the magnetic attraction from the pole applied to the farther particle compresses the thin film between the two adjacent iron particles. Similar to the piezoresistivity, the increment of the conductive area is the main cause for the conductivity increasing.

Thus, the radius r_i can be updated as

$$r_i = r_{i0} + r_p\left(\left(\sigma_0 + \sigma_1 + \sigma_2\right)^{1/3} - \sigma_0{}^{1/3}\right)\cdot\left(\frac{3\pi\left(1-\nu^2\right)}{2E_p}\right)^{1/3} \tag{8}$$

in which, σ_1 is the stress due to pressure, σ_2 is the stress from the magnetic attraction.

So the dependence of the conductivity of MREs on electric field intensity and the stress due to pressure is

$$\sigma_m = 3\phi\cdot\left[\frac{2\alpha}{h^2}E\exp\left(-\frac{h\beta}{2r_pE}\right)+\frac{\sigma_f}{r_ph}\right]\cdot\left[r_{i0} + r_p\left(\left(\sigma_0 + \sigma_1 + \sigma_2\right)^{1/3} - \sigma_0{}^{1/3}\right)\cdot\left(\frac{3\pi\left(1-\nu^2\right)}{2E_p}\right)\right]^2 \tag{9}$$

When the initial condition σ_0 and r_{i0} are set, Apart from E and σ, the other parameters in this equation are all constants. So the conductivity of MREs σ_m is dependent on the intensity of the electric field E and the stress due to pressure σ.

In response to the effects of graphite and carbonyl iron on the resistance model, two parameters λ_g and λ_i are introduced to show the effects of graphite volume fraction ϕ_g and carbonyl iron volume fraction φ_i to the conductivity of new MREs. Two parameters λ_g and λ_i show the contribution of the carbonyl iron particles and graphite powder to the resistance respectively. The higher are values for λ_g and λ_i, the less resistance the Gr-MREs have.

In the RVU, the iron particle volume fraction is set as ϕ = 0.4, which means that the volume of two hemispheres is 40% of the whole volume of RVU.

The shape of Gr MRE samples is fixed. The thickness l is l= 1mm=0.001m, the diameter D is D=0.021m, so the cross section area is $A=\pi\cdot(D/2)^2$= 0.000346 m^2.

In the tunneling equation $\left[\frac{2\alpha}{h^2}E\exp\left(-\frac{h\beta}{2r_pE}\right)+\frac{\sigma_f}{r_ph}\right]$, α and β are pre-exponential and exponential terms of the standard Fowler-Nordheim Tunneling which are both constants. In this case, value α is set as 2, value β is set as 1. h is the height of the RVU as two times of iron particle's radius. So h=0.000004 m. The iron particle's radius r_p is 0.000002 m. σ_f is the conductivity of the polymer film namely silicone rubber. Because of the high resistance of silicone rubber, the value of conductivity of silicone rubber σ_f is set as 1*10^{-10}. The electric field E is from the function file of the multimeter used in the test. The value of E is 9V. So the whole equation can be calculated as $\frac{2\alpha}{h^2}E\exp\left(-\frac{h\beta}{2r_pE}\right)+\frac{\sigma_f}{r_ph}\approx 2.01*10^{12}$. At the normal condition, the MRE Young's modulus E_p is set as 4000 Pa. The Poisson's ratio of particles ν is set as 0.1. Substitute these parameters into Equation (9), the final resistance of Gr-MREs is

$$R_g=\frac{4.78*10^{-13}\times\lambda_g\lambda_i}{10^{-8}+0.000002\left(\left(2000+\sigma_1+\sigma_2\right)^{1/3}-2000^{1/3}\right)\times 0.001166} \tag{10}$$

where σ_1 is the compressive stress from the external force and σ_2 is the compressive stress from the external magnetic field [25]. In this study, the weight of plastic plate is 10 g and the weight increment is 110 g. Thus, the force increment is 1 N. The stress, σ_1, is then calculated by dividing the force with the area A (0.000346 m^2), and shown in Table 2.

External load(N)	Real load(N)	Compressive stress σ_1 (Pa)
0	0.1	288.7165
1	1.1	3175.881
2	2.1	6063.045
3	3.1	8950.21
4	4.1	11837.37
5	5.1	14724.54
6	6.1	17611.7
7	7.1	20498.87
8	8.1	23386.03
9	9.1	26273.2
10	10.1	29160.36

Table 2. Relation of weight, load and compressive stress

The anisotropic MRE with graphite weight fraction 21.95% (contains 10g iron particles, 3g silicone rubber, 3g silicone oil and 4.5g graphite powder) can be used as an example. Table 3 shows the volume fractions of all the ingredients for this sample.

	Mass(g)	Density(g/cm³)	volume(cm³)	Volume fraction
Iron particle	10	7.86	1.272265	16.50%
Silicone rubber	3	3.18	0.943396	12.24%
Silicone oil	3	0.96	3.125	40.54%
Graphite powder	4.5	1.9	2.368421	30.72%

Table 3. Volume fractions of all the ingredients (anisotropic MRE Gr 21.95%)

The factors showing the efforts of graphite and iron particles to the resistance in Gr-MREs are λ_g and λ_i respectively and λ_g and λ_i depend on the graphite volume fraction ϕ_g and iron particles volume fraction ϕ_i . For each sample there are different values for λ_g and λ_i. Fig. 25 shows the product of λ_g and λ_i versus Gr-MREs (Gr 20%, 21.95% and 23.81%). The data is from the ratio of experimental result and theoretical prediction.

If we set $\lambda_g = \exp(a \cdot \phi_g)$ and $\lambda_i = \exp(b \cdot \phi_i)$ and substitute the data in Table 4 and Fig. 8 to λ_g and λ_i, the parameters λ_g and λ_i can be obtained and so the parameter a and b were derived as -65 and 250, respectively. So the relationship of λ_g and λ_i to ϕ_g and ϕ_i are $\lambda_g = e^{-65*\varphi_g}$, $\lambda_i = e^{250\varphi_i}$

	Gr 20%	Gr 21.95%	Gr 23.81%
Iron particle (ϕi)	17.09%	16.5%	15.96%
Graphite powder(ϕg)	28.27%	30.7%	33.01%

Table 4. Volume fractions of iron and graphite (anisotropic MRE Gr 20%, 21.95% & 23.81%)

After having determined all parameters, Equation 10 is ready to calculate the final resistance R_g of the anisotropic MRE with graphite weight fraction 20%.

5.2.2. Comparison and analysis

To show the comparison between the experimental result and the theoretical prediction, anisotropic MRE with graphite weight fraction 21.95% is considered. Comparison between experimental and theoretical result is provided in Fig. 26.

Figure 26 shows that the experimental and theoretical result do not match each other perfectly. However, the trends of both experimental result and theoretical result are the same.

Similarly some other comparison can be made from the data. At the fixed external force (which is expressed as load) when the magnetic field intensity increases the resistance of the sample decreases. The data under three external loads such as 1 N, 5 N and 10 N were chosen to compare the experimental and theoretical result in Fig. 27.

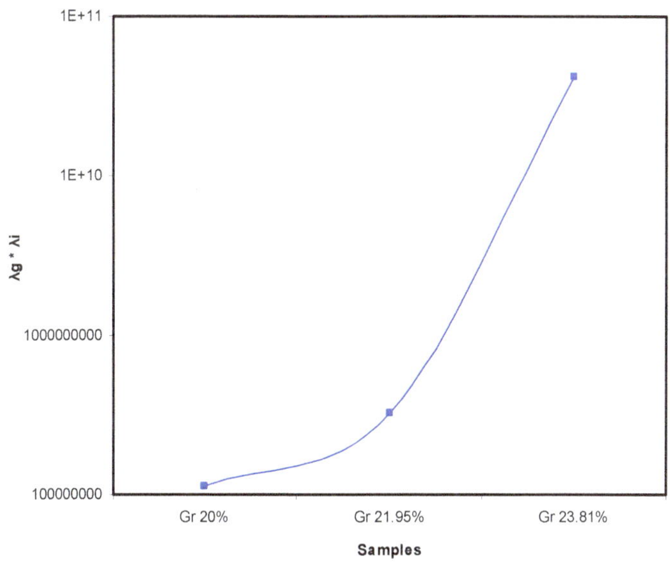

Figure 25. $\lambda_g * \lambda_i$ versus graphite weight fraction

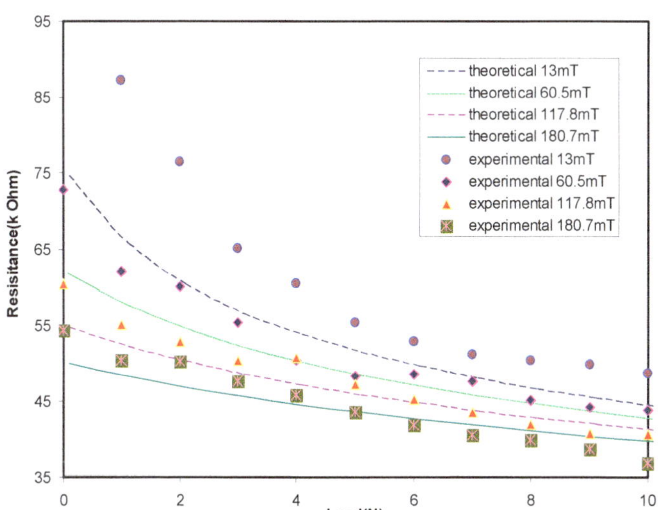

Figure 26. Comparison between experimental result and theoretical result (anisotropic MRE Gr 21.95%)

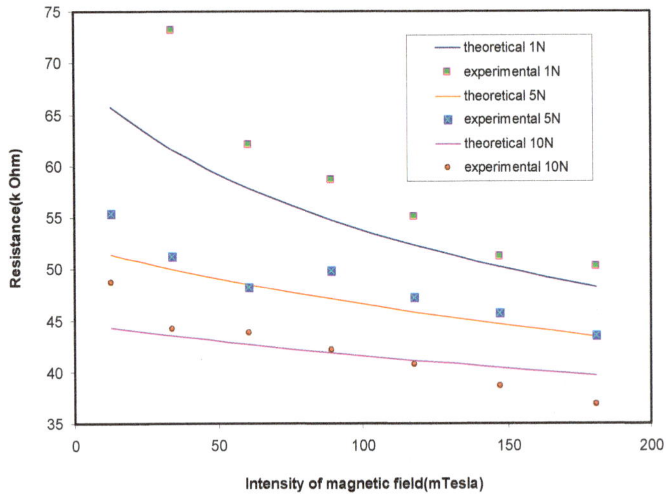

Figure 27. Resistance changing at a fixed load (anisotropic MRE Gr 21.95%)

The Fig. 27 shows the change in the resistance for given external forces. Along with the raising of magnetic field intensity, the sample's resistance decreases. The higher external load applied leads to lower resistance of Gr-MREs.

The next two figures Fig. 28 and 29 show the resistance variation between anisotropic MREs with graphite weight fraction 20%, 21.95%, and 23.81%. The magnetic field is fixed in Fig. 28 and the external load is fixed in Fig. 29.

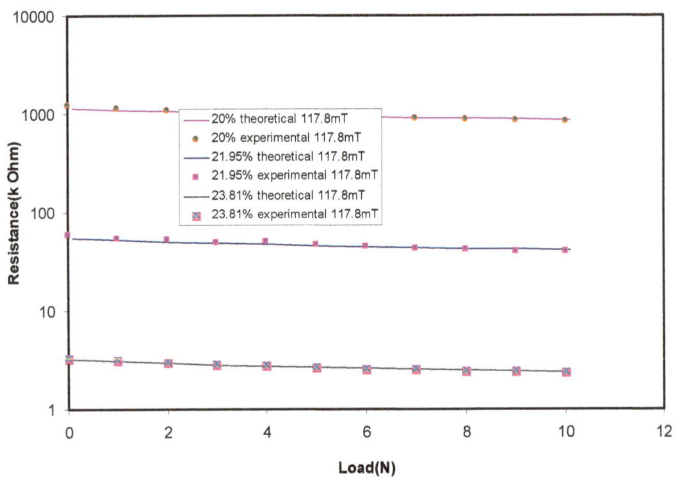

Figure 28. Resistance between different sample at 117.8mT magnetic field (anisotropic MRE with graphite weight fraction 20%, 21.95% & 23.81%

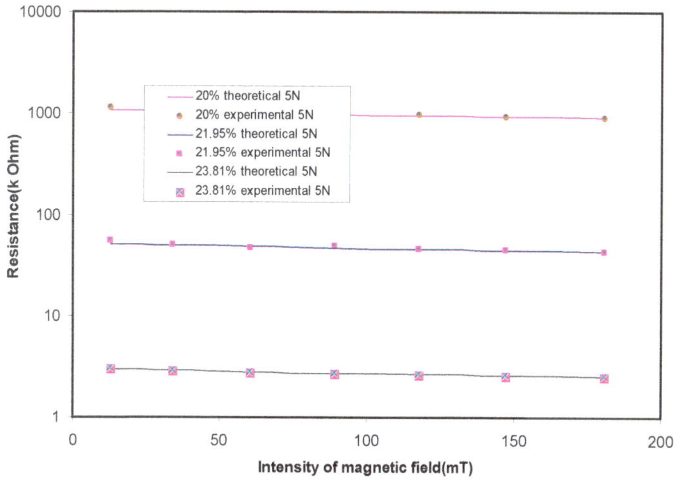

Figure 29. Resistance between different samples at 5.1N external load (anisotropic MRE with graphite weight fraction 20%, 21.95% & 23.81%)

Because of the order of values for the three samples it is easy to see from the Fig. 28 and 29 that theoretical and experimental results can match each other very well. This again provides a proof of the theory and when either external load or magnetic field intensity increases the resistance decrease.

Therefore when the magnetic field intensity and resistance are measured, Equation 10 can be used to calculate the value of the external load which means the Gr-MREs have the potential to work as the key component in a force sensor.

6. Conclusion

Both isotropic and anisotropic samples of graphite-based magnetorheological elastomers (Gr-MREs) with various graphite weight fractions ranging from 0% to 23.8% were fabricated in this study. The microstructural observation of these samples shows that the presence of graphite powder affects the forming of carbonyl iron chains. The sample with less graphite shows better-aligned carbonyl iron chains which influence the magnetorheology of MREs. In addition by connecting two iron chains in parallel and connecting the disconnected iron chains, the graphite contributes to the conductivity of MREs.

Steady state and dynamic tests such as strain amplitude sweep and angular frequency sweep were used to test the magnetorheology of Gr-MREs. With the help of graphite in MREs, the Storage and Loss Moduli are both changed. The steady state tests showed that the graphite can diminish the viscoelastic linear range of MREs. The dynamic tests proved that the samples with higher graphite weight fraction show higher initial Storage and Loss Moduli and lower MR effects. Additionally, the resistance of each MRE sample exhibits a decreasing trend with the graphite weight fraction.

Based on a Dipole model, a representative volume unit was presented to show the resistance of ideal anisotropic MREs. The current flowing through the ideal chain structure were derived by taking into account both the tunnel current and conductivity current. In the mathematical model, two parameters, λ_i and λ_g, were introduced to reflect the effect of the iron particles and graphite, which are both exponential functions of particle volume fractions. The exponential parameters were identified and then used to reconstruct modelling predictions. The comparison between experimental results with modelling predictions indicate that the proposed mathematical model can well investigate the sensing capabilities of the graphite based MRE elastomers.

Author details

Weihua Li and Tongfei Tian
School of Mechanical, Materials & Mechatronic Engineering, University of Wollongong, Wollongong, NSW, Australia

Haiping Du
School of Electrical, Computer and Telecommunications Engineering, Wollongong, NSW, Australia

Acknowledgement

This work is partially supported by the University of Wollongong through a UIC funding scheme.

7. References

[1] G. Y. Zhou and Z. J. Jiang, "Deformation in magnetorheological elastomer and elastomer-ferromagnet composite driven by a magnetic field", *Smart Mater. Struct.*, 13: 309-316 (2004)
[2] X. L. Gong, X. Z. Zhang and P. Q. Zhang, "Fabrication and Characterization of Isotropic Magnetorheological Elastomers", *Polymer Testing*, 24(5): 669–676 (2005)
[3] H. X. Deng, X. L. Gong and L. H. Wang, "Development of an adaptive tuned vibration absorber with magnetorheological elastomer", *Smart Materials & Structures*, 15 (5): N111-N116 (2006)
[4] Zou, H., Zhang, L.Q., Tian, M., Wu, S.Z. and Zhao, S.H., 2009. Study on the Structure and Properties of Conductive Silicone Rubber Filled with Nickel-Coated Graphite. Journal of Applied Polymer Science, 115(5): p. 2710-2717.
[5] Li, W.H., Kostidis, K., Zhang, X.Z., Zhou, Y. and Ieee. 2009. 'Development of a Force Sensor Working with MR Elastomers'. in 2009 Ieee/Asme International Conference on Advanced Intelligent Mechatronics, Vols 1-3. New York: Ieee.
[6] Ginder, J.M., Clark, S.M., Schlotter, W.F. and Nichols, M.E., 2002. Magnetostrictive phenomena in magnetorheological elastomers. International Journal of Modern Physics B, 16(17-18): p. 2412-2418.

[7] Lokander, M. and Stenberg, B., 2003. Performance of isotropic magnetorheological rubber materials. Polymer Testing, 22(3): p. 245-251.
[8] Lokander, M. and Stenberg, B., 2003. Improving the magnetorheological effect in isotropic magnetorheological rubber materials. Polymer Testing, 22(6): p. 677-680.
[9] Zhou, G.Y., 2003. Shear properties of a magnetorheological elastomer. Smart Materials and Structures, 12(1): p. 139.
[10] Ginder, J.M., Schlotter, W.F. and Nichols, M.E. 2001. 'Magnetorheological elastomers in tunable vibration absorbers'. in Smart Structures and Materials 2001: Damping and Isolation. Newport Beach, CA, USA: SPIE.
[11] Gong, X.L., Zhang, X.Z. and Zhang, P.Q., 2005. Fabrication and characterization of isotropic magnetorheological elastomers. Polymer Testing, 24(5): p. 669-676.
[12] Kchit, N. and Bossis, G., 2009. Electrical resistivity mechanism in magnetorheological elastomer. Journal of Physics D-Applied Physics, 42(10): p. 5505-5505.
[13] Wang, X.J., Gordaninejad, F., Calgar, M., Liu, Y.M., Sutrisno, J. and Fuchs, A., 2009. Sensing Behavior of Magnetorheological Elastomers. Journal of Mechanical Design, 131(9): p. 6.
[14] Bica, I., 2009. Influence of the transverse magnetic field intensity upon the electric resistance of the magnetorheological elastomer containing graphite microparticles. Materials Letters, 63(26): p. 2230-2232.
[15] Li, W.H., Kostidis, K., Zhang, X.Z., Zhou, Y. and Ieee, Development of a Force Sensor Working with MR Elastomers, in 2009 Ieee/Asme International Conference on Advanced Intelligent Mechatronics, Vols 1-3. 2009, Ieee: New York. p. 233-238.
[16] Zhao, Y., Maietta, D.M. and Chang, L., 2000. An Asperity Microcontact Model Incorporating the Transition From Elastic Deformation to Fully Plastic Flow. Journal of Tribology, 122(1): p. 86-93.
[17] McLachlan, D.S., 2000. Analytical functions for the dc and ac conductivity of conductor-insulator composites. Journal of Electroceramics, 5(2): p. 93-110.
[18] Woo, L.Y., Wansom, S., Hixson, A.D., Campo, M.A. and Mason, T.O., 2003. A universal equivalent circuit model for the impedance response of composites. Journal of Materials Science, 38(10): p. 2265-2270.
[19] Weinberg, Z.A., 1982. On tunneling in metal-oxide-silicon structures. Journal of Applied Physics, 53(7): p. 5052-5056.
[20] Serdouk, S., Hayn, R. and Autran, J.L., 2007. Theory of spin-dependent tunneling current in ferromagnetic metal-oxide-silicon structures. Journal of Applied Physics, 102(11): p. 113707-1-113707-5.
[21] Dahlke, W.E. and Sze, S.M., 1967. Tunneling in metal-oxide-silicon structures. Solid State Electronics, 10(8): p. 865-873.
[22] Zhupanska, O.I. and Ulitko, A.F., 2005. Contact with friction of a rigid cylinder with an elastic half-space. Journal of the Mechanics and Physics of Solids, 53(5): p. 975-999.
[23] Etsion, I., Levinson, O., Halperin, G. and Varenberg, M., 2005. Experimental investigation of the elastic-plastic contact area and static friction of a sphere on flat. Journal of Tribology, 127(1): p. 47-50.

[24] Maouche, N., Maitournam, M.H. and Dang Van, K., 1997. On a new method of evaluation of the inelastic state due to moving contacts. Wear, 203-204: p. 139-147.
[25] Chang, W.R., Etsion, I. and Bogy, D.B., 1988. Static Friction Coefficient Model for Metallic Rough Surfaces. Journal of Tribology, 110(1): p. 57-63.

Chapter 13

Novel Self-Oscillating Polymer Actuators for Soft Robot

Yusuke Hara, Shingo Maeda,Takashi Mikanohara, Hiroki Nakagawa, Satoshi Nakamaru and Shuji Hashimoto

Additional information is available at the end of the chapter

1. Introduction

Intelligent materials with changing properties and functions have been studied in many fields [1-9]. In particular, stimuli-responsive polymer systems have been considerably investigated for the purpose of the many types of possible applications, such as soft actuators, microfluidics and medical devices, *etc.* due to their light weight, flexibility and low noise, *etc.* [10-18]. Polymer gels have the flexible bodies allowing volume changes and sensors properties for chemical substances at the same time. By utilizing these properties, the gel actuators have been considered to be applied for the biomimetic robots. In particular, a thermoresponsive poly(N-isopropylacrylamide) (PNIPAAm) gel have been significantly investigated especially for medical devices. [19, 20] However, in order to drive these polymer materials, external device for controlling the external stimulus must be needed. On the other hand, organic systems can generate autonomous motion without external stimuli. In recent years, self-oscillating gels and polymer chains were developed and researched in order to understand and mimic the spontaneous mechanical works. [21-27] The energy source of self-oscillating polymer systems is chemical energy like a living organism, that is, the polymer system cause the self-oscillation induced by the Belouzov-Zhabotinsky (BZ) reaction or the pH oscillating reaction. [28-43] The Self-oscillating reactions have been widely studied for their relevance to nonlinear, chaotic dynamics and biological signals. In previous investigations, the self-oscillating behaviors of polymer systems induced by the Belousov-Zhabotinsky (BZ) or the pH oscillating reactions were realized.[44-60] The BZ reaction is well known as an oscillating reaction accompanying spontaneous redox oscillations to generate a wide variety of nonlinear phenomena, e.g., a target or spiral pattern in an unstirred solution, and periodicity, multi-periodicity, or chaos in a stirred solution. The overall process of the BZ reaction is the oxidation of an organic substrate by an oxidizing agent in the presence of the catalyst under strong acidic conditions. In the BZ reaction, the metal catalyst undergoes spontaneous redox self-oscillation, and it changes the solubility from hydrophilic to hydrophobic at the same time. In order to cause the self-oscillation under temperature constant condition, the different solubility of the Ru catalyst in the reduced and oxidized state is utilized. In previous investigations, poly(Nisopropylacrylamide) (poly(NIPAAm)) was covalently bonded to (ruthenium (4-vinyl-4'-methyl-2,2-bipyridine) bis(2,2'-bipyridine)bis(hexafluorophosphate)) ($Ru(bpy)_3$) and a negatively charged acrylamide-2-methylpropanesulfonic acid (AMPS) as a pH and solubility control site [47-50]. The AMPS-containing polymer systems caused self-oscillations originating from the different solubility of the $Ru(bpy)_3$ moiety in the oxidized and reduced state. However, the conventional-type self-oscillating polymer system has the temperature limitation for causing the self-oscillation. That is because the conventional-type self-oscillating polymer system causes the aggregation or the contraction above the LCST. In this chapter, we introduce novel type self-oscillating polymers systems that are constituted the non-thermoresponsive and biocompatible poly-vinylpyrrolidone (PVP) as a polymer main-chain in order to expand the application fields. In our previous investigations, we first succeeded in causing the aggregation-disaggregation self-oscillation of the novel polymer chain and polymer gel under the constant condition induced by the BZ reaction. As for the VP-based self-oscillating polymer chain, the influence of the concentration of the three BZ substrates (sodium bromate, malonic acid and nitric acid) other than the metal catalyst on the waveform and period of the self-oscillation was investigated. As a result, it was demonstrated that the amplitude of the self-oscillation is hardly affected by the initial concentration of the BZ substrates. Furthermore, we firstly succeeded in causing the swelling-deswelling self-oscillation at high temperature condition by adapting the VP-main chain. We studied the influence of the initial concentration of the three BZ substrates other than the metal catalyst and the temperature on the period of the self-oscillation. As a result, it was clarified that the period of the self-oscillation for the VP-based polymer gel can be controllable by the selection of the initial concentration of the three BZ substrates (malonic acid (MA), sodium bromated ($NaBrO_3$) and nitric acid) and the temperature. In addition, by optimizing the initial concentration of the BZ substrates and the temperature, it was demonstrated that the maximam frequency of the swelling-deswlling self-oscillation is 0.5 Hz. This value was 20 times as large as that of the conventional-type self-oscillating polymer gel (poly(NIPAAm-co-$Ru(bpy)_3$ gel) with the thermoresponsive nature.

Next, in this chapter, we introduce the challenge of mimicking the peculiar locomotion of living organism, such as mollusks and apods, by utilizing polymer gel that cause the swelling-deswelling induced by the pH oscillating chemical reaction. Gastropods, like snails and slugs, can obtain the peristaltic movement by forming contraction waves, which is the propagation of the shrinking part of the body [61]. In this research, we focused on producing a functional gels obtaining peristaltic movements by the chemical energy directly.

In addition, we introduce a novel-tpye nanofiber gel actuator that was manufactured by the electrospining method. The nanofiber gel actuator can cause bending-streching motion synchronized with the pH oscilltion. Polymer nanofibers have been much studied for the application to various fields, such as chemical detector, scaffolds, wound dressing, and multifunctional membranes, sensors, filtration media and biomaterials, *etc.* [62-65]. That is because the nanofiber gel membranes have a high surface-to-volume ratio and a relatively small pore size compared to normal-type polymer hydrogel. The merit of the electrospining method to manufacutre nanofiber gel actutors is low cost, relatively high production rate, and applicability to many types of polymers. Moreover, by utilizing the electrospinning method, we can construct various shapes of nanofiber gel actuators because it does not require a mold to synthesize the gel. By controlling the external conditions, we can construct more complex structures such as porous fiber, core sheathe fiber, *etc.* [66-69]. Recently, some researchers succeeded in manufacturing functional nanofiber gels by utilizing electrospining method. For example, Hsieh and coworkers have reported ultrafine gel fibrous membranes that consist of poly(acrylic acid) (PAAc), poly(N-isopropylacrylamide-*co*-acrylic acid) and poly(vinyl alcohol)(PVA) and PVA/PAAc mixtures [70-72]. However, there are a few challenges to realize the nanofiber gel actuator [73,74]. In this study, in order to construct an autonomous nanofiber gel actuator, we utilized Landolt pH-oscillator based on a bromated/ sulfite/ferrocyanide reaction. By coupling with this pH-oscillator, we realized a nanofiber gel actuator that shows the bending and stretching motions with a constant period and displacement.

2. Experimental section

2.1.1. Synthesis of the poly(VP-co-Ru(bpy)$_3$)

The polymer chain was prepared as follows. 0.5g of ruthenium(4-vinyl-4′-methyl-2,2′-bipylridine)bis(2,2′-bipyridine)bis(hexafluorophosphate) (Ru(bpy)$_3$) as a metal catalyst for the BZ reaction, 9.5 g of vinylpyrrolidone (VP) and 0.35 g of 2,2′-azobis(isobutyronitrile) (AIBN) as an initiator were dissolved in the methanol solution (31 g) under total monomer concentration of 20 wt%. These polymerizations were carried out at 60 °C for 24 h *in vacuo*. These resulting reaction mixtures were dialyzed against graded series of water/methanol mixtures, for 1 day each in 0, 25, 50, 75, and 100 wt% of water, and then freeze-dried.

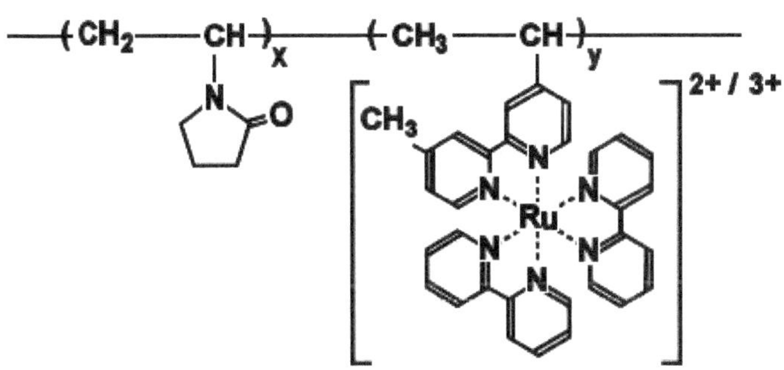

Figure 1. Chemical structure of poly(VP-*co*-Ru(bpy)$_3$).

2.1.2. Measurement of Lower Critical Solution Temperature (LCST) for the Poly(VP-co-Ru(bpy)3) solution

The lower critical solution temperature (LCST) of the polymer solution was measured under the reduced and oxidized states, by using $Ce(SO_4)_2$ as an oxidizing agent and $Ce_2(SO_4)_3$ as a reducing agent, respectively. The polymer solutions (0.5 wt%) of poly(VP-*co*-$Ru(bpy)_3$) were prepared by dissolving the polymer in a 0.3 M HNO_3 aqueous solution and adding 5 mM $Ce(SO_4)_2$ or 5 mM $Ce_2(SO_4)_3$, respectively. The LCST measurements were carried out with a spectrophotometer (JASCO, Model V-630) equipped with magnetic stirrers and a thermostatic controller. In this measurement, the 570 nm wavelength was used because it is the isosbestic point for the reduced and oxidized states of $Ru(bpy)_3$. The transmittance (%) of the polymer solution at 570 nm was then recorded by raising the temperature at a rate of 0.5 °C/min.

2.1.3. Measurement of optical oscillations for the poly(VP-co-Ru(bpy)3) solution

The poly(VP-*co*-$Ru(bpy)_3$) solutions were prepared by dissolving the polymer (0.5 wt%) into an aqueous solution containing the three BZ substrates (malonic acid (MA) and sodium bromate ($NaBrO_3$), nitric acid (HNO_3)). The transmittance self-oscillations of the polymer solutions were measured under constant temperature and stirring. In order to detect the transmittance change which is based on the autonomous transmittance change, 570-nm wavelength was used. The time course of the transmittance at 570 nm was monitored by a spectrophotometer.

2.2.1. Synthesis of the poly(VP-co-Ru(bpy)3) gel

The gel was prepared as follows. 0.110g of $Ru(bpy)_3$ as a metal catalyst for the BZ reaction was dissolved in 0.877g of vinylpyrrolidone (VP). 0.012g of N,N′-methylenebisacrylamide (MBAAm) as a cross-linker, and 0.020g of 2,2′-azobis(isobutyronitrile) (AIBN) as an initiator were dissolved in the methanol solution (3ml) (Figure 1). The two solutions were mixed together well, and then the mixed solution purged with dry nitrogen gas. The monomer solution was injected between Teflon plates separated by silicone rubber as a spacer (thickness: 0.5mm), and then polymerized at 60°C for 18 hours. After gelation, the gel strip was soaked in pure methanol for a week to remove unreacted monomers. The gel was carefully hydrated through dipping it in a graded series of methanol-water mixtures, for 1 day each in 75, 50, 25 and 0%.

2.2.2. Measurements of the equilibrium swelling ratio of poly(VP-co-Ru(bpy)3) gels

The equilibrium swelling ratio of the gels was measured under the reduced and the oxidized state, by using the oxidizing and the reducing agents. Gels were put into two solutions of Ce(III) and Ce(IV) under the same acidity, $[Ce_2(SO_4)_3]$=0.001M and $[HNO_3]$=0.3M; $[Ce_3(SO_4)_2]$=0.001M and $[HNO_3]$=0.3M, respectively. The equilibrium swelling ratio of the gels was observed and recorded by using a microscope (Fortissimo Corp.WAT-

250D), a LED light (LEDR-74/40 W), and a video recorder (Victor corp. SR-DVM700). The analysis was conducted by using the image processing software (Image J 1.38x). Measurements of the equilibrium swelling ratio were performed in a water-jacketed cell made of acrylic plates.

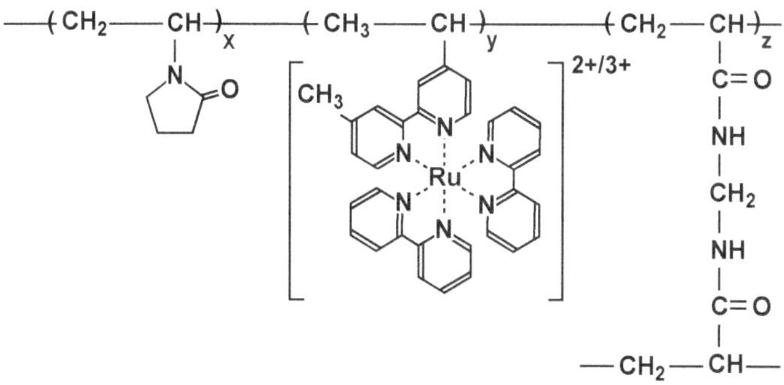

Figure 2. Chemical structure of poly(VP-co-Ru(bpy)$_3$) gel.

2.2.3. *Measurements of the oscillating behavior for poly(Vp-co-Ru(bpy)$_3$) gels*

The gel membrane was cut into rectangles(side length, about 2 × 20 mm) and immersed into 8mL of an aqueous solution containing malonic acid (MA), sodium bromate ($NaBrO_3$), and nitric acid (HNO_3). Shape changes of the gel strip were observed and recorded using a microscope (OLYMPUS.IX71), and a video recorder (Victor Corp. SR-DVM700). The analysis was conducted by using the image processing software (Image J 1.38x). A one-pixel line along the length of recorded gel image was stored at regular time intervals (0.05 s). The stored pixel line images were sequentially lined up as a function of time by the computer. This image processing procedure constructs a spatio-temporal diagram. From the obtained diagram, the time-dependent change in the gel edge was traced for observing the behavior of the volume change.

2.3.1. *Synthesis of the poly(AAm-co-AAC) gel*

The pH-responsive cylindrical hydrogels were prepared as follows. 0.533 g of acrylamide (AAm), 54.0 mg of acrylic acid (AAc), 12.8mg of N,N′-methylenebisacrylamide (MBAAm) as a cross-linker, and 22.6 mg of 2,2'-Azobis (2-methylpropionamidine) dihydrochloride as an initiator were dissolved in the O_2 free pure water (3.0 mL). The gelation was carried out in glass tubes (internal diameters from 1.0 mm to 2.0 mm) at 50 °C for 21 h. After removing from their molds, the cylindrical gels were cut in pieces 3.0 mm long and hydrated in pure water.

The pH-responsive cylindrical microphase-separated gels were prepared as follows. The composition of monomers is same as the cylindrical hydrogels, and they were dissolved in the solutions of O_2 free water/acetone mixtures (80/20, 70/30, 60/40, and 50/50 wt %). The polymerization was carried out in glass tubes (the size is same as above) at 50 °C for 21 h.

After removing from their molds, the cylindrical gels were also cut in pieces 3.0 mm long and hydrated in a graded series of methanol-water mixtures for 1 day each in 75, 50, 25, and 0 vol % methanol.

The pH-responsive microphase-separated tubular gel was prepared as follows. The composition of monomers is same as the cylindrical microphase-separated gels, and they were dissolved in the water/acetone mixture solution (70/30 wt %). The mold was made by coupling different size of glass tubes (external/internal diameters are 2.6/2.0, 1.0/0.6 mm), and the monomer solution was flowed into the tube-shaped space (external/internal diameters are 2.0/1.0 mm). The tubular gel was cut in pieces 4.0 cm long and hydrated in a graded series of methanol-water mixtures for 1 day each in 75, 50, 25, and 0 vol % methanol.

2.3.2. SEM Observation of the gels

Two samples were prepared in the following way. The microphase-separated gel cylinders were cut into pieces and soaked in acid (pH= 2) and base (pH = 11) solutions for 1 day until reaching the equilibrium. Then the gels were quickly frozen in liquid nitrogen and freeze-dried under vacuum for 1 day. The freeze-dried samples were fixed on the aluminum stubs and coated with the gold for 30 s under vacuum. The gel network structures were observed by the SEM (scanning electron microscope, HITACHI S2500CX).

2.3.3. Measurement of phase transition kinetics

In our measurement of the gels kinetics, two acid-base solutions with low and high pH (pH = 2 and pH = 11) were used. A piece of cylindrical gel under the equilibrium swelling in one solution was transferred quickly into the other. The changes of the cylindrical gel diameters were monitored by using a microscope equipped with a CCD camera controlled by a computer. The obtained images were analyzed by image processing software.

2.3.4. Experimental apparatus for causing the CT reaction in the pH-responsive tubular gel

The feeding solutions were stored in three separated reservoirs containing an alkaline sodium chlorite solution ($[NaClO_2] = 1.2 \times 10^{-1}$ M, $[NaOH] = 1.5 \times 10^{-4}$ M), an alkaline potassium tetrathionate solution ($[K_2S_4O_6] = 3.0 \times 10^{-2}$ M, $[NaOH] = 1.5 \times 10^{-4}$ M), and a sodium hydroxide solution ($[NaOH] = 1.3 \times 10^{-2}$ M). The solutions were pumped by peristaltic pumps (ATTO SJ-1211L) and premixed by a magnetic stirrer (AS-ONE CT-3A) before entering the tubular gel. The flows of chlorite and tetrathionate solutions were maintained at 30 mL/h, and the mixed solution flowed into the tubular gel by a peristaltic pump (ATTO SJ-1211H). The tubular gel was soaked in the mixed CT solution for 30 min in advance. After entering the flow of the mixed solution, an acid perturbation was applied by touching the gel at a particular spot with a paper soaked in a 1 M H_2SO_4 solution. The behavior of the tubular gel was observed by using a microscope equipped with a CCD camera controlled by a computer. Methyl red, a color indicator that changes from yellow (basic) to red (acidic), was added to the solutions to directly visualize the propagation of acid regions. Fig. 3 shows the schematic illustration of experimental apparatus.

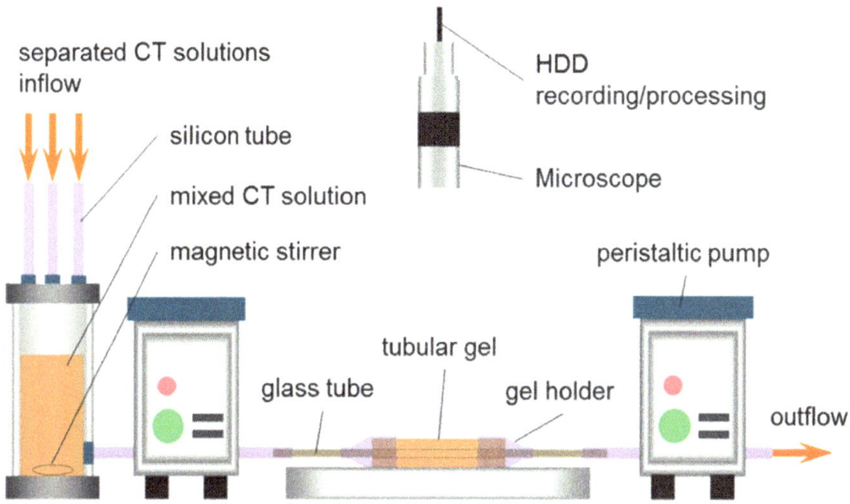

Figure 3. Schematic illustration of experimental apparatus for causing the CT reaction in the pH-responsive tubular gel. (Reprinted ref. 58, Copyright *IEEE*. Reproduced with permission.)

2.4.1. *Preperation of Poly(AAc-co-tBMA)*

Using AAc (25.9 g), tert- butyl methacrylate (tBMA) (34.1 g), and 2,2′-azobisiso-butyronitrile (AIBN) (0.49 g) as an initiator, *poly(AAc-co-tBMA)* (Figure 4) was synthesized by the radical polymerization in ethanol (139.5 g) using a total monomer concentration of 30 wt%. The molar ratio of tBMA incorporated into the copolymer was 30 mol%. The polymerization was carried out at 60 °C under nitrogen flow for 24 h. The resulting reaction mixture was dialyzed against ethanol for seven days.

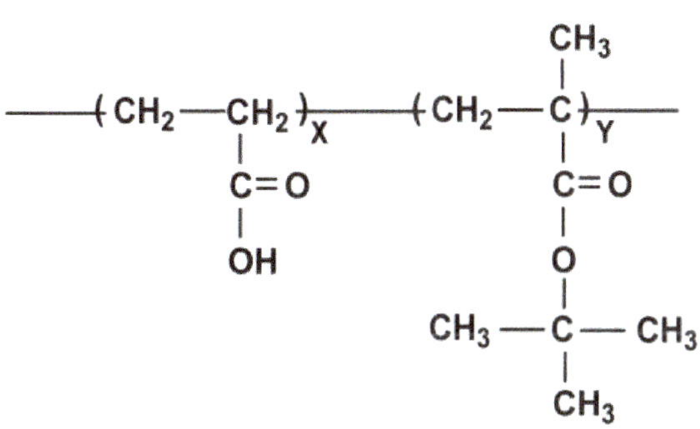

Figure 4. The chemical structure of poly(AAc-*co*-tBMA) polymer.

2.4.2. *Electrospinning*

The polymer solution (18 wt%) was poured into a 2.5 mL syringe. A potential of 10 kV was applied by connecting the power supply (GT80 GREEN TECHNO) to the syringe tip (Figure 5). In order to introduce anisotropic structure into the nanofiber gel, the 1.0 mL of the polymer solution in the syringe was sprayed at a flow rate of 2.0 mL/hour (sprayed for 30 minutes), and then the flow rate was changed to 1.0 mL/hour (sprayed for 60 minutes). The fibers were collected on the grounded glass substrate as a collector. The distance between the collector and the syringe tip was 15 cm. The temperature and humidity were 25 °C and 70%, respectively. After the electrospinning, the obtained sheet, with a thickness of about 200 μm, was dried overnight at 50 °C.

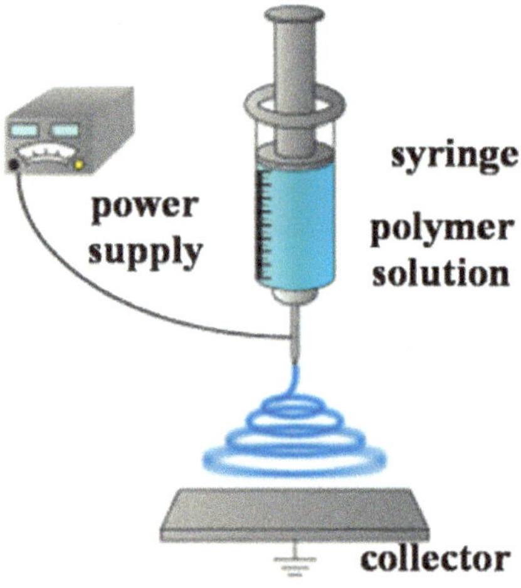

Figure 5. Schematic illustration of electrospinning set-up.

2.4.2. *Measurement of motion of the nanofiber gel actuator*

The open continuously stirred tank reactor (CSTR) (40 mL) was designed using an acrylic cell with a water jacket in order to control the solution temperature in the cell. Potassium bromated (0.26 mol/L), sodium sulfite (0.3 mol/L), potassium ferrocyanide (0.08 mol/L), and sulfuric acid (0.04 mol/L), solutions were pumped into the reactor at a flow rate of 50 mL/hour. The pH changes in the reactor were monitored continuously by utilizing a pH meter (F-55 HORIBA) held in the reactor, and its electronic output was directly recorded by a computer. The nanofiber gel (length 15 mm, width 3 mm) was set at the bottom of the water jacket. One end of the gel strip was sandwiched in the incision of the silicone rubber. Shape changes of the gel strip were recorded by a fixed microscope (Fortissimo Corp. WAT-250D) and a video recorder (Victor Corp. SR-DVM700). The temperature in the reactor was controlled at 25 °C by utilizing the water bath equipment.

3. Results and discussion

3.1. Self-oscillating behavior of poly(VP-*co*-Ru(bpy)$_3$) solution

The measurement of the solubility for the poly(VP-*co*-Ru(bpy)$_3$) solution in the reduced and oxidized state was conducted by changing the temperature from 10 to 50 °C. As shown in Figure 6, there are no the lower critical solution temperature (LCST) for the polymer solutions in the reduced and oxidized state. The polymer solutions in the reduced and oxidized state have different transmittance values. That is, this result indicates that the polymer solution has different solubility in the reduced and oxidized state, respectively. The solubility of the polymer solution in the reduced state is higher than that in the oxidized state. Figure 7 shows self-oscillating behaviors of the poly(VP-*co*-Ru(bpy)$_3$) solution in the different concentrations of malonic acid ([MA] = 0.04, 0.05, 0.06, 0.07, 0.08 and 0.09 M) under the fixed concentration of sodium bromate and nitric acid ([$NaBrO_3$] = 0.3 M and [HNO_3] = 0.3 M). As shown in Figure 7, the base line of the transmittance self-oscillation gradually decreased with time in all malonic acid concentrations. The damping behavior originates from the change in the ionic strength of the polymer solution when the transmittance measurement starts [47-49]. In order to cause the BZ reaction in the polymer solution, the self-oscillating polymer solution and the other solution of the BZ substrates are mixed just before the transmittance measurements. In general, the solubility of the polymer chain is significantly affected by the ionic strength of the solution. Therefore, when the ionic strength increased at the start point of the self-oscillation, the solubility of the polymer chain decreased. In the solution condition of this study, the ionic strength of the polymer solution was very high because the BZ reaction required a significant high concentration of the BZ substrates. Therefore, the damping behaviors occurred from the start point of the self-oscillation. In addition, as shown in Figure 7, the width of the waveform increased with decreasing the concentration of malonic acid. Basically, the width of the waveform of the transmittance self-oscillation depends on the rate of the BZ reaction because the self-oscillation was induced by the BZ reaction. As the concentration of the BZ substrates decreased, the rate of the BZ reaction decreased due to decrease in the collision rate among the BZ substrates. Therefore, the width of the waveform increased with decreasing the concentration of the BZ substrates. This tendency was observed in the transmittance self-oscillation of the AMPS-containing polymer solutions [48].

Figure 8 shows the transmittance self-oscillations of the novel polymer solution in the different concentration of sodium bromate ([$NaBrO_3$] = 0.1, 0.2, 0.3 and 0.4 M) at 14 °C under the fixed concentration of malonic acid and nitric acid ([MA] = 0.1M and [HNO_3] = 0.3 M). As shown in Figure 8, the amplitude of the self-oscillation gradually decreased with time in the same manner as in Figure 7. Moreover, the width of the waveform decreased with the increase in the concentration of sodium bromate due to the increase in the reaction rate of the BZ reaction. In addition, as shown in Figure 8, the amplitudes of the transmittance self-oscillations were hardly affected by the initial concentration of sodium bromate. In additon, Figure 9 showed the amplitude of the transmittance self-oscillation for the polymer solution under the different concentrations of the BZ substrates. As shown in Figure 9, all the BZ substrate concentrations hardly influence the amplitude of the transmittance self-oscillation. That is, the amplitude values were almost the same in all BZ substrate conditions. In our previous investigations, we studied the effect of the concentration of the BZ substrates on the waveform of the transmittance self-oscillation for the AMPS-containing polymer solution. As a result, we clarified that the amplitude of the self-oscillation is significantly affected by the initial concentration of the BZ substrates [47-49]. This is because the AMPS-containing polymer chain caused damping, that is, the amplitude of the self-oscillation gradually decreased with time. The damping behavior of the polymer solution originates from the change in the size of the polymer aggregation with time. In the case of the NIPAAm-based polymer chains, the reduced Ru moiety in the polymer chain strongly interacts with the other reduced Ru one. Once the reduced Ru moiety strongly interacts with the other Ru one, the interaction hardly dissociates [48]. In the BZ reaction, the time in the reduced state is much longer than in the oxidized state. Therefore, the hydrophobic $Ru(bpy)_3^{2+}$ moiety in the polymer chain dominantly behaved for the determination of the polymer aggregation state in the self-oscillating behavior. For this reason, the mole fraction of the $Ru(bpy)_3^{2+}$ moiety in the polymer chain significantly affect the waveform of the transmittance self-oscillation. This influence can be explained by the overall process of the BZ reaction based on the Field-koros-Noyes (FKN) mechanism [33-36, 48]. On the other hand, in the case of the VP-based polymer chain, the bipyridine ligands interacted with the VP based main-chain. The strength of the interaction of the bipyridine ligands in the oxidized state is higher than in the reduced state. That is, the polymer aggregation increased in the oxidized state. However, the time in the oxidized state is much shorter than in the reduced state. Therefore, the size of the polymer aggregation changed very slowly compared to the AMPS-containing polymer solution. Consequently, the degree of the damping for the Vp-based polymer solution is considerably small. Therefore, the amplitude is hardly affected by the initial concentration of the BZ substrates.

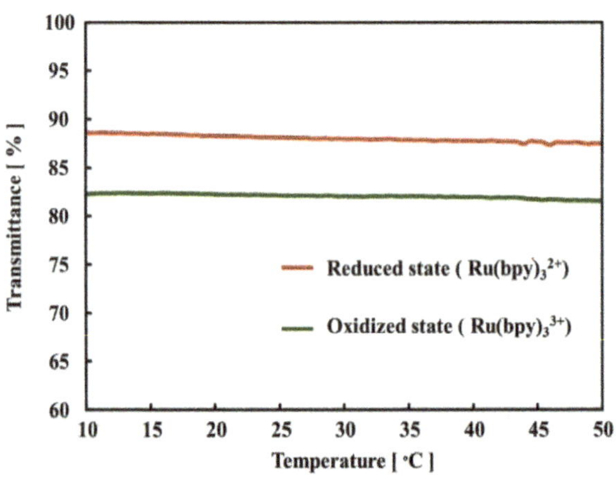

Figure 6. Temperature dependence of transmittance for poly(VP-*co*-Ru(bpy)3) solutions under the different conditions of reduced Ru(II) (in Ce(III) solution) and Ru(III) (in Ce(IV) solution) states.

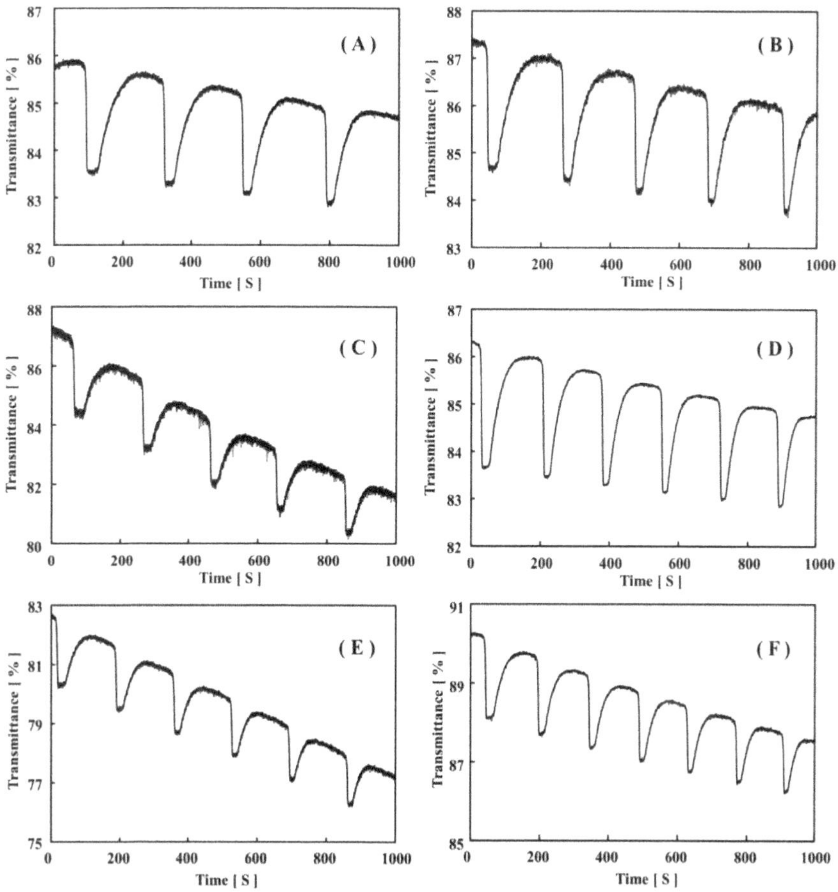

Figure 7. Oscillating profiles of transmittance at 14 °C for 0.5 wt% poly(VP-*co*-$Ru(bpy)_3$) solution in the fixed nitric acid and sodium bromate conditions ([HNO_3] = 0.3 M and [$NaBrO_3$] = 0.3 M): (**A**) [MA] = 0.04 M, (**B**) [MA] = 0.05 M, (**C**) [MA] = 0.06 M, (**D**) [MA] = 0.07 M, (**E**) [MA] = 0.08 M, (**F**) [MA] = 0.09 M.

3.2. Self-oscillating behavior of poly(VP-*co*-$Ru(bpy)_3$) gel

Figure 10 shows the equilibrium swelling behaviors of the poly(VP-*co*-$Ru(bpy)_3$) gels in the Ce(III) and Ce(IV) solutions under the same acidic condition. In the Ce(III) solution, the gel kept a tinge of orange, which indicated that the copolymerized $Ru(bpy)_3$ moiety in the gel was in the reduced state. On the other hand, in the Ce(IV) solution, the gel quickly turned from orange to green, which showed the $Ru(bpy)_3$ moiety in the gel had changed the reduced state to the oxidized state. In the oxidized state, the equilibrium volume of the gel was larger than that in the reduced state in all temperature condition. This is because the solubility of the $Ru(bpy)_3$ moiety has significantly different properties in the oxidized and

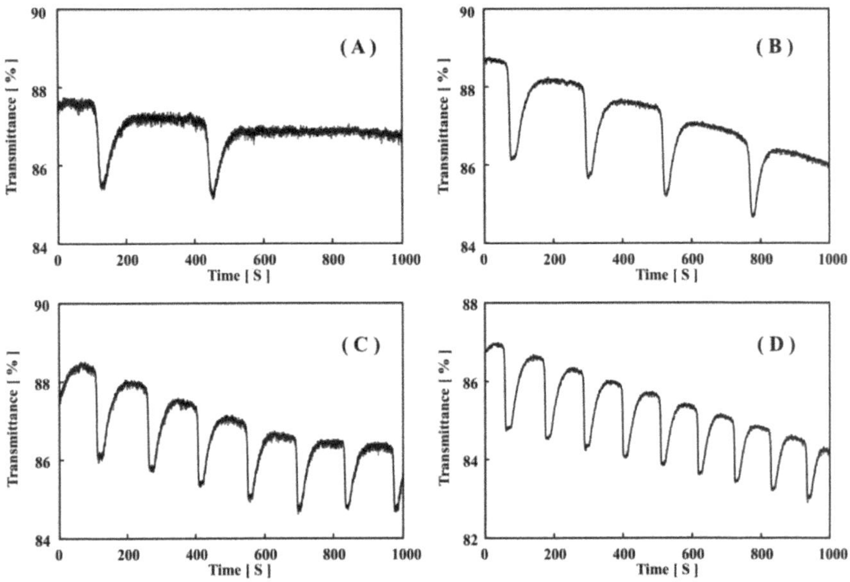

Figure 8. Oscillating profiles of transmittance at 14 °C for 0.5 wt% poly(VP-*co*-Ru(bpy)$_3$) solution in fixed nitric acid and malonic acid conditions ([HNO_3] = 0.3 M, [MA] = 0.1 M) (**A**) [$NaBrO_3$] = 0.1 M, (**B**) [$NaBrO_3$] = 0.2 M, (**C**) [$NaBrO_3$] = 0.3 M, (**D**) [$NaBrO_3$] = 0.4 M.

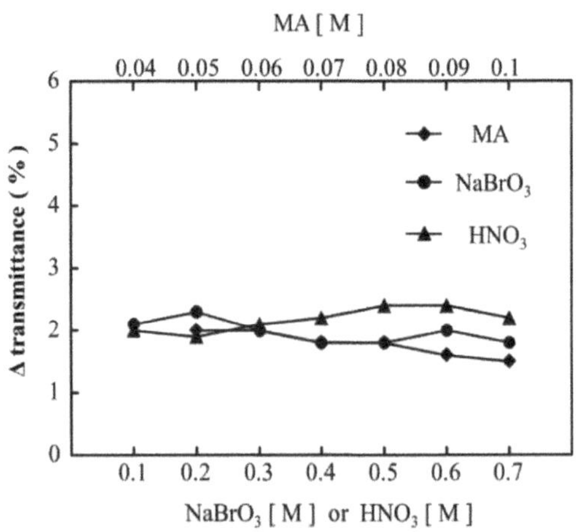

Figure 9. Dependence of amplitude of transmittance self-oscillation for polymer solution at 14 °C in the change in one BZ substrate under fixed concentrations of the other two BZ substrates: **MA** ([MA] = 0.04, 0.05, 0.06, 0.07, 0.08 and 0.09 M, fixed [$NaBrO_3$] = 0.3 M and [HNO_3] = 0.3 M); **$NaBrO_3$** ([$NaBrO_3$] = 0.1, 0.2, 0.3, 0.4, 0.5 and 0.6 M, fixed [MA] = 0.1 M and [HNO_3] = 0.3 M); **HNO_3** ([HNO_3] = 0.1, 0.2, 0.3, 0.4, 0.5 and 0.6 M, fixed [$NaBrO_3$] = 0.1 M and [HNO_3] = 0.3 M).

reduced state. The reduced $Ru(bpy)_3$ moiety in the gel has an extreme hydrophobic property in the VP-based polymer gel. This property is attributed to the conformation of the bipyridine ligands surrounding the Ru ion, which induces the deswelling behavior. That is, in the VP-based gel, the bipyridine ligands surrounding the Ru ion exert a greater influence on the solubility of the polymer chain in the reduced state as compared with the ionization effect of the Ru ion.[47-52] On the contrary, the oxidized $Ru(bpy)_3$ parts in the gel has a great hydrophilic property. The driving force of the swelling-deswelling self-oscillation is originated in the different solubility of the $Ru(bpy)_3$ moiety in the reduced and oxidized states as shown in Figure 10. In the reduced and oxidized state, there is no observation of the volume phase transition because of the PVP main chain of the gel without LCST.

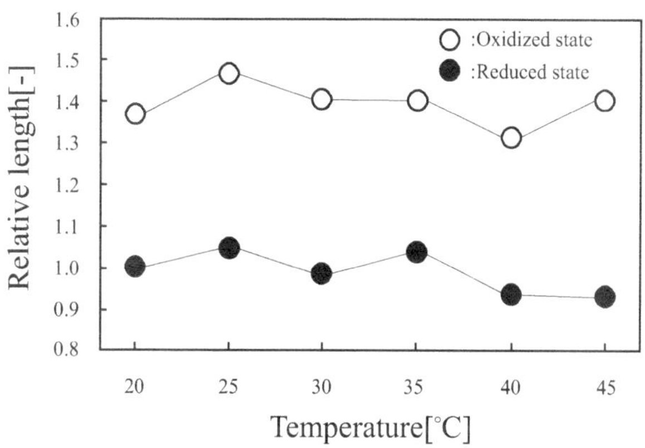

Figure 10. Equilibrium swelling ratio of poly(VP-*co*-$Ru(bpy)_3$) gel in cerium sulfate solutions as a function of temperature. (●) $[Ce_2(SO_4)_3]$ = 0.001M and $[HNO_3]$ = 0.3M; (○) $[Ce(SO_4)_2]$ = 0.001M and $[HNO_3]$ = 0.3M. The relative length is defined as the ratio of characteristic diameter at the initial state at 20°C. (Reprinted ref. 57, Copyright American Chemical Society. Reproduced with permission.)

Figure 11 showed the logarithmic plots of the period against the initial concentration of one BZ substrate under fixed concentration of the other two BZ substrates at a constant temperature (T=20 °C). As shown in Figure 11, all the logarithmic plots had a good linear relationship. Therefore, the period [T(s)] of the swelling-deswelling self-oscillation can be expressed as $a[\text{substrate}]^b$ where a and b are the experimental constants, and the brackets indicate the initial concentration. Moreover, as shown in Figure 11, the period of the self-oscillation have the saturation point at the following initial concentration: [MA] = 0.07M (Figure 11(a)), $[NaBrO_3]$ = 0.5M (Figure 11(b)) and $[HNO_3]$ = 0.7M (Figure 11(c)). The period at the saturated point in Figure 11(a) was significantly higher than that in Figure 11(b) and Figure 11(c). This tendency can be explained by considering the mole fraction of the reduced $Ru(bpy)_3$ moiety in the gel. This is because the reduced $Ru(bpy)_3$ moiety in the gel has the significantly high hydrophobic property. Therefore, the number of the hydrophobic reduced $Ru(bpy)_3$ moiety in the gel exert influence of the self-oscillating behavior. The Field-koros-Noyes (FKN) mechanism explained the overall process of the BZ reaction. [33-36, 48] According to the FKN mechanism, the overall reaction is divided into the following three

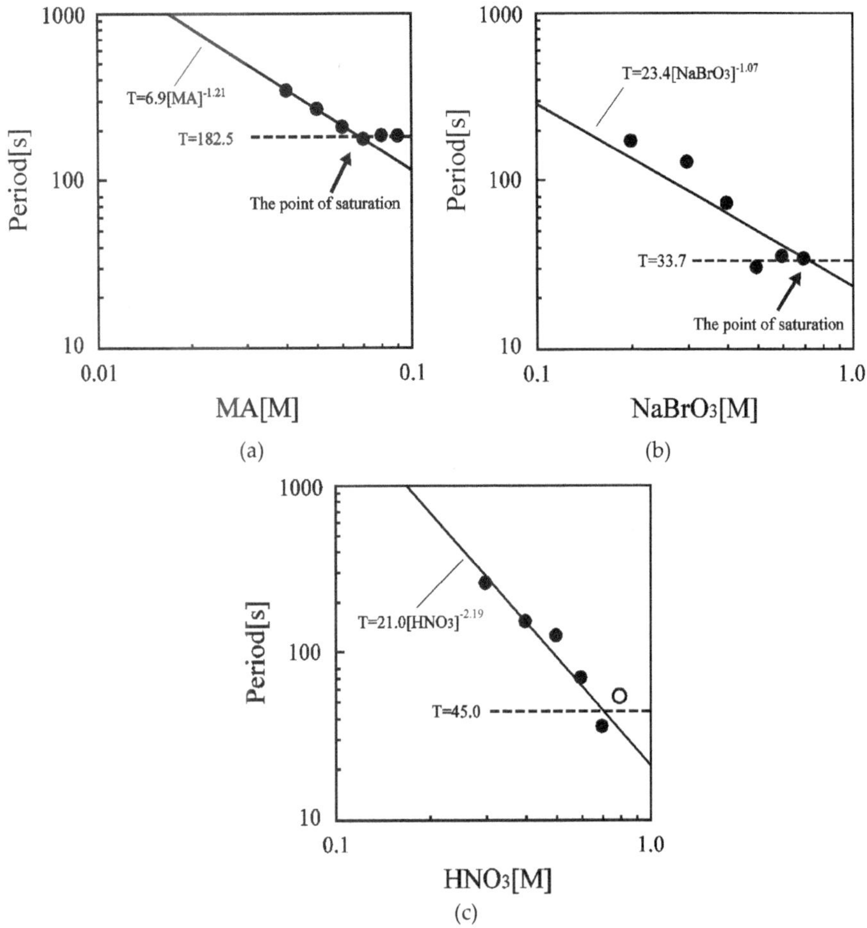

Figure 11. Logarithmic plots of period (T in s) vs initial molar concentration of one BZ substrate at a constant temperature (T=20 °C) under fixed concentrations of the other two BZ substrates; (a) [$NaBrO_3$] = 0.084M and [HNO_3] = 0.3M, (b) [MA] = 0.0625M and [HNO_3] = 0.3M, (c) [MA] = 0.0625M and [$NaBrO_3$] = 0.084M. (●) plots and (○) plots show the linear relation and the saturated line vs initial concentration of one BZ substrate, respectively. (Reprinted ref. 57, Copyright American Chemical Society. Reproduced with permission.)

main processes: consumption of Br^- ions (process A), autocatalytic formation of $HBrO_2$ (process B), and formation of Br^- ions (process C).

A: $BrO_3^- + 2Br^- + 3H^+ \rightarrow 3HOBr$

B: $BrO_3^- + HBrO_2 + 2Mred + 3H^+ \rightarrow 2HBrO_2 + 2Mox + H_2O$

C: $2Mox + MA + BrMA \rightarrow fBr^- + 2Mred + \text{other products}$

In the process of B and C, the $Ru(bpy)_3$ moiety in the gel works as the catalyst: the reduced $Ru(bpy)_3$ moiety is oxidized (process B), and the oxidized one is reduced (process C). Therefore, as the initial concentration of the MA increased, the mole fraction of the reduced $Ru(bpy)_3$ moiety in the gel increased in accordance with the FKN mechanism. With increasing in the mole fraction of the reduced $Ru(bpy)_3$ in the gel, the shrinking force originating in the hydrophobic reduced $Ru(bpy)_3$ greatly increased as well. Generally, as for a polymer gel, deswelling speed is faster than the swelling one. Once the gel collapsed, it takes a lot of time for the aggregated polymer domain in the gel to recover the elongated state. This is because the polymer aggregation state is thermodynamically more stable in the polymer gel. Therefore, as the shrinking force increased, the swelling speed of the poly(VP-*co*-$Ru(bpy)_3$) gel significantly decreased. As a result, in the higher MA condition, the period at the saturated point was long (T=182.5) compared with the other condition (T=33.7 (Figure 11(b)) and T=45.0 (Figure 11(c)). On the other hand, in the case of Figure 11(b), the period at the saturated point was greatly shorter than that in Figure 11(a). In the condition of Figure 11(b), the swelling force originating in the hydrophilic oxidized $Ru(bpy)_3$ moiety increased due to the increase in the mole fraction of the oxidized $Ru(bpy)_3$ moiety in the gel in accordance with the FKN mechanism. Therefore, the gel can cause the swelling-deswelling self-oscillation at the high speed due to the strong recovering force originating in the higher mole fraction of the hydrophilic oxidized $Ru(bpy)_3$ moiety in the gel. Moreover, in the condition of Figures 11(c), the period for the poly(VP-*co*-$Ru(bpy)_3$) gel had the different aspect from that of the conventional-type poly(NIPAAm-*co*-$Ru(bpy)_3$) gel.[75] The period of the self-oscillation decreased with increasing the initial concentration of the BZ substrates because of the increase in the collision frequency among the BZ substrates. Therefore, we considered that the relationship between the period and the [HNO_3] for the poly(VP-*co*-$Ru(bpy)_3$) gel is of a more natural tendency. In addition, in the condition of Figure 11(c), the control range of the period by changing the initial concentration of the HNO_3 for the gel was much wider than that for the poly(NIPAAm-*co*-$Ru(bpy)_3$ gel. [75]

Moreover, as shown in Figure 12, the period of the swelling-deswelling self-oscillation decreased with increasing the temperature because the temperature affects the BZ reaction rate in accordance with the Arrenius equation.[30] The period of the swelling-deswelling self-oscillation to the temperature for the poly(VP-*co*-$Ru(bpy)_3$) gel has the linear relationship. The period (2 second) reached the saturation at 46 °C in the BZ condition ([MA] = 0.08M, [$NaBrO_3$] = 0.48M and [HNO_3] = 0.48M). This is because the swelling-deswelling speed of the gel is slower than the changing rate of the redox states of the $Ru(bpy)_3$ in the gel. That is, the self-oscillating behavior of the gel cannot follow the changing the redox state of the $Ru(bpy)_3$ moiety. The maximum frequency (0.5Hz) of the poly(VP-*co*-$Ru(bpy)_3$) gel was 20 times as large as that of poly(NIPAAm-*co*-$Ru(bpy)_3$ gel. [75] The self-oscillating behaviors of the poly(VP-*co*-$Ru(bpy)_3$) gel at 20°C and 50°C were shown in the Figure 12(b) and 12(c), respectively. The displace of the volume change self-oscillation at 20°C and 50°C were about 10 μm and 4 μm, respectively. These results clarified that the displacement of the swelling-deswelling self-oscillation for the gel has the trade-off relationship against the period of the self-oscillation, that is, the length of the volume change decreased with increase in the period.

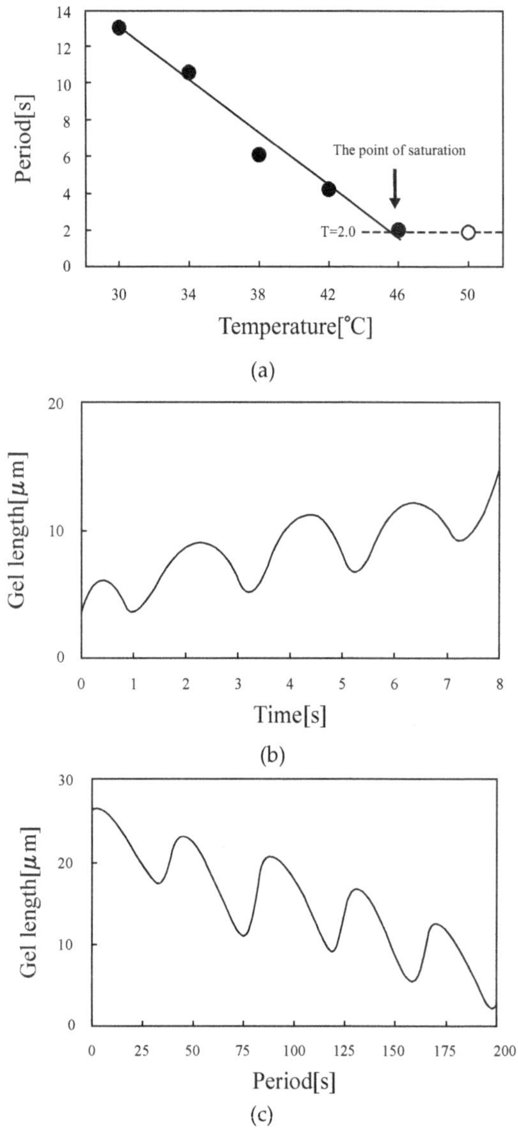

Figure 12. (a) Dependence of the self-oscillation period on the temperature. (●) plots and (○) plots show the linear relation and the saturated line vs temperature, respectively. (b) Self-oscillating profile of cubic poly(VP-*co*-Ru(bpy)$_3$) gel at 50°C (MA = 0.08M, $NaBrO_3$ = 0.48M and HNO_3 = 0.48M). (c) Self-oscillating profile of cubic poly(VP-*co*-Ru(bpy)$_3$) gel at 20°C (MA = 0.08M, $NaBrO_3$ = 0.48M and HNO_3 = 0.48M). Cubic gel (each side length is about 2mm and 20mm) was immersed in 8ml of the mixture solution of the BZ substrates. (Reprinted ref. 57, Copyright American Chemical Society. Reproduced with permission.)

3.3. Tubular gel motility driven by chemical reaction networks

In previous studies, De Kepper *et al.* showed a method to form the contraction waves by coupling the pH-responsive hydrogel and the acid autocatalytic chlorite tetrathionate (CT) reaction [42, 43]. They presented a system specially designed to show the chemomechanical instabilities, in other words, dynamical deformations of the functional gels, such as contraction waves and complex spatio-temporal volume oscillations. However, in this system, the pH responsive cylindrical gel is fixed to continuous stirred tank reactors (CSTR) because it must be permanently fed with constant flows of fresh CT reactants. Therefore, this system is unsuitable for the development of a locomotive gel robot. It is necessary to develop a novel concentration tuning mechanism without the stirred reactor for the coupled system of the functional gel and the CT reaction.

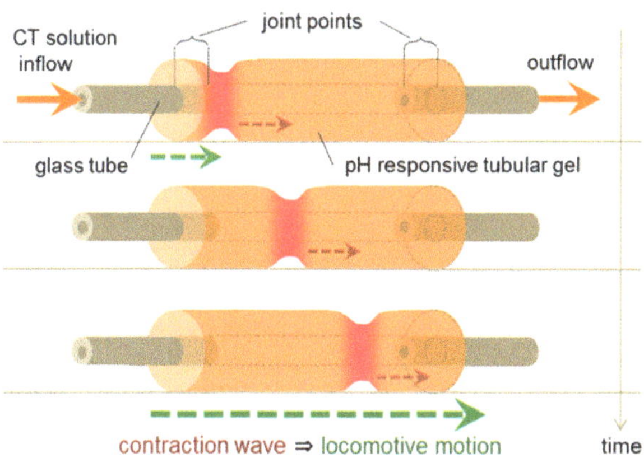

Figure 13. Schematic representation of the principles of the tubular gel feeding system (Reprinted ref. 58, Copyright *IEEE*. Reproduced with permission.)

In this chapter, we introduce a novel pH-responsive tubular gel and propose a new method of generating contraction waves on the functional gel. Figure 13 shows the schematic representation of the principle of the feeding system of the tubular gel. We controlled the reaction diffusion system by pouring the reactant solution into the hollow of the tubular gel, and attempted to achieve the peristaltic movement of the tubular gel by formulating contraction wave. This system enables us to feed with constant flow of fresh reactants, and also to be free from tank reactors.

The CT reaction is known as the acid autocatalytic reaction with the concentration changes of various sorts of reactive substrates, including the chlorite and tetrathionate ion, and this reaction exhibits a bi-stability of an acid steady state and an alkaline steady state. The CT reaction shows very complex kinetics which is still not completely understood [76]. However, it can be described by the following chemical reaction,

$$7ClO_2^- + 2S_4O_6^{2-} + 6H_2O \rightarrow 7Cl^- + 8SO_4^{2-} + 12H^+.$$

It runs in slight chlorite excess, and is associated with the following autocatalytic empirical rate law,

$$r = -\frac{1}{7}\frac{d[ClO_2^-]}{dt} = k[ClO_2^-][S_4O_6^-][H^+]^2,$$

r indicates the concentration per unit mol of the reactive substrates, and k is a proportionally coefficient. The changes between the steady states of the CT reaction take place depending on the feeding condition of the proton. When an alkali is constantly supplied, the CT reaction solution can maintain an alkaline steady state. As an acid perturbation is added or an alkali supply decreases below a certain concentration, the acid autocatalytic reaction develops and acid region spreads by the reaction-diffusion process, finally the solution becomes acid steady state. When autocatalytic reactions are operated in a pH responsive spherical gel, the gel shrinks following the one-way reaction, regularly. However, in some conditions, the relatively faster diffusivity of the proton can lead to the oscillatory reaction diffusion instability [42,43,77,78]. The volume oscillation of spherical gel is generated by coupling the CT reaction with the volume phase transition of the pH-responsive gel, and we call the concatenation of the two phenomena "chemical reaction networks."

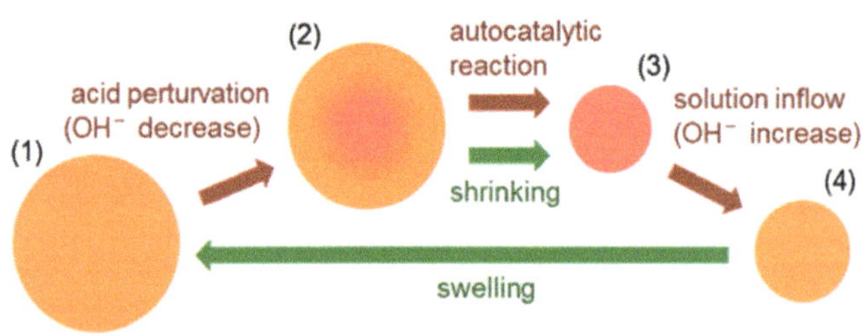

Figure 14. Flowchart of the chemical reaction networks. Each disk shows the condition of a spherical gel. The red arrows indicate changes of the interior CT solution and the green arrows indicate volume phase transition of the pH-responsive gel, that is, the gel causes shrinking or swelling. (Reprinted ref. 58, Copyright *IEEE*. Reproduced with permission.)

Figure 14 shows the flowchart of the chemical reaction networks. The red areas indicate the acid steady state and the yellow areas indicate the alkaline steady state. There are four steps in the networks:

1. The spherical gel is in the uniform alkaline steady state.
2. When the gel radius grows and exceeds a certain threshold value, an acid perturbation rises in the core of the gel.
3. The acid region fills the gel entirely, and the gel shrinks.
4. When the gel radius shrinks below a certain threshold value, it swells again by inflowing of alkali solution.

As described above, the stabilities of such steady states depend on the gel radius. It is shown theoretically and experimentally that the switching between the chemical reaction networks occurs with hysteresis as a function of such geometric parameters. In this case, the reaction diffuses symmetrically from the center of the gel. When the reaction is applied to the cylindrical gels or tubular gels, it diffuses symmetrically to the diameter direction and also diffuses to the length direction. The disks shown in Figure 14 indicate the changes of the cross-section surface at one location of the cylindrical gel.

For coupling pH-responsive gels and the CT reaction, it is quite important to achieve large volume changes of the gels, because the stabilities of the steady states depend on the gel sizes. Also, the speed of volume change has a high correlation with the contraction wave shape. However, in general, the reactivities of hydrogels composed of chemically cross-linked polymer networks are low because the polymer chains are molecularly restricted by a large number of cross-links. An *N*-isopropylacrylamide (NIPAAm) gel with a microscale phase separation that underwent a quick response has previously been reported [79]. By preparing a NIPAAm gel above the LCST, the NIPAAm gel forms a porous structure with polymer-rich domains and aggregations in the matrix of loosely bound network structures [80]. Consequently, polymer-rich domains inside the gel aggregate or disperse rapidly through the porous structure within the gel by an effluent pathway of the solvent. Therefore, in this investigation, we prepared the microphase-separated AAm-based gels. The microphase-separation in the gel depends strongly on the methods of gel preparation. Meanwhile, a method for preparing the microphase-separated gels by altering the composition of the solvent has been reported [55]. Therefore, we synthesized the poly (AAm-*co*-AAc) gels in water/acetone solvents. Also, we investigated the gels kinetics at the different mixture proportions of the solvents.

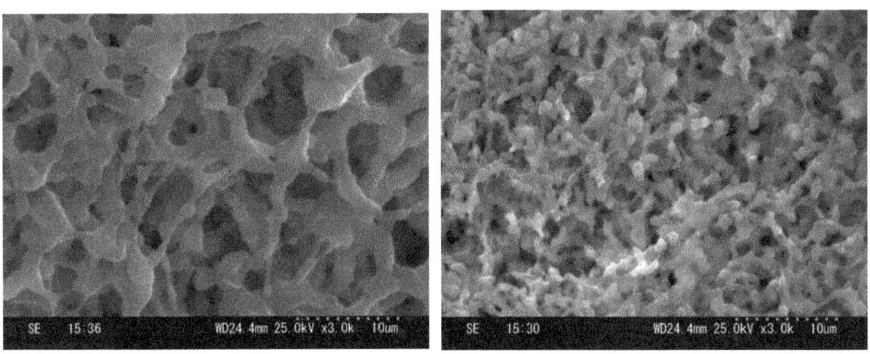

(a) (b)

Figure 15. SEM images of the interior morphologies of poly (AAm-*co*-AAc) microphase-separated gel with the mixing ratio of water/acetone solvents at 50/50 ((a) swelling state and (b) shrinking state at equal magnification). (Reprinted ref. 58, Copyright *IEEE*. Reproduced with permission.)

First, we observed the structures of the gels. Figure 15 shows the SEM images of the interior morphologies of the poly(AAm-*co*-AAc) microphase-separated gels with the mixing ratio of water/acetone solvents at 50/50. There are polymer rich domains inside the both gels (swelling state and shrinking state) and their sizes change in proportion as the volume phase transitions of the gels. This microscale structures are quite different from those of ordinary poly(AAm-*co*-AAc) gels with water solvents.

Next, we investigated the phase transition kinetics of the poly(AAm-*co*-AAc) cylindrical gels. When the external pH was rapidly changed, the gel diameters gradually changed to approach the equilibrium state. Figure 16 shows the plots of the gel diameters as a function of time on the shrinking process. Here, let us define L_f, $L(t)$ and L_i, as the final gel diameter, the gel diameter at t = t and the initial gel diameter (t = 0), respectively. The time evolution was found to be well described by a single exponential:

$$L(t) = L_i - \Delta L(1 - e^{-(t/\tau)}) \tag{1}$$

where ΔL and τ represents the total gel length change (=$L_i - L_f$) and the characteristic time of shrinking. When τ and L_f are determined, we can estimate the diffusion constant D by using the following relation for cylindrical gels [81]:

$$D = \frac{L_f^{\ 2}}{24\tau} \tag{2}$$

From the results, the corrective diffusion constant of the poly(AAm-*co*-AAc) cylindrical gels can be estimated as follows; (A) 4.39×10^{-5} mm²/s, (B) 5.38×10^{-5} mm²/s, (C) 1.75×10^{-3} mm²/s, (D) 1.99×10^{-3} mm²/s, (E) 2.07×10^{-3} mm²/s. In the case of sample (A) and (B), the corrective diffusion constants are almost the same as those of the normal type hydrogels. On the other hand, in the case of sample (C), (D) and (E), they are two orders of magnitudes larger than those of sample (A) and (B). These results indicate that the mixture fractions of acetone to the solvents have the threshold values for the microphase-separation. Also, these results confirm that the microphase-separation plays an important role for increasing speed of the volume phase transition.

We also investigated the detailed pH-responsiveness of the gels. Figure 17 shows the equilibrium swelling of the poly (AAm-*co*-AAc) cylindrical gels at various pH values. From this results, all samples caused volume phase transition between the pH variation range of the CT reaction (from pH = 2 to pH = 11). Also, the diameter change rates of sample (A) and (C) are larger than that of sample (E). It is notable that the stiffness of the gels depends on the mixture proportions of water/acetone solvents. The stiffness of the gels tends to get lower as the mixture fractions of acetone increase. In this experiment, sample (D) and (E) do not stand up under their own weight and became easily deformed in air. Therefore, for this reason, we chose sample (C) as the material of the tubular gel because of its stiffness and phase transition kinetics.

We succeeded in coupling the pH-responsive hydrogel and the CT reaction. Figure 18 shows the propagation of the acid region in the swollen part of the tubular gel. The mixture proportion of the solvent of the tubular gel was same as sample (C). As shown in Figure 18, the CT solution was colored with methyl red. The yellow area corresponds to alkaline composition and the red area corresponds to acid composition. The red allows indicate the forefront of the acid region and the green allows indicate the forefront of the gel contraction. When the acid perturbation was applied to the left part of the tubular gel, the red area propagated to the right, and finally covered the entire gel. Also, the gel diameter shrank following the propagation of the acid region, as shown in Figure 18. These results suggest that the poly(AAm-*co*-AAc) tubular gel can be coupled with the CT solution, in terms of the acid propagation and the volume change of the gel.

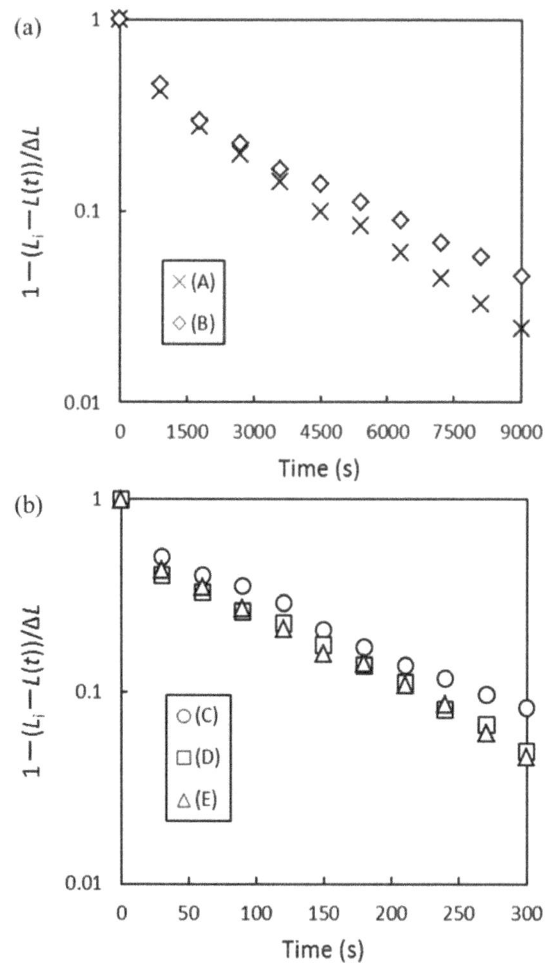

Figure 16. Shrinking behaviors of poly(AAm-*co*-AAc) cylindrical gels at the different mixture proportions of water/acetone solvents ((a): (A) 100/0, (B) 80/20, (b): (C) 70/30, (D) 60/40, and (E) 50/50 wt %). (Reprinted ref. 58, Copyright *IEEE*. Reproduced with permission.)

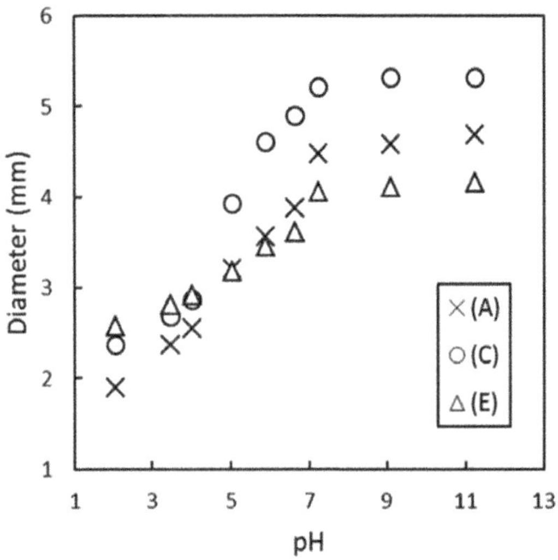

Figure 17. Equilibrium swelling of poly(AAm-*co*-AAc) cylindrical gels at various pH values adjusted using HCl and NaOH solutions, at the different mixture proportions of water/acetone solvents ((A) 100/0, (C) 70/30, and (E) 50/50 wt %). (Reprinted ref. 58, Copyright *IEEE*. Reproduced with permission.)

Fig. 19 shows the spatiotemporal diagram constructed from sequential images of the acid propagation. The extractive line is the black bar in Figure 18 (a). As shown in Figure 19, the acid region propagates at the constant speed, and the velocity of the acid propagation was 16 μm/s. It is 10 times faster than that of the previous research [42]. We also measured the velocity of the acid propagation in the mixed CT solution in the glass tube, and it was 200 μm/s. This is much larger value than that in the case of Figure 18. This result indicates that the propagation occurred inside the gel. These results confirm that a part of the chemical reaction networks, from step (1) to step (3) in Figure 14, was realized experimentally. In order to achieve autonomous swelling of the gel, from step (3) to step (4) in Figure 14, the gel diameter needs to shrink below a certain threshold value. It means that the minimum and maximum sizes of the tubular gels should be regulated for realizing oscillatory volume changes by inflowing of alkali in the CT solutions. The concentrations or mixing ratios of the CT solutions also affect the forming of the chemical reaction networks. The regulation of the gel sizes and the CT solutions will take an important role in the generation of the contraction waves in the tubular gels. This research will be the first step for realizing the biomimetic chemical robot which causes peristaltic locomotion.

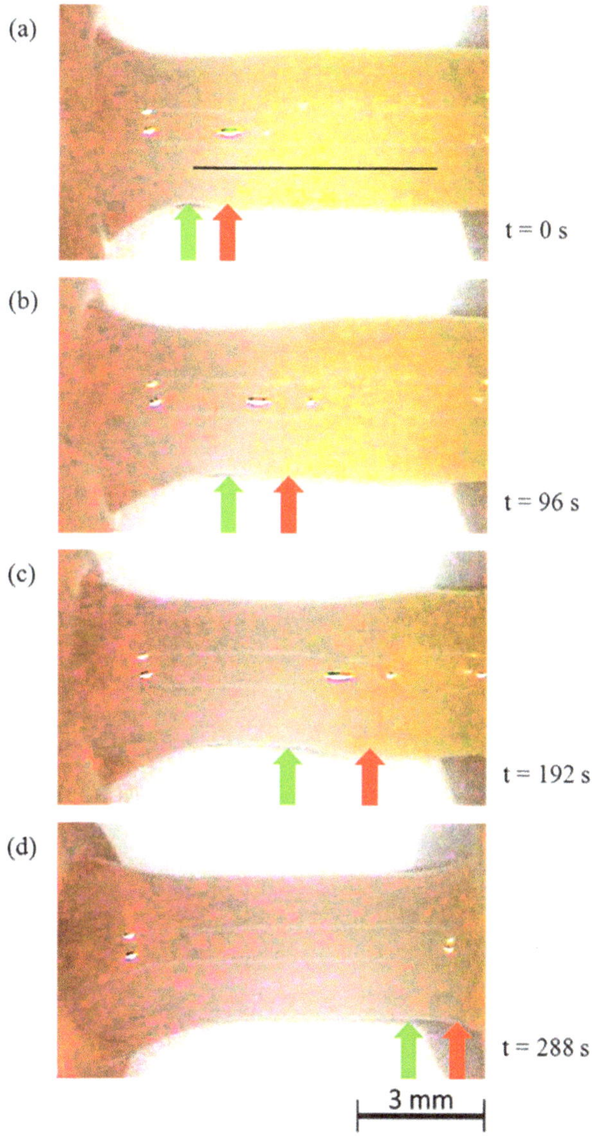

Figure 18. Propagation of the acid front in the swollen part of the tubular gel. An acid perturbation was applied to the left part of the gel. The yellow area corresponds to alkaline composition and the red area corresponds to acid composition. The red allows indicate the forefronts of the acid regions and the green allows indicate the forefronts of the gel contraction, which are the rightmost points obtained locally minimum diameters. (a) t = 0 indicates the criterion time after applying the acid perturbation. (Reprinted ref. 58, Copyright *IEEE*. Reproduced with permission.)

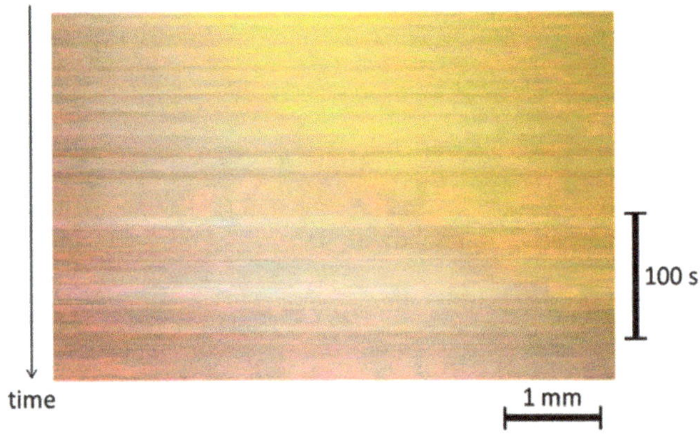

Figure 19. Spatiotemporal diagram constructed from sequential images of the acid propagation. The extracted horizontal lines correspond to the black bar in Fig. 18 (a). (Reprinted ref. 58, Copyright *IEEE*. Reproduced with permission.)

3.4. A pendulum-like motion of nanofiber gel actuator synchronized with external periodic pH oscillation

In this study, in order to drive the nanofiber gel actuator in response to the external pH changes, we selected the pH responsive poly(AAc) (PAAc) as a main polymer chain. The PAAc is protonated when the pH is below the pKa. When the pH of the solution is below pKa, the nanofiber gel collapses due to hydrogen bonding among the polymer chains. On the other hand, when the pH is above the pKa, the PAAc polymer chain is ionized, that is, the solubility of the polymer chain changes from a hydrophobic to a hydrophilic nature. As a result, the nanofiber gel expands because of the electrostatic repulsion force among the charged PAAc polymer chains. However, the fiber gel that consists of the only PAAc polymer chain, finally dissolves into the aqueous solution because the gel does not have the cross-linkage among the polymer chains into the nanofiber. In order to avoid the nanofiber gel dissolving, especially when it is above the pKa, we adopted the tBMA domain into the PAAc as a cross-linkage and a solubility control site, due to the hydrophobic interaction among the tBMA in the nanofiber. As a result, the poly(AAc-*co*-nBMA) nanofiber gel does not dissolve in the aqueous solution.

Figure 20 shows distributions of diameter of the poly(AAc-*co*-nBMA) nanofibers electrospun at two flow rates (2.0 mL/h and 1.0 mL/h) (See Figure 21). In general, the fiber diameters depend on the flow rate. In our experiment, the average diameter at the flow rate 2.0 mL/h (302 nm) was thicker than at 1.0 mL/h (233 nm).

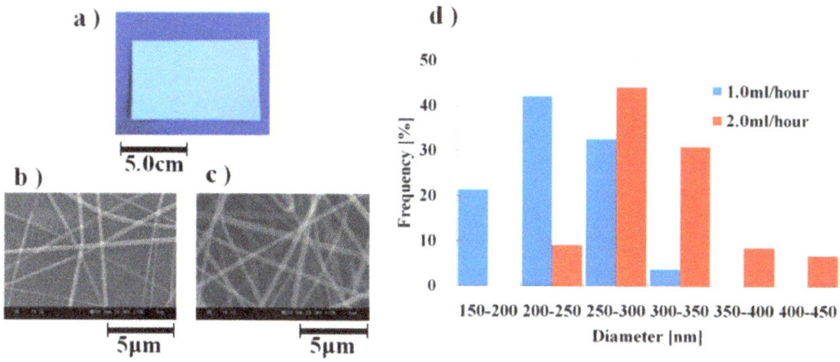

Figure 20. (**a**) An electrospun fiber sheet. (**b**) The SEM image of nanofibers at flow rate 1.0 mL/h. (**c**) The SEM image of nanofibers at flow rate 2.0 mL/h. (**d**) Distribution of fiber diameters of poly(AAc-*co*-nBMA) nanofibers electrospun at different flow rates.

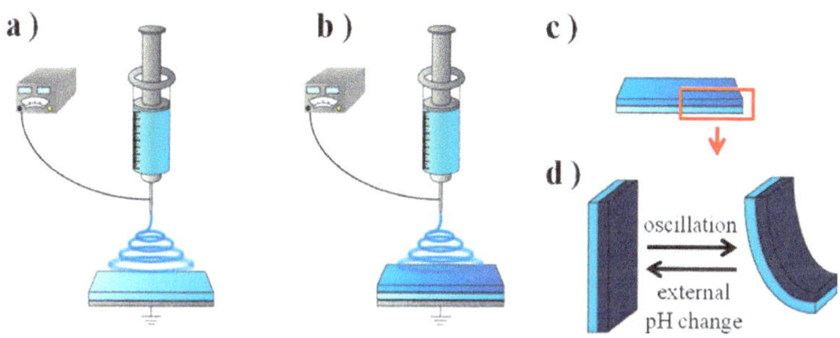

Figure 21. The method of introducing the anisotropic structure into the nanofiber gel. (**a**) Electrospinning at a flow rate of 2.0 mL/hour (sprayed for 30 minutes). (**b**) Electrospinning at a flow rate of 1.0 mL/hour (sprayed for 60 minutes). (**c**) Drying in 50 °C over night. (**d**) Cutting into 15 mm × 3 mm × 200 μm.

In order to drive the nanofibrous gel actuator synchronized with autonomous pH oscillation, we focused on the Landolt pH-oscillator, based on a bromated/ sulfite/ ferrocyanide reaction discovered by Edblom *et al.* [39,40]. This reaction causes the autonomous cyclic pH changes with a wide range at room temperature. The reaction has many reaction steps, so we estimated the main reactions as follow [41].

$$BrO_3^- + 3HSO_3^- + H^+ \rightarrow Br^- + 3SO_4^{2-} + 4H^+ \tag{3}$$

$$BrO_3^- + 6Fe(CN)_6^{4-} + 6H^+ \rightarrow Br^- + 6Fe(CN)_6^{3-} + 3H_2O \tag{4}$$

In Process (1), H_2SO_3 is oxidized by bromate, and ferrocyanide is oxidized by bromate in Process (2). In the above two processes, the hydrogen ions produced and consumed at comparable rates. Therefore, in this reaction, the pH oscillation takes place in the CSTR. Figure 22 shows the experimental set up of the CSTR. The CSTR was constructed by using four peristaltic pumps in order to feed four solutions of potassium bromated, sodium sulfite, potassium ferrocyanide and sulfuric acid. Moreover, this system had one more peristaltic pump to drain the excess solution. The degree of changing the pH range (amplitude) and period of the oscillating reaction can be controlled by changing the feed concentration, flow rate and solution temperature.

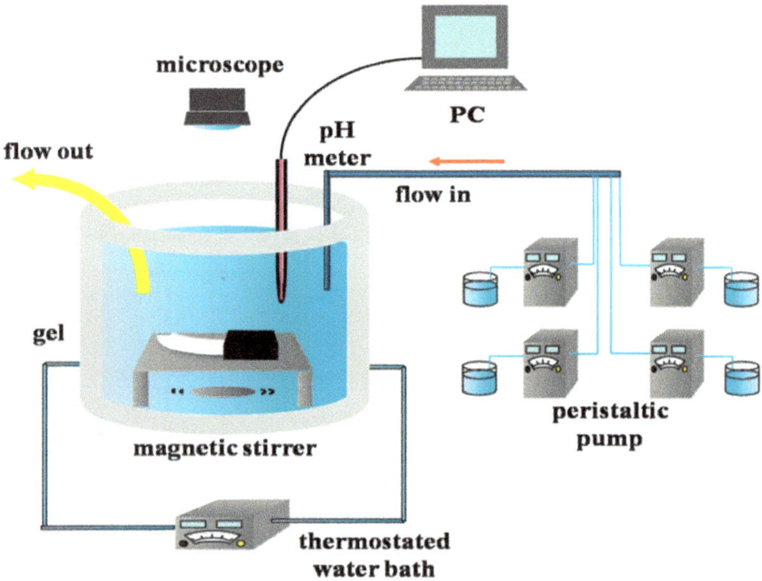

Figure 22. Continuous monitoring system for oscillation: monitoring both the medium pH and the gel motion in a CSTR.

Figure 23 shows a motion of the nanofiber gel actuator. The bending and stretching motions of the gel actuator synchronized with the pH oscillating reaction. As shown in Figure 23, we defined R as the length between two edges of the gel. Figure 24 shows the trajectory of the nanofiber gel strip. As shown in Figure 24, the gel strip caused the pendulum-like motion. As the external pH is below the pKa, the nanofiber gel stretches because of the deswelling originating from the hydrogen bonding (1→3). Next, when the pH is above the pKa, the gel bends because of the swelling originating from the repulsive force among the anionic polymer chains (4→6).

Figure 25 shows the temporal changes of R of the gel strip and the external pH, respectively. The range of the pH oscillation based on a bromate/sulfite/ferrocyanide reaction was $3.1 < pH < 7.2$, and the period was about 20 min. When the external pH changes periodically, the R of the gel strip cyclic changes synchronized with the external pH change. As shown in Figure 25, when the pH sharply decreased, the R of the gel strip starts to increase because the gel collapsed. Next, when the pH increased rapidly, R gradually decreased, due to the gel actuator swelling originating from the repulsive force of AAc domain in the polymer chain. That is because the gel has different rates at swelling and deswelling. In general, the swelling motion of the gel is slower than the deswelling motion. Therefore, when the gel actuator bended, the R value gradually decreased.

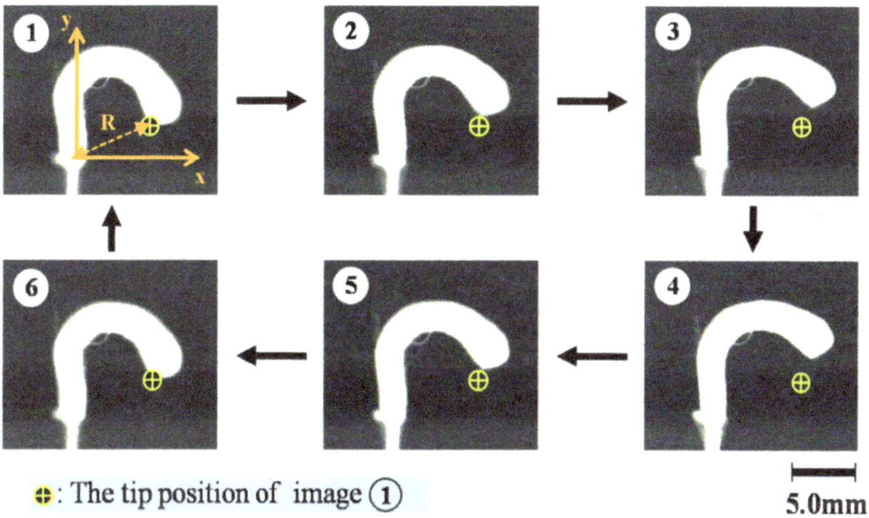

Figure 23. Periodical pendulum motion of poly(AAc-*co*-nBMA) nanofiber gel.

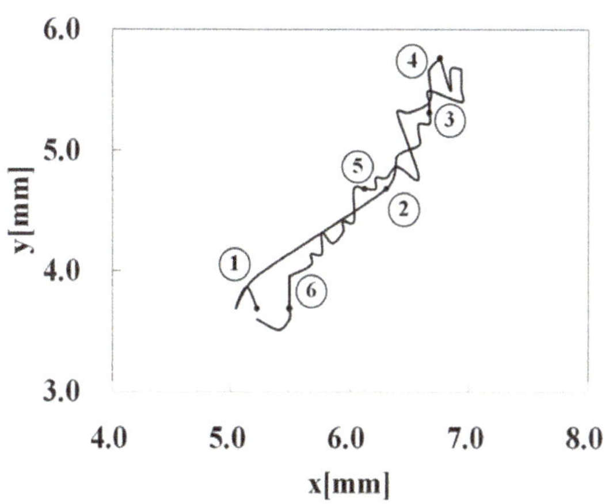

Figure 24. Trajectory of the tip of the gel relative to its attachment position during pH oscillation.

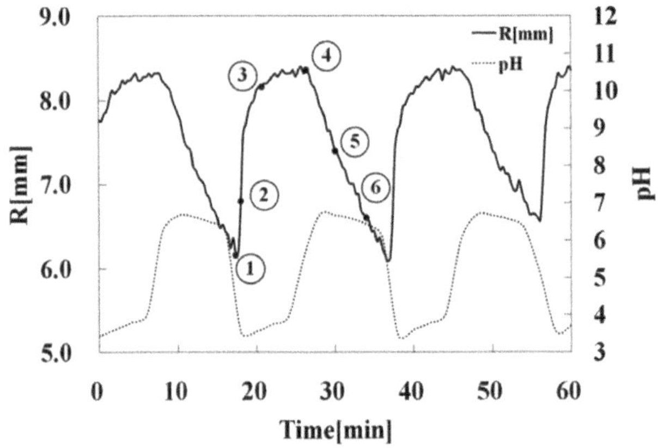

Figure 25. The time series of pH (discontinuous line) and R (solid line).

4. Conclusion

As for the VP-based self-oscillating polymer chain, we examined the influence of initial substrate concentrations of the BZ reaction on transmittance self-oscillation for the novel poly(VP-*co*-Ru(bpy)$_3$) solution. As a result we noted that the width of the self-oscillation is much affected by the initial concentration of the BZ substrates. In addition, we clarified that the amplitudes of the transmittance self-oscillation is hardly affected by the initial concentrations of the BZ substrates. This tendency was not observed in the case of the AMPS-containing polymer solution. Furthermore, we demonstrated that the period of the transmittance self-oscillation can be controlled by the selection of the initial concentration of the BZ substrates.

Furethermore we succeeded in clarifying the influence of the initial concentration of the BZ substrates and the temperature on the period of the swelling-deswelling self-oscillation for the novel poly(VP-*co*-Ru(bpy)$_3$) gel. The logarithmic plots of the period against the initial concentration of one BZ substrate under fixed the other two BZ substrates showed the good linear relationships. Moreover, the period of the self-oscillation increased with the increasing the temperature in accordance with Arrenius equation. The maximum frequency (0.5Hz) of the poly(VP-*co*-Ru(bpy)$_3$) gel was 20 times as large as that of poly(NIPAAm-*co*-Ru(bpy)$_3$ gel. Therefore, the period of the swelling-deswelling self-oscillation for the gel can be controllable in wide range by optimizing the initial concentration of the three BZ substrates and the temperature. In addition, we showed that the displacement of the self-oscillation for the gel has the trade-off relationship against the period of the self-oscillation.

Next, we introduced the new method of generating contraction waves in the functional gel, which controls the reaction diffusion system by flowing the CT solution into the hollow tubular gel. In order to realize this system, the poly(AAm-*co*-AAc) microphase-separated gels were synthesized with various mixture proportions of the water/acetone solvents. From the view of the swelling kinetics, we experimentally clarified that the suitable mixture proportion for the fast volume changes and the adequate robustness for solution sending. Also, we coupled the poly (AAm-*co*-AAc) microphase-separated tubular gel and the CT solution. Moreover, we successfully demonstrated the propagation of the acidic contraction region of the gel. In the next step, we will accomplish the entire chemical reaction networks for realizing a swelling from the acid contraction region, which leads to a contraction wave in the tubular gel. This research will be the first step for realizing the biomimetic chemical robot which causes a peristaltic locomotion.

Finally, we introduced the fabrication of nanofiber gel actuator with anisotropic internal structure by changing the flow rate in electrospinning. The developed gel generates bending and stretching motion according to the external pH. In order to drive the nanofiber gel actuator automatically, we focused on the pH oscillating reaction based on a bromate/sulfite/ferrocyanide reaction. As a result, we succeeded in causing the bending-stretching motion of the nanofiber gel actuator synchronized with the external pH oscillation. By analyzing the motion of the gel, we found that the gel actuator caused the pendulum-like motion. Moreover, we clarified that the displacement and period of the nanofiber gel were stable, which makes it promising as a molecular device for potential applications.

Author details

Yusuke Hara*
Nanosystem Research Institute (NRI), National Institute of Advanced Industrial Science and Technology (AIST), Japan

Shingo Maeda
Shibaura Institute of Technology, Japan

Takashi Mikanohara, Hiroki Nakagawa, Satoshi Nakamaru and Shuji Hashimoto
Department of Applied Physics, Graduate School of Science and Engineering, Waseda University, Japan

5. References

[1] Fukushima, T.; Asaka, K.; Kosaka, A.; Aida, T. (2005). Fully plastic actuator through layer-by-layer casting with ionic-liquid-based bucky gel, *Angewandte Chemie International Edition*, 44, pp.2410-2413, ISSN 1433-7851

[2] Oguro, K.; Kawami, Y.; Takenaka, H. (1992). An actuator element of polyelectrotyte gel membrane-electrode composite, *Bulletin of the Government Industrial Research Institute, Osaka*, 43, pp. 21–24.

* Corresponding Author

[3] Otake, M.; Kagami, Y.; Inaba, M.; Kim, B. & Inoue, H. (2002). Motion design of a starfishshaped gel robot made of electro-active polymer gel. *Robotics and Autonomous Systems,* Vol.40, No.2-3, (August 2002), pp. 185-191, ISSN 0921-8890

[4] Steinberg, I, Z.; Oplatka, A.; Kachalsky, A. (1996). Mechanochemnical engines, *Nature,* 210, pp. 568–571, ISSN 0028-0836

[5] Feinberg, A. W. Feigel, A. Shevkoplyas, S. S. Sheehy, S. Whitesides, G, M. & Parker, K. K (2007). Muscular thin films for building actuators and powering devices, *Science,* Vol.317, No.5843, (September 2007), pp.1366-1370, ISSN 0036-8075

[6] Harada, A. ; Kataoka, K. (1999). Chain length recognition: core-shell supramolecular assembly from oppositely charged block copolymers. *Science,* 283, pp. 65-67 ISSN 1095-9203

[7] Kim, J.; Nayak, S.; Lyon, L. A. (2005). Bioresponsive hydrogel microlenses, *Journal of the American Chemical Society,* 127, pp. 9588–9592 ISSN 0002-7863

[8] Kwon, G.H.; Park, J.Y.; Kim, J.Y.; Frisk, M.L.; Beebe, D.J.; Lee, S.H. Biomimetic Soft Multifunctional Miniature Aquabots. *Small* 2008, 4, 2148-2153.

[9] Hoffman, A. S. (2002) Hydrogels for biomedical applications, *Advanced Drug Delivery Reviews,* 43, pp. 3-12, ISSN 0169-409X

[10] Tanaka, T. (1981). Gels, Scientific American.,244, pp.110-116, ISSN 0036-8733

[11] Hirokawa, Y.; Tanaka, T. (1984). Volume phase transition in a non ionic gel, *Journal of chemical Physics,* 81, pp.6379-6380, ISSN *0021-9606*

[12] Ilmain, F.; Tanaka, T., Kokufuta, E. (1991). Volume transition in a gel by hydrogen bonding, *Nature,* 349, pp.400-401, ISSN 0028-0836

[13] Tanaka, T. (1978). Collapse of gels and critical endpoint, *Physical Review Letters,* 40, pp.820-823, ISSN 0031-9007

[14] Suzuki, A.; Tanaka, T. (1990). Phase transition in a polymer gels induced by visible-light, *Nature,* 346, pp. 345-347, ISSN 0028-0836

[15] Kuhn, W. ; Hargity, B. ; Katchalsky, A. & Eisenberg, H. (1950). Reversible Dilation and Contraction by Changing the State of Ionization of High-Polymer Acid Networks, *Nature,* Vol.165, No.4196, (April 1950), pp. 514-516, ISSN 0028-0836

[16] Hu, Z.; Zhang, X. & Li, Y. (1995). Synthesis and Application of Modulated polymer gels. *Science,* Vol.269, No.5223, (April 1995), pp. 525-527, ISSN 0036-8075

[17] Yeghiazarian, L. Mahajan, S. Montemagno, C. Cohen, C. & Wiesner, U. (2005). Directed Motion and Cargo Transport Through Propagation of Polymer-Gel Volume Phase Transitions, *Advanced Materials,* Vol.17, No.15, (August 2005), pp. 1869-1873, ISSN 0935-9648

[18] Huber, D. L.; Manginell, R. P.; Samara, M. A.; Kim, B. & Bunker, B. C. (2003). Programmed Adsorption and Released of proteins in a Microfluidic device, *Science,* Vol.301, No.5631, (July 2003), pp. 352-354, ISSN 0036-8075

[19] Beebe, D. J.; Moore, J. S.; Bauer, J. M.;Yu, Q.;Liu, R. H.;Devadoss, C.; Jo, B. H. (2000). Functional hydrogel structures for autonomous flow control inside microfluidic channels, *Nature,* 404, pp. 588-590, ISSN 0028-0836

[20] Asoh, T.; Matsusaki, M.; Kaneko, T.; Akashi, M. (2008). Fabrication of Temperature-Responsive Bending Hydrogels with a Nanostructured Gradient, *Advanced Materials*, Vol.20, No.11, pp. 2080-2083, ISSN 0935-9648.
[21] Ross, J.; Muller, S. C.; Vidal, C. (1988). Chemical Waves, Science, Vol. 240, pp.460-465, ISSN 1095-9203.
[22] Epstein, I. R.; Showalter, K.(1996). Nonlinear Chemical Dynamics: Oscillations, Patterns, and Chaos, *The Journal of Chemcal Chemistry*, Vol. 100, No. 31, pp.13132-13147, ISSN 1932-7447.
[23] Epstein, I. R. (1995). The Consequences of Imperfect Mixing in Autocatalytic Chemical and Biological Systems, Nature Vol. 374, pp. 321-327, ISSN 0028-0836
[24] Lee, K. J.; Cox, E. C.; Goldstein, R. E. (1996). Competing Patterns of Signaling Activity in Dictyostelium Discoideum. *Physical Review Letters*, Vol. 76, pp. 1174-1177, ISSN 0031-9007.
[25] Shibayama, M.; Tanaka, T. (1993). Volume Phase Transition and Related Phenomena of Polymer Gels, *Advances in Polymer Science*, Vol. 109, pp. 1-62, ISSN 0065-3195.
[26] Landrot, V.; De Kepper, P.; Boissonade, J.; Szalai, I.; Gauffre, F. (2005). Wave Patterns driven bu chemomechanical instabilities in responsive gels, *The Journal of Physical Chemistry B*, Vol. 109, pp. 21476-21480, ISSN 1520-6106.
[27] Boissonade, J. (2003). Efficient Method for Predicting Crystal Structures at Finite Temperature: Variable Box Shape Simulations, *Physical Review Letters*, Vol. 90, pp. 188302 (4pages), ISSN 0031-9007.
[28] Yashin, V, V. & Balazs, A, C. (2006). Pattern formation and shape changes in self-oscillating polymer gels. *Science*, Vol. 314, No.5800, (September 2006) pp.798-801, ISSN 0036-8075
[29] Yashin, V, V. & Balazs, A, C. (2007). Theoretical and computational modeling of self-oscillating polymer gels, The *Journal of Chemical Physics*, Vol.126, No.12, (March2007) pp.124707-1-124707-17 ISSN 0021-9606
[30] Zaikin, A.N.; Zhabotinsky, A.M. (1970). Concentration Wave propagation in two-dimensional liquid-phase self-oscillating system, *Nature*, 225, pp. 535–537, ISSN 0028-0836
[31] Reusser, E.J.; Field, R.J. (1979). The transition from phase waves to trigger waves in a model of the Zhabotinskii reaction, *Journal of the American Chemical Society*, 101, pp. 1063–1071, ISSN 0002-7863
[32] Nicolis, G.; Prigogine, I. (1997). Self Orgainization in Nonequilibrium Systems; *Wiley: New York, NY, USA.*
[33] Field, R.J.; Burger, M. (1985). Oscillations and Traveling Waves in Chemical Systems; *John Wiley & Sons: New York, NY, USA.*
[34] Field, R.J.; Noyes, R. M. (1974). Oscillations in chemical systems. IV. Limit cycle behavior in a model of a real chemical reaction, *Journal of Chemical Physics*, 60, pp. 1877–1884, ISSN 0021-9606
[35] Gyorgyi, L.; Turanyi, T.; Field, R J. (1990). Mechanistic details of the oscillatory Belousov-Zhabotinskii reaction, *Journal of Chemical Physics*, 94, pp. 7162–7170, ISSN 0021-9606

[36] Turanyi, T.; Gyorgyi, L.; Field, R. J. (1993), Analysis and simplification of the GTF model of the Belousov-Zhabotinsky reaction, *Journal of Chemical Physics*, 97, pp. 1931–1941, ISSN 0021-9606.

[37] Scott, S. K. (1991), Chemical Chaos, *1st ed.; Oxford University Press: Oxford, UK.*

[38] Mori, H. ; Kuramoto, Y. (1997). Dissipative Structures and Chaos, *Springer-Verlag, Berlin.*

[39] Edblom, E.C.; Orban, M.; Epstein, I.R. (1986). A New Iodate Oscillator: The Landolt Reaction with Ferrocyanide in a CSTR. *Journal of the American Chemical Society*, Vol. 108, pp. 2826-2830, ISSN 0002-7863.

[40] Edblom, E.C.; Luo, Y.; Orban, M.; Kustin, K.; Epstein, I.R. (1989) Kinetics and Mechanism of the Oscillatory Bromate-Sulfite-Ferrocyanide Reaction. *The Journal of Physical Chemistry*, Vol. 93, pp. 2722-2727, ISSN 1089-5639.

[41] Crook, J.; Smith, A.; Jones, A.; Ryan, J. (2002). Chemically Induced Oscillations in a pH-Responsive Hydrogel. *Physical Chemistry Chemical Physics*, Vol. 4, pp. 1367-1369, ISSN 1463-9076.

[42] V. Labrot, P. De Kepper, J. Boissonade, I. Szalai, and F. Gauffre, (2005). Wave Patterns Driven by Chemomechanical Instabilities in Responsive Gels, *Journal of Chemical Physics*, vol. 109, No. 46, pp. 21476-21480, ISSN 0021-9606.

[43] J. Boissonade, P. De Kepper, F. Gauffre, and I. Szalai, (2006) Spatial bistability: A source of complex dynamics. From spatiotemporal reaction-diffusion patterns to chemomechanical structures, *Chaos*, Vol. 16, No. 3, pp. 037109 (8pages), ISSN 1054-1500.

[44] Ishiwatari, T.; Kawaguchi, M.; Mitsuishi, M. (1984). Oscillatry reactions in polymer systems, *Journal of Polymer Science Part A: Polymer Chemistry*, 22, pp. 2699-2704 ISSN 0887-624X

[45] Yoshida, R.; Takahashi, T.; Yamaguchi, T.; Ichijo, H. (1996). Self-oscillating gel, *Journal of the American Chemical Society*, 118, pp. 5134-5135, ISSN 0002-7863

[46] Yoshida, R.; Sakai, T.; Ito, S.; Yamaguchi, T. (2002). Self-oscillation of polymer chains with rhythmical soluble-insoluble changes, *Journal of the American Chemical Society*, 124, pp. 8095-8098, ISSN 0002-7863

[47] Hara, Y.; Yoshida, R. Self-oscillation of polymer chains induced by the Belousov-Zhabotinsky reaction under acid-free conditions. *Journal of Physical Chemistry B* 2005, 109, pp. 9451–9454, ISSN 1089-5647

[48] Hara, Y.; Yoshida, R. (2009). Damping behavior of aggregation-disaggregation self-oscillation for a polymer chain, *Macromolecular Rapid Communications*, 30, pp. 1656–1662, ISSN 1521-3927

[49] Hara, Y.; Yoshida, R. (2005). Control of oscillating behavior for the self-oscillating polymer with pH-control site, *Langmuir*, 21, pp. 9773–9776, ISSN 0743-7463

[50] Hara, Y.; Yoshida, R. (2008). A viscosity self-oscillation of polymer solution induced by the BZ reaction under acid-free condition, *Journal of Chemical Physics*, 128, 224904, ISSN 0021-9606

[51] Hara, Y.; Sakai, T.; Maeda, S.; Hashimoto, S.; Yoshida, R. (2005).b Self-oscillating soluble-insoluble changes of polymer chain including an oxidizing agent induced by the Belousov-Zhabotinsky reaction, *Journal of Physical Chemistry B*, 109, pp. 23316–23319, ISSN 1089-5647

[52] Hara, Y.; Yoshida, R. (2008). Self-oscillating polymer fueled by organic acid. *Journal of Physical Chemistry B*, 112, pp. 8427–8429, ISSN 1089-5647

[53] Maeda, S.; Hara, Y.; Yoshida, R.; Hashimoto, S. (2008). Control of the dynamic motion of a gel actuator driven by the Belousov-Zhabotinsky reaction, *Macromolecular Rapid Communications*, 29, pp. 401–405, ISSN 1521-3927

[54] Maeda, S.; Hara, Y.; Sakai. T.; Yoshida, R.; Hashimoto, S. (2007). Self-walking gel, *Advanced Materials*, 19, pp. 3480–3484, ISSN 1521-4095

[55] Maeda, S.; Hara, Y.; Yoshida, R.; Hashimoto, S. (2008). Peristaltic motion of polymer gels, *Angewandte Chemie International Edition*, 47, pp. 6690-6693, ISSN 1433-7851

[56] Hara, Y.; Yoshida, R. (2009). Influence of a positively charged moiety on aggregation-disaggregation self-oscillation induced by the BZ reaction, *Macromolecular Chemistry and Physics*, 210, pp. 2160–2166, ISSN 1022-1352

[57] Nakamaru, S,; Maeda, S. ; Hara, Y.; Hashimoto, S. (2009). Control of Autonomous Swelling-Deswelling Behavior for a Polymer gel. *Journal of Physical Chemistry B*, 2009, 113, pp. 4609-4613, ISSN 1089-5647

[58] T. Mikanohara, S. Maeda, Y. Hara, S. Hashimoto (2011). Tubular Gel Motility Driven by Chemical Reaction Networks", *Proceeding of 2011 IEEE International Conference on Robotics and Biomimetics*, pp.2008-2013, ISBN 978-1-4577-2136-6.

[59] Nakagawa, H. Hara, Y. Maeda, S. & Hashimoto, S. (2010). A Novel Design of Nanofibrous Gel Actuator by Electrospinning, *Proceedings of IEEE Nano* 2010. ISBN 978-4244-7031-0

[60] Nakagawa, H.; Hara, Y.; Maeda, S. & Hashimoto, S (2010). A Pendulum-Like Motion o Nanofiber Gel Actuator Synchronized with External Periodic pH Oscillation, *Polymers*, Vol.3, No.1, (December 2010) pp 405-412, ISSN 2073-4360

[61] R. McNeill Alexander, (2006). Principles of Animal Locomotion, *Princeton University Press*, ISBN: 9780691126340

[62] Stevens, M.M.; George, J.H. Exploring and Engineering the Cell Surface Interface. *Science* 2005, 310, 1135-1138.

[63] Cui, Y.; Wei, Q.; Park, H.; Lieber, C.M. (2001). Nanowire Nanosensors for Highly Sensitive and Selective Detection of Biological and Chemical Species. *Science*, Vol. 293, pp. 1289-1292, ISSN 0036-8075

[64] Gu, S.Y.; Wang, Z.M.; Li, J.B.; Ren, J. (2010). Switchable Wettability of Thermo-Responsive Biocompatible Nanofibrous Films Created by Electrospinning. *Macromolecular Materials and Engineering*, Vo. 295, pp. 32-36 ISSN 1438-7492.

[65] Adam, D. (2001). Fine Set of Threads, *Nature*, Vol. 411, No. 236, pp. 236-236, ISSN 0028-0836

[66] Yu, J.H.; Sergey, V.F.; Gregory, C.R. (2004). Production of Submicrometer Diameter Fibers by Two-Fluid Electrospinning. *Advanced Materials* Vol. 16, pp. 1562-1566, ISSN 1521-4095.

[67] Dan, L.; Gong, O.; Jesse, T.M.; Younan, X. (2005). Collecting Electrospun Nanofibers with Patterned Electrodes, *Nano Letter*, Vol. 5, pp. 913-916, ISSN 1530-6984.

[68] Donzhi, Y.; Jianfeng, Z.; Jing, Z.; Jun, N. Aligned (2008). Electrospun Nanofibers Induced by Magnetic Field, *Journal of Applied Polymer Science*, Vol. 110, pp. 3368-3372, ISSN 1097-4628.

[69] Bogntizki, M.; Frese, T.; Steinhart, M.; Greiner, A.; Wendoreff, J.H. (2001). Preparation of Fibers with Nanoscaled Morphologies: Electrospinning of Polymer Blends, *Polymer Engineering & Science*, Vol. 41, pp. 982-989, ISSN 1548-2634.

[70] Li, L.; Hsieh, Y. (2005). Ultra-fine Polyelectrolyte Fibers from Electrospinning of Poly(acrylic acid), *Polymer*, Vol. 46, pp. 5133-5139, ISSN 0032-3861.

[71] Chen, H.; Hsieh, Y. (2004). Ultrafine Hydrogel Fibers with Dual Temperatureand pH-Responsive Swelling Behaviors. *Journal of Polymer Science Part A: Polymer Chemistry* ,Vol. 42, pp. 6331-6339, ISSN 1099-0518.

[72] Jin, X.; Hsieh, Y. (2005). pH-Responsive swelling behavior of poly(vinyl alcohol)/poly(acrylic acid) bi-component fibrous hydrogel membranes, *Polymer*, Vol. 46, pp. 5149-5160, ISSN 0032-3861.

[73] Okuzaki, H.; Kobayashi, K.; Yan, H. (2009). Thermo-Responsive Nanofiber Mats, *Macromolecules* Vol. 42, pp. 5916-5918, ISSN 0024-9297.

[74] Jin, X.; Hsieh, Y. Electrospinning (2007). pH-Responsive Block Copolymer Nanofibers. *Advanced Materials*, Vol. 19, pp. 3544-3548, ISSN 1521-4095.

[75] Yoshida, R.; Tanaka, M.; Onodera, S.; Yamaguchi, T.; Kokufuda, E. (2000). In-phase synchronization of chemical and mechanical oscillations in self-oscillating gels, *The Journal of Physical Chemistry A*, Vol. 104, pp. 7549–7555, ISSN 1089-5639.

[76] AK. Horvath, I. Nagypal, G. Peintler, and IR. Epstein (2004). Autocatalysis and self-inhibition: Coupled kinetic phenomena in the chlorite-tetrathionate reaction, *Journal of the American Chemical Society*, vol. 126, No. 20, pp. 6246-6247, ISSN 0002-7863.

[77] J. Boissonade (2005). Self-oscillations in chemoresponsive gels: A theoretical approach, *Chaos*, vol. 15, No. 2, pp. 023703, ISSN 1054-1500.

[78] J. Boissonade (2003) Simple chemomechanical process for self-generation of rhythms and forms, *Physical Review Letters*, Vol. 90, No. 18, pp. 188302, ISSN 0031-9007.

[79] B. G. Kabra, and S. H. Gehrke (1991). Synthesis of fast response, temperature-sensitive poly(N-isopropylacrylamide) gel, *Polymer Communications*, vol. 32, no. 11, pp. 322-323, ISSN 0263-6476.

[80] T. Norisuye, Y. Kida, N. Masui, and Q. Tran-Cong-Miyata (2003). Studies on two types of built-in inhomogeneities for polymer gels: Frozen segmental concentration fluctuations and spatial distribution of cross-links, *Macromolecules*, Vol. 36, No. 16, pp. 6202-6212, ISSN 0024-9297.

[81] A. Onuki (1993). Theory of phase-transition in polymer gels, *Advances in Polymer Science*, Vol. 109, pp. 63-121, ISSN 0065-3195.

Permissions

The contributors of this book come from diverse backgrounds, making this book a truly international effort. This book will bring forth new frontiers with its revolutionizing research information and detailed analysis of the nascent developments around the world.

We would like to thank Giovanni Berselli, Ph.D., for lending his expertise to make the book truly unique. He has played a crucial role in the development of this book. Without his invaluable contribution this book wouldn't have been possible. He has made vital efforts to compile up to date information on the varied aspects of this subject to make this book a valuable addition to the collection of many professionals and students.

This book was conceptualized with the vision of imparting up-to-date information and advanced data in this field. To ensure the same, a matchless editorial board was set up. Every individual on the board went through rigorous rounds of assessment to prove their worth. After which they invested a large part of their time researching and compiling the most relevant data for our readers. Conferences and sessions were held from time to time between the editorial board and the contributing authors to present the data in the most comprehensible form. The editorial team has worked tirelessly to provide valuable and valid information to help people across the globe.

Every chapter published in this book has been scrutinized by our experts. Their significance has been extensively debated. The topics covered herein carry significant findings which will fuel the growth of the discipline. They may even be implemented as practical applications or may be referred to as a beginning point for another development. Chapters in this book were first published by InTech; hereby published with permission under the Creative Commons Attribution License or equivalent.

The editorial board has been involved in producing this book since its inception. They have spent rigorous hours researching and exploring the diverse topics which have resulted in the successful publishing of this book. They have passed on their knowledge of decades through this book. To expedite this challenging task, the publisher supported the team at every step. A small team of assistant editors was also appointed to further simplify the editing procedure and attain best results for the readers.

Our editorial team has been hand-picked from every corner of the world. Their multi-ethnicity adds dynamic inputs to the discussions which result in innovative outcomes. These outcomes are then further discussed with the researchers and contributors who give their valuable feedback and opinion regarding the same. The feedback is then collaborated with the researches and they are edited in a comprehensive manner to aid the understanding of the subject.

Apart from the editorial board, the designing team has also invested a significant amount of their time in understanding the subject and creating the most relevant covers. They scrutinized every image to scout for the most suitable representation of the subject and create an appropriate cover for the book.

The publishing team has been involved in this book since its early stages. They were actively engaged in every process, be it collecting the data, connecting with the contributors or procuring relevant information. The team has been an ardent support to the editorial, designing and production team. Their endless efforts to recruit the best for this project, has resulted in the accomplishment of this book. They are a veteran in the field of academics and their pool of knowledge is as vast as their experience in printing. Their expertise and guidance has proved useful at every step. Their uncompromising quality standards have made this book an exceptional effort. Their encouragement from time to time has been an inspiration for everyone.

The publisher and the editorial board hope that this book will prove to be a valuable piece of knowledge for researchers, students, practitioners and scholars across the globe.

List of Contributors

Adelaide Nespoli, Carlo Alberto Biffi, Riccardo Casati, Francesca Passaretti, Ausonio Tuissi and Elena Villa
Consiglio Nazionale delle Ricerche, Istituto per l'Energetica e le Interfasi (CNR-IENI), Lecco, Italy

A. Spaggiari, G. Scirè Mammano and E. Dragoni
Department of Science and Methods for Engineering, University of Modena and Reggio Emilia, via Amendola, Campus S. Lazzaro Reggio Emilia, Italy

William Coral, Claudio Rossi, Julian Colorado, Daniel Lemus and Antonio Barrientos
Centre for Automation and Robotics (CAR) of the Technical University of Madrid, Spain
CSIC (National Council for Scientific research)

Simone Pittaccio and Stefano Viscuso
National Research Council of Italy, Institute for Energetics and Interphases, Lecco, Italy

Daniele Davino, Alessandro Giustiniani and Ciro Visone
University of Sannio - Engineering Department, 82100 Benevento (BN), Italy

Maletta Carmine
Assistant professor of Machine Design at University of Calabria, Department of Mechanical Engineering, Rende (CS), Italy

Furgiuele Franco
Full professor of Machine Design at University of Calabria, Department of Mechanical Engineering,
Rende (CS), Italy

Thomas G. McKay and Benjamin M. O'Brien
The Biomimetics Laboratory of the Auckland Bioengineering Institute, The University of Auckland,
Level 6, Auckland, New Zealand

Iain A. Anderson
The Biomimetics Laboratory of the Auckland Bioengineering Institute, The University of Auckland,
Level 6, Auckland, New Zealand
Department of Engineering Science, School of Engineering, The University of Auckland, Level 3,
Auckland, New Zealand

Ahn Kyoung Kwan
School of Mechanical and Automotive Engineering, University of Ulsan, Korea

Doan Ngoc Chi Nam
Graduate School of Mechanical and Automotive Engineering, University of Ulsan, Korea

P.Maiolino and G.Cannata
Department of Communication, Computer and System Sciences (DIST), University of Genova,
Genova, Italy

A.Ascia, M.Maggiali, L.Natale and G.Metta
Fondazione Istituto Italiano di Tecnologia (Italian Institute of Technology), Genova, Italy

Zheng Chen, T. Um and Hilary Bart-Smith
University of Virginia, USA

José G. Martínez, Joaquín Arias-Pardilla and Toribio F. Otero
Universidad Politécnica de Cartagena, Spain

Weihua Li and Tongfei Tian
School of Mechanical, Materials & Mechatronic Engineering, University of Wollongong, Wollongong, NSW, Australia

Haiping Du
School of Electrical, Computer and Telecommunications Engineering, Wollongong, NSW, Australia

Yusuke Hara
Nanosystem Research Institute (NRI), National Institute of Advanced Industrial Science and
Technology (AIST), Japan

Shingo Maeda
Shibaura Institute of Technology, Japan

Takashi Mikanohara, Hiroki Nakagawa, Satoshi Nakamaru and Shuji Hashimoto
Department of Applied Physics, Graduate School of Science and Engineering, Waseda University,
Japan